AF604393

# HEINEMANN CHEMISTRY 2

## SKILLS AND ASSESSMENT

Penny Commons

**VCE Units 3 and 4**

Written for the VCE Chemistry Study Design 2024–2027

**Pearson Australia**
(a division of Pearson Australia Group Pty Ltd)
459–471 Church Street
Level 1, Building B
Richmond, Victoria 3121
www.pearson.com.au

Copyright © Penny Commons and Pearson Australia 2025
(a division of Pearson Australia Group Pty Ltd)
First published 2023 by Pearson Australia
2026 2025 2024 2023
10 9 8 7 6 5 4 3 2 1

**Reproduction and communication for educational purposes**
The Australian *Copyright Act 1968* (the Act) allows a maximum of one chapter or 10% of the pages of this work, whichever is the greater, to be reproduced and/or communicated by any educational institution for its educational purposes provided that that educational institution (or the body that administers it) has given a remuneration notice to the Copyright Agency under the Act. For details of the copyright licence for educational institutions contact the Copyright Agency (www.copyright.com.au).

**Reproduction and communication for other purposes**
Except as permitted under the Act (for example any fair dealing for the purposes of study, research, criticism or review), no part of this book may be reproduced, stored in a retrieval system, communicated or transmitted in any form or by any means without prior written permission. All enquiries should be made to the publisher at the address above.

This book is not to be treated as a blackline master; that is, any photocopying beyond fair dealing requires prior written permission.

PHOTOCOPYING OF BOOKS IS RESTRICTED UNDER LAW

VCE Heinemann Project Leads: Fiona Cooke, Bryonie Scott, Misal Belvedere, Malcolm Parsons
Schools Programme Manager: Michelle Thomas
Production Editor: Chris Woods
Editor: Catherine Greenwood
Series Designer: Anne Donald
Senior Rights and Permissions Analyst: Amirah Fatin Binte Mohamed Sapi'ee
Desktop Operator: JitPin Chong
Illustrators: DiacriTech and QBS Learning
Proofreader: Marta Veroni
Printed in Malaysia (CTP-VVP)

ISBN 978 0 6557 0027 2

Pearson Australia Group Pty Ltd ABN 40 004 245 943

**Disclaimer**
The selection of internet addresses (URLs) provided for this book was valid at the time of publication and was chosen as being appropriate for use as a secondary education research tool. However, due to the dynamic nature of the internet, some addresses may have changed, may have ceased to exist since publication, or may inadvertently link to sites with content that could be considered offensive or inappropriate. While the authors and publisher regret any inconvenience this may cause readers, no responsibility for any such changes or unforeseeable errors can be accepted by either the authors or the publisher.

**Indigenous Australians**
We respectfully acknowledge the traditional custodians of the lands upon which the many schools throughout Australia are located. We acknowledge traditional Indigenous Knowledge systems are founded upon a logic that developed from Indigenous experience with the natural environment over thousands of years. This is in contrast to Western science, where the production of knowledge often takes place within specific disciplines. As a result, there are many cases where such disciplinary knowledge does not reflect Indigenous worldviews, and can be considered offensive to Indigenous peoples.

Some of the images used in *Heinemann Chemistry 2 Skills and Assessment* might have associations with deceased Indigenous Australians. Please be aware that these images might cause sadness or distress in Aboriginal or Torres Strait Islander communities.

**Practical activities**
All practical activities, including the illustrations, are provided as a guide only and the accuracy of such information cannot be guaranteed. Teachers must assess the appropriateness of an activity and take into account the experience of their students and facilities available. Additionally, all practical activities should be trialled before they are attempted with students and a risk assessment must be completed. All care should be taken and appropriate personal protective clothing and equipment should be worn when carrying out any practical activity. Although all practical activities have been written with safety in mind, Pearson Australia and the authors do not accept any responsibility for the information contained in or relating to the practical activities, and are not liable for loss and/or injury arising from or sustained as a result of conducting any of the practical activities described in this book.

**Attributions**
We thank the following for their contributions to our skills and assessment book:
The following abbreviations are used in this list: t = top, b = bottom, l = left, r = right, c = centre.

**Cover: Science Photo Library:** Sambraus, Daniel.
**Shutterstock:** Smile Fight, pp. 1-2, 63-4; Zzcapture, pp. 119, 171-2, 234-5.
**Victorian Curriculum and Assessment Authority (VCAA):** Selected examination questions and extracts from the VCE Chemistry Study Design (2023–2027) are copyright Victorian Curriculum and Assessment Authority (VCAA), reproduced by permission. VCE® is a registered trademark of the VCAA. The VCAA does not endorse this product and makes no warranties regarding the correctness or accuracy of its content. To the extent permitted by law, the VCAA excludes all liability for any loss or damage suffered or incurred as a result of accessing, using or relying on the content. Current VCE Study Designs and related content can be accessed directly at www.vcaa.vic.edu.au.
**Wikimedia:** The Incredulity of Saint Thomas by Cima da Conegliano, p. 198. Image in the National Gallery Collection, object number NG816.

# Contents



## Unit 3 How can design and innovation help to optimise chemical processes?

### AREA OF STUDY 1

### What are the current and future options for supplying energy?



### AREA OF STUDY 2

### How can the rate and yield of chemical reactions be optimised?

# Contents

## Unit 4 How are carbon-based compounds designed for purpose?

# How to use this book

The *Heinemann Chemistry 2 Skills and Assessment* book provides the opportunity to practise, apply and extend your learning through a range of supportive and challenging activities. These activities reinforce key concepts and skills, and enable a flexible approach to learning. There are also regular opportunities for reflection and self-evaluation in the final worksheet in each area of study.

This resource has been written to the VCE Chemistry Study Design 2023–2027 and is divided into five areas of study—two in Unit 3 and three in Unit 4. Areas of study 1 and 2 in each unit consist of four main sections:

- key knowledge
- worksheets
- practical activities
- exam questions.

## CHEMISTRY TOOLKIT

The Chemistry toolkit supports development of the skills and techniques needed to undertake practical and secondary-sourced investigations, and covers examination techniques and study skills. It also includes checklists, models, exemplars and scaffolded steps. The toolkit can serve as a reference tool and be consulted as needed.

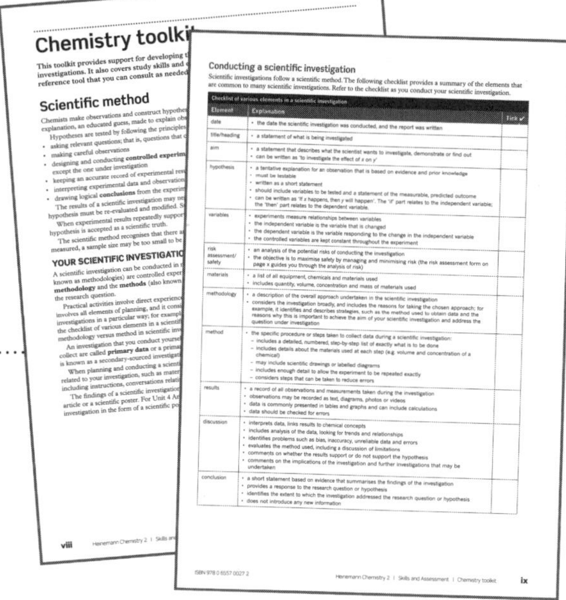


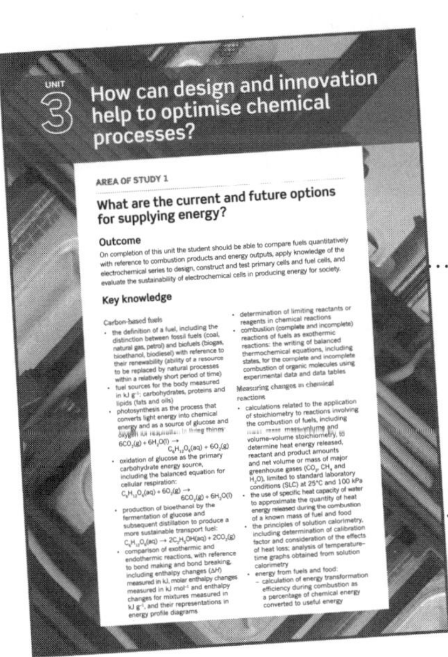


## AREA OF STUDY OPENER

*Heinemann Chemistry 2 Skills and Assessment* is structured to follow the study design units and areas of study. The area of study opening page lists the study design key knowledge for easy reference to the activities that follow.

## KEY KNOWLEDGE

Each area of study begins with a key knowledge section. This consists of a set of summary notes that cover the key knowledge for that area of study. Key terms are in bold and are included in the glossary of the student book. The section also serves as a ready reference for completing the worksheets and practical activities.

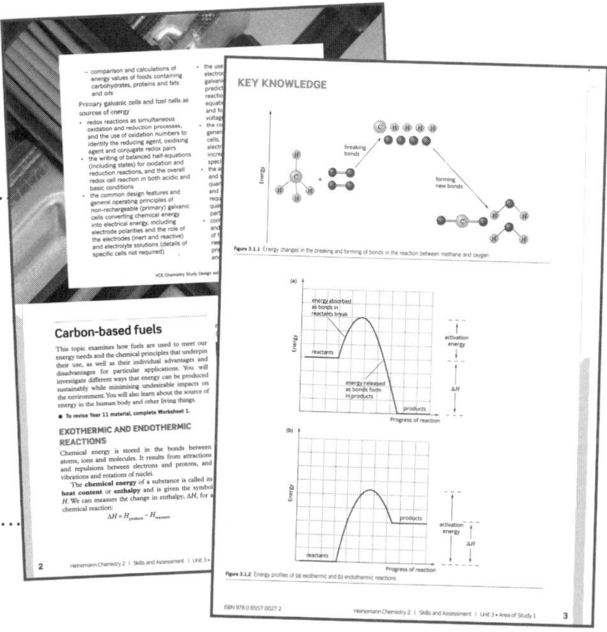

## WORKSHEETS

The worksheets feature questions that allow you to practise and apply your knowledge and skills. Each area of study includes a 'Knowledge review' worksheet to activate prior knowledge, a 'Literacy review' worksheet that provides opportunities for vocabulary and literacy support, and a 'Reflection' worksheet, which you can use for self-assessment. Other worksheets provide opportunities to revise, consolidate and further your understanding. All worksheets function as formative assessment and are clearly aligned with the study design. A range of questions building from foundation to challenging is included in each worksheet.

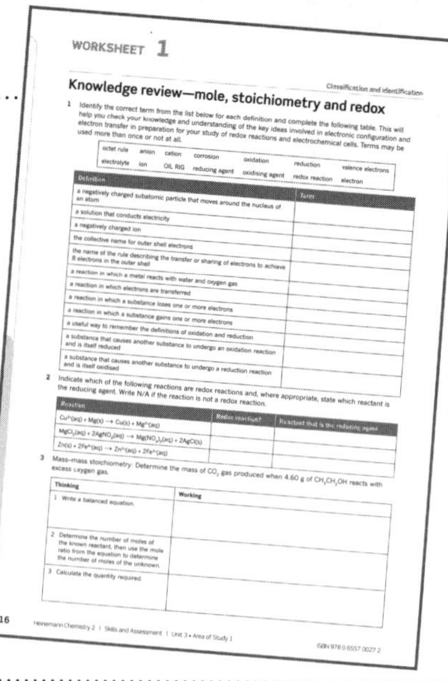
WORKSHEET 1

Knowledge review—mole, stoichiometry and redox

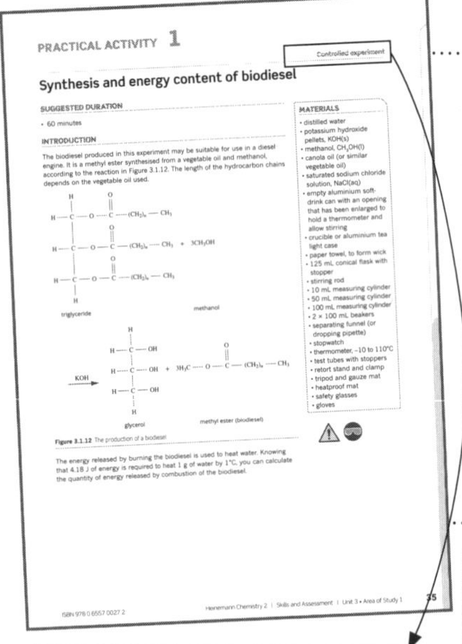
PRACTICAL ACTIVITY 1

Controlled experiment

Synthesis and energy content of biodiesel

## PRACTICAL ACTIVITIES

Practical activities offer you the chance to complete practical work related to the various themes covered in the study design. You have the opportunity to design and conduct scientific investigations, generate, evaluate and analyse data, appropriately record results, and prepare evidence-based conclusions. Where relevant, you will also need to conduct risk assessments to identify any potential hazards.

Each practical activity includes a suggested duration. Together with the Area of Study 3 scientific investigation, the practical activities meet the 34 hours of practical work mandated for Units 3 and 4 in the study design.

Controlled experiment

## METHODOLOGIES

Each worksheet and practical activity is mapped to one or more of the scientific investigation methodologies outlined in the study design. Completing these activities gives you experience in applying the methodologies in a wide variety of contexts and prepares you for designing and conducting your own scientific investigation in Unit 4 Area of Study 3.

## EXAM QUESTIONS

Each area of study finishes with a selection of VCAA exam questions. These give you the opportunity to gain valuable experience applying your knowledge and understanding to exam questions.

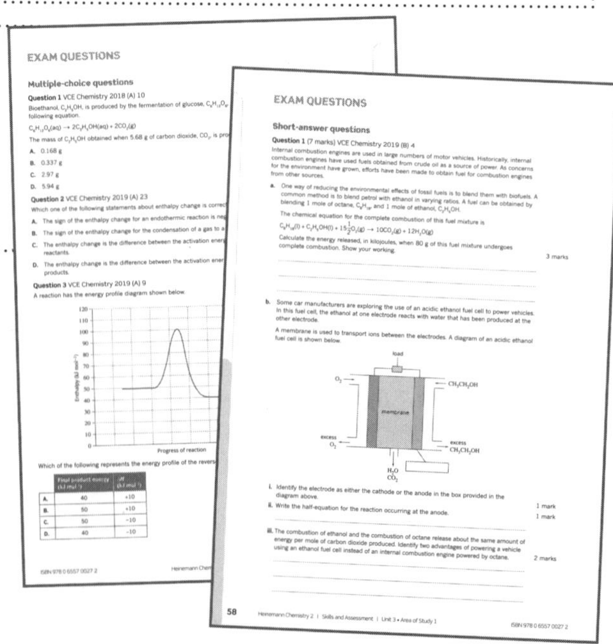
EXAM QUESTIONS

Multiple-choice questions

EXAM QUESTIONS

Short-answer questions

## TEACHER SUPPORT

Comprehensive answers and fully worked solutions for all worksheets, practical activities and exam questions are provided via the *Heinemann Chemistry 2 eBook + Assessment* or Pearson Places. In-depth support for Unit 4 Area of Study 3 in the form of samples, templates and teacher notes is also included, along with an interactive SPARKlab for every practical activity.

 ISBN 978 0 6557 0027 2

# Your one-stop shop for VCE Chemistry success

**Heinemann Chemistry 6th edition, written to the new VCE Chemistry Study Design 2023–2027, offers a seamless experience for teachers and learners.**

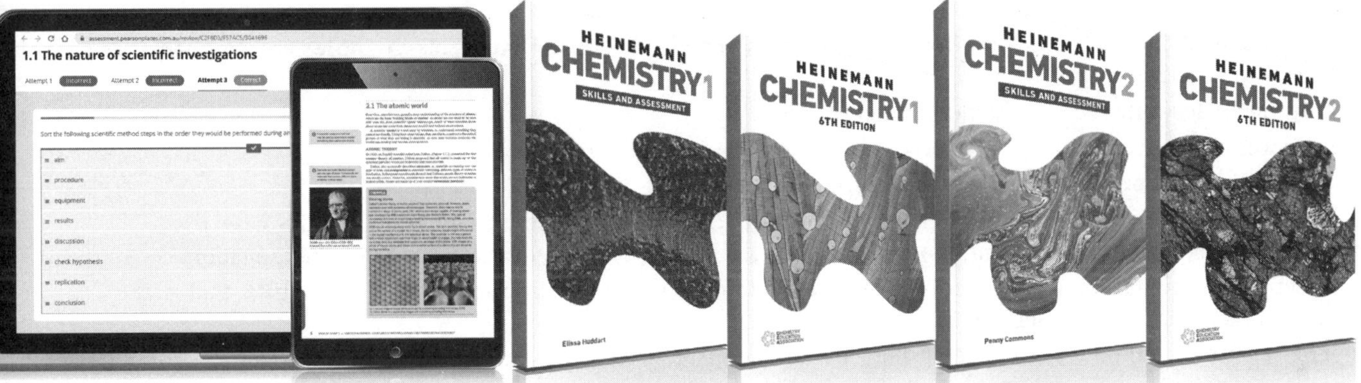


## Enhanced eBook with assessment

Save time by accessing content and assessment in one simple-to-use platform.

### Teacher access

- digital assessment tool for informing planning, assigning work, tracking progress and identifying gaps
- comprehensive sample teaching plans
- suite of practice exam materials
- rich support for Area of Study 3
- solutions and support for the skills and assessment book

### Learner access

- digital assessment questions with scaffolded hints, auto-correction and instant feedback
- student book fully worked solutions
- videos by VolkScience
- SPARKlab versions of the skills and assessment book practical activities, developed in collaboration with PASCO

## Student book

The **student book** addresses the latest developments and applications in chemistry. It also incorporates best-practice literacy and learning design to ensure the content and concepts are fully accessible to all learners.

Key features include:

- case studies with real-world data and analysis questions
- a smooth progression from low to high order questions in section, chapter and area of study reviews.

## Skills and assessment book

The **skills and assessment book** gives students the edge in applying key science skills and preparing for all forms of assessment.

Key features include:

- a skills toolkit
- key knowledge study notes
- worksheets
- practical activities
- VCAA exam and exam-style questions for each area of study
- sample Area of Study 3 investigations.

Find out more at **pearson.com.au/vce-science**

# Chemistry toolkit

**This toolkit provides support for developing the skills required to undertake scientific investigations. It also covers study skills and examination preparation. The toolkit can serve as a reference tool that you can consult as needed throughout the year.**

## Scientific method

Chemists make observations and construct hypotheses to account for their observations. A **hypothesis** is a possible explanation, an educated guess, made to explain observations.

Hypotheses are tested by following the principles of the **scientific method**. These include:

- asking relevant questions; that is, questions that can be tested
- making careful observations
- designing and conducting **controlled experiments**; in controlled experiments all **variables** are kept constant, except the one under investigation
- keeping an accurate record of experimental results
- interpreting experimental data and observations in a logical manner
- drawing logical **conclusions** from the experimental results.

The results of a scientific investigation may negate, or refute, the hypothesis being tested. In this case, the hypothesis must be re-evaluated and modified. Such results are useful in redirecting scientific investigation.

When experimental results repeatedly support a hypothesis, it may become a **theory** or **principle**; that is, the hypothesis is accepted as a scientific truth.

The scientific method recognises that there are limitations in investigations. For example, some factors cannot be measured, a sample size may be too small to be representative, or unknown factors may influence investigations.

### YOUR SCIENTIFIC INVESTIGATION

A scientific investigation can be conducted in many ways. Examples of scientific investigation approaches (also known as methodologies) are controlled experiments, case studies and modelling. The scientific investigation **methodology** and the **methods** (also known as procedures) selected will depend on the aim of the investigation and the research question.

Practical activities involve direct experiences or hands-on activities. The methodology for a scientific investigation involves all elements of planning, and it considers the focus of the investigation and the rationale for approaching investigations in a particular way; for example, through controlled experiments, case studies or modelling. See the checklist of various elements in a scientific investigation on the following page for further information about methodology versus method in scientific investigations.

An investigation that you conduct yourself is known as a primary investigation, and the data and information you collect are called **primary data** or a primary source. An investigation in which you analyse data collected by others is known as a secondary-sourced investigation.

When planning and conducting a scientific investigation, you must maintain a logbook to record information related to your investigation, such as materials and methods, raw data, data analysis and sources of information, including instructions, conversations relating to the investigation.

The findings of a scientific investigation may be presented in a variety of formats, such as a scientific report, an article or a scientific poster. For Unit 4 Area of Study 3, VCAA requires you to present the results of a scientific investigation in the form of a scientific poster (see pages xii and xiii).


ISBN 978 0 6557 0027 2

## Conducting a scientific investigation

Scientific investigations follow a scientific method. The following checklist provides a summary of the elements that are common to many scientific investigations. Refer to the checklist as you conduct your scientific investigation.

| Checklist of various elements in a scientific investigation | | |
|---|---|---|
| **Element** | **Explanation** | **Tick ✔** |
| date | • the date the scientific investigation was conducted, and the report was written | |
| title/heading | • a statement of what is being investigated | |
| aim | • a statement that describes what the scientist wants to investigate, demonstrate or find out<br>• can be written as 'to investigate the effect of *x* on *y*' | |
| hypothesis | • a tentative explanation for an observation that is based on evidence and prior knowledge<br>• must be testable<br>• written as a short statement<br>• should include variables to be tested and a statement of the measurable, predicted outcome<br>• can be written as 'If *x* happens, then *y* will happen'. The 'if' part relates to the independent variable; the 'then' part relates to the dependent variable. | |
| variables | • experiments measure relationships between variables<br>• the independent variable is the variable that is changed<br>• the dependent variable is the variable responding to the change in the independent variable<br>• the controlled variables are kept constant throughout the experiment | |
| risk assessment/ safety | • an analysis of the potential risks of conducting the investigation<br>• the objective is to maximise safety by managing and minimising risk (the risk assessment form on page x guides you through the analysis of risk) | |
| materials | • a list of all equipment, chemicals and materials used<br>• includes quantity, volume, concentration and mass of materials used | |
| methodology | • a description of the overall approach undertaken in the scientific investigation<br>• considers the investigation broadly, and includes the reasons for taking the chosen approach; for example, it identifies and describes strategies, such as the method used to obtain data and the reasons why this is important to achieve the aim of your scientific investigation and address the question under investigation | |
| method | • the specific procedure or steps taken to collect data during a scientific investigation:<br>– includes a detailed, numbered, step-by-step list of exactly what is to be done<br>– includes details about the materials used at each step (e.g. volume and concentration of a chemical)<br>– may include scientific drawings or labelled diagrams<br>– includes enough detail to allow the experiment to be repeated exactly<br>– considers steps that can be taken to reduce errors | |
| results | • a record of all observations and measurements taken during the investigation<br>• observations may be recorded as text, diagrams, photos or videos<br>• data is commonly presented in tables and graphs and can include calculations<br>• data should be checked for errors | |
| discussion | • interprets data, links results to chemical concepts<br>• includes analysis of the data, looking for trends and relationships<br>• identifies problems such as bias, inaccuracy, unreliable data and errors<br>• evaluates the method used, including a discussion of limitations<br>• comments on whether the results support or do not support the hypothesis<br>• comments on the implications of the investigation and further investigations that may be undertaken | |
| conclusion | • a short statement based on evidence that summarises the findings of the investigation<br>• provides a response to the research question or hypothesis<br>• identifies the extent to which the investigation addressed the research question or hypothesis<br>• does not introduce any new information | |

## LOGBOOK

Your teacher will talk with you about how to set up and use your logbook. Scientists use logbooks constantly and your logbook is where you will write everything about the investigation (including both negative and positive results), dating each entry. It is your evidence of the work you have done and enables your investigation to be authenticated and audited if necessary.

## RISK ASSESSMENT

Safety should be considered before conducting a scientific investigation. The school and your teacher are responsible for reducing most risks. As a student scientist, you can take measures to reduce the risks by following your teacher's instructions, wearing a lab coat, safety glasses and closed shoes, tying back loose hair and clothing, and following the safety protocols for handling the chemicals and equipment you use.

To determine the safety protocols involved in an investigation, you should complete a risk assessment beforehand. This will determine, assess and provide ways to manage the risks. The following risk assessment form can be copied into your logbook to help you complete your risk assessment.

| Risk assessment form | | |
|---|---|---|
| **List ...** | **Identify any risks** | **State how you will ...** |
| equipment you will be using: | | use each piece: |
| safety data sheets (SDSs) required for each chemical (and its concentration) you will be using: | | use each chemical:<br><br>dispose of each chemical safely: |
| ethical issues you need to consider: | | honestly and accurately record and report your data and any secondary data from a literature review you use: |
| any other possible risks: | | reduce these risks: |

## STUDENT-DESIGNED INVESTIGATION

You will be required to design and conduct a scientific investigation based on the concepts you have learnt in Unit 3 or Unit 4, or across both Units 3 and 4. This assessment task gives you the opportunity to apply key science skills and to pursue an area of interest to you, within the directions provided by your teacher, and based on the key knowledge addressed during the course.

You will be required to develop a question that drives your investigation and then conduct the investigation, completing the elements listed in the table on page ix and recording all information in your logbook. You will then present your investigation as a scientific poster.

### Determining your research question

As you develop the question, consider the resources available to you in the school lab and your ability to perform the investigation. Your research question should allow you to:

- analyse the topic
- consider the relationship between the dependent and independent variables
- record data accurately and precisely for the dependent variable
- keep other variables constant.

### Sourcing information

The resources you refer to in investigations may be primary and/or secondary.

**Primary sources** of information are created by a person directly involved in an investigation or study. Examples of primary sources are results from experiments (such as raw data or photographs), reports of scientific investigations, and peer-reviewed scientific articles reporting the results of an investigation.

**Secondary sources** of information are summaries, reviews or interpretations of primary sources. Secondary sources include textbooks, biographies, documentaries, newspaper articles and websites.

Refer to the following two checklists. The first shows features to look for when assessing and selecting the best sources of information. The second shows how to set out the information that is required for the references section of your report in American Psychological Association (APA) style. This is only one of several referencing systems that you might be required to use in your studies.

 ISBN 978 0 6557 0027 2

| Selecting resources for the investigation | Tick ✔ |
|---|---|
| The resource is: | |
| • credible and I can identify the author, author's expertise and publisher | |
| • current because the date of publication of the material is provided and is recent | |
| • factual and I know that it is objective material and not biased | |
| • accurate and all information is correct | |
| • relevant and covers the area I am investigating | |
| • readable and neither too simple nor too complex in its coverage of the material. | |

| Examples of information required for references and bibliographies (APA style) |
|---|
| **Article in scientific magazine**<br>Author, initials. (year). Title of article. *Journal title, volume number* (issue number), page numbers. Digital object identifier (DOI) or URL<br>Lee, M. L. (2017). Materials science: Crystals aligned through graphene. *Nature 544*(7650), 301–302. |
| **Book**<br>Author, initials. (year). *Title of book* (edition, if not first). Publisher.<br>Chan, D., Commons, C., Commons, P., Derry, L., Freer, E., Huddart, E., Lennard, L., MacEoin, M., Moylan, M., O'Shea, P., Ross, R., Vanderkruk, K. & Waldron, P. (2023). *Heinemann Chemistry 2* (6th ed.). Pearson Australia.<br>Include names of all authors or editors up to 20. Special rules apply for 21 or more authors. You can find information about the rules online. |
| **Internet**<br>Author, initials/name of organisation. (year). Title of webpage or web document. URL.<br>Royal Society of Chemistry. (2017). *Periodic table.* http://www.rsc.org/periodic-table |

Always remember to record the above details for each resource you use correctly in your logbook and date the entry. It is important to accurately cite resources in your reference section as you use them, because it can be time-consuming and difficult to find this information later.

## NOTE-TAKING AND ORGANISING NOTES

Note-taking and organising requires skill. Good note-taking helps you avoid plagiarism and provides excellent information to support your scientific report writing. Plagiarism is when you take someone else's ideas and words and present them as your own work. You plagiarise if you copy sections or sentences from sources or you cut and paste from the internet. It is acceptable to use the ideas of others but you must state clearly where the information has come from in your reference section.

The following table provides examples of original text, plagiarised text and acceptably rephrased text.

| Original text | Plagiarised text | Rephrased text |
|---|---|---|
| Critical elements are elements that are heavily relied on for industry and society in areas such as energy, electronics and food production. The demand for critical elements is so high that current methods of obtaining or recovering the elements are often unsustainable. Future development of recycling methods will be crucial to allow their continued use. | Critical elements are heavily relied on for energy, electronics, food production and other areas of society. The high demand for critical elements means current methods of obtaining the elements are often unsustainable. Continued use will require future development of recycling methods. | Critical elements are important in society in many areas including production of energy, electronics and food. The way critical elements are currently obtained from mining or through recycling is not enough to supply current demand. Ways of improving recycling of critical elements is essential in the future. |

There are various approaches to effective note-taking. Whatever technique you use, try to keep your notes brief and focused on key points. Examples include:

- dot point summaries
- underlining or highlighting text
- labelled diagrams
- flow charts—to show sequences
- concept maps—to show connections between ideas
- Venn diagrams—to show similarities and differences
- tables—useful for summarising longer and more complex information that has subparts and can incorporate any of the other note-taking techniques.

## Scientific writing

Scientists have a particular writing style. You should use this distinctive style to communicate your ideas. Scientific writing is:

- objective—it describes events rather than what people think or feel and is as free as possible of bias or personal opinion
- precise—it avoids exaggeration and uses qualified language
- formal—it uses scholarly language rather than colloquial or everyday language
- concise—it conveys information in short, clearly understandable sentences without unnecessary information
- simple—it uses short sentences where possible
- predominantly written in passive voice, although sometimes active voice can be used to avoid confusion
- structured to include headings, tables, diagrams and mathematical calculations.

Examples of unscientific and scientific writing are demonstrated in the following table.

| Unscientific writing | Scientific writing |
|---|---|
| Subjective, biased:<br>• The results were fantastic.<br>• This produced a disgusting odour.<br>• The breathtakingly beautiful bowerbird ... | Objective, unbiased:<br>• The results showed ...<br>• This produced a pungent odour.<br>• The golden bowerbird ... |
| Exaggerated:<br>• The object weighed a huge amount.<br>• The magnesium burst into huge flames.<br>• Millions of ants swarmed over ... | Accurate, precise:<br>• The mass of the object was 250 kg.<br>• The magnesium burned vigorously.<br>• Ants swarmed over ... |
| Everyday, informal language:<br>• The bacteria passed away.<br>• The results don't ...<br>• We guessed that ...<br>• Previous researchers were slack and missed ... | Formal language:<br>• The bacteria died.<br>• The results do not ...<br>• It was hypothesised that ...<br>• Previous researchers have not found ... |
| Active voice:<br>• We recorded oxygen levels every hour.<br>• We put 50 g of solute in a conical flask containing distilled water, and then we slowly added 1 mol $L^{-1}$ hydrochloric acid. | Passive voice:<br>• The oxygen level was recorded hourly.<br>• 50 g of solute was placed in a conical flask containing distilled water, and then 1 mol $L^{-1}$ of hydrochloric acid was slowly added. |

## Data quality

Consider the quality of the primary and secondary data you collect. In the discussion section of your scientific investigation report, use terms such as **accuracy**, **precision**, **repeatability**, **reproducibility**, **resolution**, **measurement result**, and **true value** to support your analysis of the quality of your results; any possible improvements and modifications you might make when conducting a future similar investigation; and the **validity** of the experimental method and results.

## Types of errors

Mistakes: **Mistakes** are avoidable errors that may occur in calculations, or while performing the steps in an experiment or reading an instrument. They are not considered in the error analysis, as it is assumed the researcher is competent. If a mistake occurs, the experiment should be repeated correctly.

Systematic: A **systematic error** is reproducible and different from the true value by a fixed amount in the same direction. Systematic errors can often be corrected by calibrating instruments.

Random: **Random errors** are errors that follow no regular variations in the measurement and result in a range of readings. They cannot be eliminated but can be reduced by taking several measurements and averaging them.

## PRESENTING YOUR INVESTIGATION AS A POSTER

Scientific findings may be presented in a variety of ways. A common presentation format at science conferences is a poster. Posters can get ideas across to a large audience in an organised, concise and creative way. Other common presentation formats are essays, reports, oral presentations and articles. Each presentation format has its own conventions. A poster, such as you will use for your Unit 4 Area of Study 3 scientific investigation report, is characterised by:

- a balance of text and visuals
- a title and subheadings
- balanced layout
- captions for figures and tables
- references
- a hierarchy of font size according to subheading level
- consistent font style—no more than three fonts.


ISBN 978 0 6557 0027 2

The poster format provided by VCAA for your scientific investigation report is shown below.

<table>
<tr><td colspan="3">Title<br>Student name</td></tr>
<tr><td>Introduction<br><br>Methodology and methods<br><br>Results</td><td rowspan="1">Communication statement reporting the key finding of the investigation as a one-sentence summary</td><td>Discussion<br><br>Conclusion</td></tr>
<tr><td colspan="3">References and acknowledgements</td></tr>
</table>

VCE Chemistry Study Design extracts © VCAA (2022); reproduced by permission.

Your teacher can provide you with poster and logbook samples and templates and other support for Unit 4 Area of Study 3 to guide you through your investigation.

## Proofreading

After you have completed the investigation and prepared your presentation, it is important to proofread and check diagrams and results. Proofreading your work enables you to minimise errors and maximise effective communication of your investigation. Use the following questions as a proofreading checklist.

| Proofreading checklist | Tick ✔ |
| --- | --- |
| Have I: | |
| • investigated the question fully? | |
| • expressed myself clearly to communicate my ideas well? | |
| • used scientific writing style? | |
| • included data analysis? | |
| • checked spelling, punctuation and grammar? | |
| • included references? | |
| • met the requirements of the presentation format? | |

# Study skills

There are a variety of techniques and strategies you can use to help you study. You may find that you use different strategies in different situations. For example, you may prefer to highlight key phrases in your notebook throughout a month and make summaries of the topics at the end of the month in preparation for an examination. The strategies you choose will depend on personal preference and may not be the same as those used by your classmates.

Effective study skills involve more than the learning strategies you use. Equally important is when you use those skills. It is more effective to apply study skills throughout the year, revising and consolidating your knowledge as you progress through the course, rather than doing a rushed cram just before the examination. Revise your work regularly. Being organised and setting up a study plan is key to reducing your stress.

## GETTING ORGANISED FROM THE BEGINNING OF THE YEAR

To get yourself organised, try the following steps.

- Use a diary to write down all homework and assessment tasks as soon as you get them. Note due dates and what is required.
- Be specific about the tasks you need to do. Rather than writing 'do Chemistry', it is more effective to note things such as which questions to answer, which page to read in your student book, and what topic you plan to summarise.
- Write a list of everything you need to do each day. Tick off or cross out items as you complete them.
- Break down large tasks into smaller separate parts that are manageable.
- Make sure your lists and planners are realistic. Do not set yourself more than you can actually do.
- Plan to have one set of well-expressed notes for your study revision before the exams. You will gradually develop these notes over the entire year.

### Study techniques

Studying requires concentration. Remove any distractions and plan a 10-minute break every hour. Vary your study technique depending on the content to be learnt and your personal preference. Although you may have already found a study technique that works for you, also consider the following options.

| Study technique | Tips |
|---|---|
| **Highlighting**<br>Standard solution<br>A solution with an accurately known concentration. This illustration shows a standard solution being made from a primary standard. | • Highlight or underline key points as you read your notes or text. |
| **Summary notes**<br>*IONIC COMPOUNDS*<br>*Common properties of ionic compounds:*<br>• *brittle, which means they shatter when hit with a hammer*<br>• *hard, which means they are resistant to scratching*<br>• *high melting point*<br>• *unable to conduct electricity in solid state*<br>• *good electrical conductor in liquid or aqueous states* | • Create a list of key headings and add some dot points about each heading.<br>• Write your own summary of the key ideas in each chapter.<br>• Use headings and subheadings.<br>• Underline key words and key phrases.<br>• Use simple diagrams.<br>• Remember, the most effective chapter summaries are clear, concise and uncluttered. |
| **Diagrams**<br>lowest energy orbit where electron is normally found<br>higher energy orbits | • Diagrams can be used as a summary of key concepts.<br>• Diagrams are useful memory triggers.<br>• Diagrams present a lot of information in a visual way, with minimal text. |
| **Concept maps and flow charts**<br>types of chemical bonds<br>ionic bond<br>covalent bond<br>hydrogen bond<br>metallic bond<br>polar covalent bond<br>non-polar covalent bond | • Concept maps are a great way of connecting key terms and ideas in a simple and ordered way.<br>• Use the lines that connect ideas to note the relationships between the words and phrases.<br>• They may include text and images.<br>• They may be a simple or more complex summary tool.<br>• Concept maps and other graphic organisers such as Venn diagrams and flow charts show how information is connected and help deepen understanding. |


ISBN 978 0 6557 0027 2

<table>
<tr><th>Study technique</th><th>Tips</th></tr>
<tr><td>Tables<br>
<table>
<tr><th colspan="4">Subatomic particles</th></tr>
<tr><td>Particle</td><td>Relative mass</td><td>Charge</td><td>Location</td></tr>
<tr><td>neutron</td><td>1</td><td>neutral</td><td>nucleus</td></tr>
<tr><td>proton</td><td>1</td><td>positive</td><td>nucleus</td></tr>
<tr><td>electron</td><td>$\frac{1}{1800}$</td><td>negative</td><td>orbiting nucleus</td></tr>
</table></td>
<td>• Tables are useful to show relationships between different factors.<br>• Information is uncluttered.</td></tr>
<tr><td>Mnemonic devices<br>AN OILRIG CAT<br>At the ANode, Oxidation Is Loss of electrons and Reduction Is Gain of electrons at the CAThode.</td>
<td>• Mnemonic tools help you remember information, e.g. the essential concepts in galvanic cells.</td></tr>
<tr><td>Glossary<br>Electron—negatively charged particle in an atom<br>Functional group—atom or group of atoms attached to a molecule whose presence determines the properties of the molecule</td>
<td>• Compile your own glossary—writing the terms and definitions will make them easier to remember.<br>• Add images to help you remember.<br>• Write each term in a sentence.<br>• Memorise these terms and definitions.</td></tr>
<tr><td>Trigger words<br>chiral centre—4 different groups<br>mirror image is not superimposable<br>enantiomers</td>
<td>• Write down the key words associated with a topic or theme.<br>• Trigger words are useful in helping to remember other related words and ideas.</td></tr>
<tr><td>Repeating information aloud</td>
<td>• Repetition is a good way to remember information.<br>• Helps you to slow down and absorb the information.</td></tr>
<tr><td>Practising</td>
<td>• Complete as many review questions and past exams as possible.</td></tr>
<tr><td>Flash cards<br>What is the Schrödinger model?<br>Identify the states of matter and changes between them.</td>
<td>• Making flash cards helps your understanding.<br>• Make your own cards.<br>• Write a question on one side and the answer on the back.<br>• Cards can include definitions, brief explanations, diagrams, equations and graphs.<br>• Be aware of the time involved.</td></tr>
<tr><td>Teaching someone</td>
<td>• Teach friends or family members.<br>• Teaching a difficult concept to someone means you must first understand the concept yourself.</td></tr>
<tr><td>Handwriting notes</td>
<td>• Handwrite rather than type summary notes for a small part of a topic.<br>• Remember, the examination requires you to write answers.<br>• Be aware—this is time-consuming and should not be the way you plan to produce all your summary notes.</td></tr>
<tr><td>Responding to feedback and self-correcting</td>
<td>• Check through all feedback from your teacher.<br>• Highlight what was right or wrong.<br>• Attempt to identify where you have made errors and rework the answer to get it right.<br>• Find the misunderstood concept in your summary notes and highlight or annotate it to help you remember.</td></tr>
</table>

## EXAMINATION PREPARATION

In the weeks before the examination, begin your exam preparation. The earlier you begin revising, the easier it will be. It is also helpful to begin practising exam-style questions throughout the year.

Like most skills, your ability to do exams and to handle different types of exam questions will improve with practice. Doing practice exams is vital because you gain experience in:

- using reading time effectively
- allocating the right amount of time to each question
- working to a time limit
- reading and interpreting questions
- understanding what is required for each question
- planning answers to use the language of chemistry effectively and concisely
- deciding on relevant information
- proofreading/checking your answers
- writing efficiently for the duration of the exam.

Use the following checklist as a reminder of your study program.

| Study program checklist | Tick ✔ |
|---|---|
| Have I: | |
| • revised all areas of the course? | |
| • highlighted important points? | |
| • made a summary of the important points in each topic? | |
| • read over my revision notes? | |
| • looked at and worked through practice exam papers? | |
| • answered practice exam questions within the appropriate time limit? | |

## Getting ready for the exam

Once you have finished revising and feel confident with the material, use the following checklist to make sure you are prepared before the exam day arrives.

| The exam itself | Tick ✔ |
|---|---|
| Have you: | |
| • checked the exact exam requirements? | |
| • obtained and carefully read the instruction cover sheet from previous exams? (Check that no changes are expected for your exam.) | |
| • planned how to best use your reading time? | |
| **Practical steps** | **Tick ✔** |
| Have you: | |
| • assembled all materials required and allowed? This includes: | |
| – black pens to write your answers | |
| – pencils of the correct lead for the multiple-choice answer sheet | |
| – sharpeners, erasers, rulers and other permissible materials such as scientific calculators and spare batteries | |
| – a water bottle (check that your bottle meets the requirements—it may need to be clear with no label) | |
| – a watch to keep track of the time (you are not permitted to look at mobile phones) | |
| • double-checked dates, times and locations for each exam? | |
| • remembered your student number? | |
| • packed a form of photo ID (if required)? | |
| • arranged transport, allowing ample time to avoid rushing and to manage possible delays? (Be sure to arrive early to settle your nerves and gather your thoughts.) | |
| • planned for appropriate meals and drinks because exam times vary? (Avoid heavy meals before exams because they make you sleepy.) | |


ISBN 978 0 6557 0027 2

## Examination strategies

Familiarise yourself with the conditions of the examination well before the day you sit the exam. You should know:

- the amount of reading time allowed in the exam before writing begins
- the amount of writing time allocated
- any particular equipment allowed and/or required, such as a calculator, pencils, pens and ruler
- strategies to tackle the exam.

  Exam strategies are listed in the following table.

| Exam strategies |
| --- |
| **Reading time**<br>• You will be given reading time at the beginning of the exam (usually 15 minutes).<br>• Remember that no writing at all is allowed during this time—no note-taking, highlighting or underlining.<br>• Read the instructions.<br>• Read through the short-answer questions first—this will give you an overall sense of the themes of questions that require written responses.<br>• Read the multiple-choice questions next.<br>• Decide the order in which you will answer the short-answer questions. Start with what you consider to be the easiest question to build confidence. |
| **Writing time**<br>• You will be given 2.5 hours of writing time.<br>• Teachers often suggest you begin with the multiple-choice questions.<br>• Answer every multiple-choice question in order, even if you can only make an educated guess.<br>• If you are unsure of an answer to a multiple-choice question, mark it so you can come back to it if time allows.<br>• Attempt the short-answer questions next.<br>• Attempt the easiest short-answer questions first and work your way to the more challenging questions. |
| **Tips for answering questions**<br>• Carefully read each question, underlining key words.<br>• Be aware that most questions are structured so they become more challenging towards the end. You may not be able to answer the last part of a question, but you will have earned most of the marks by answering earlier parts.<br>• Look carefully at any diagrams, pictures, tables and graphs and make sure you understand their relevance to the questions involved.<br>• For questions with graphs, read the graph title and the labels (units) on the axes carefully so that you can establish the relationship the graph is showing.<br>• For questions with tables, read the headings on the columns and rows carefully so that you can analyse the content of the table effectively.<br>• Check for key words in a question. Highlight them but do not colour the whole question.<br>• Use correct spelling—it can mean the difference between scoring a point and not scoring a point.<br>• Use correct significant figures in calculated answers and include units.<br>• When writing chemical equations, include states of matter.<br>• Plan your answers before you write, remembering to address the exam criteria. Be sure to write down key words and phrases; perfect sentences are not so important.<br>• For questions with parts, read the whole question first. This gives you an overall picture of the question. It will also help to ensure that you do not repeat yourself in subsequent parts of the question.<br>• Make sure you actually answer the question that is asked.<br>• Once you have answered the question, re-read your answer and then re-read the question to ensure that you have actually answered all of the question.<br>• When writing a definition, avoid using the word you are defining in your definition.<br>• If giving values from a graph, use a ruler to line up points with the axes so you can be accurate, and always include units in your answer.<br>• Be sure to attempt all questions.<br>• Read over your answers to pick up careless errors—the mind is faster than the hand and you may not always write what you intend (especially when you have limited time).<br>• Write legibly. Exams are scanned and marked online. If the assessor can't read your answer, they can't mark it as correct.<br>• Keep an eye on the time.<br>• Never leave an exam early. Use any spare time to re-read and check your answers. |
| **Exam cues**<br>• The number of marks allocated to a question provides a clue about how much you are expected to write. Two marks usually means you need to make a minimum of two key points.<br>• The number of lines allowed for the answer indicates the length of the expected answer. If your writing is large, you may need to turn the page over and continue on the back. Make sure you indicate to the examiner that they must turn to the back of the page for the rest of the answer. Where the answer continues, clearly state that it is the continuation of the question and state the question number. |

## Handling exam questions

The exam is divided into two sections: Section I consists of multiple-choice questions and Section II contains a series of short-answer questions.

| Multiple-choice questions | Short-answer questions |
|---|---|
| • In many cases, you can use your basic knowledge of the subject to cancel out one or two options.<br>• Mark an answer to every question on your answer sheet as you go through them. This avoids leaving any questions unanswered if you run out of time, and avoids getting out of order by leaving blank rows. You can always go back and change the answers to questions you were unsure about.<br>• Circle the letter of your answer to the question on the exam paper as well, so you can easily check if you get out of sync on the answer sheet.<br>• If you are finding a question very difficult, don't spend too much time on it as it is only worth 1 mark. Guess an answer if necessary, then mark the question on your paper and come back to it at the end if you have time.<br>• If you have time at the end of the exam, redo the questions with your original answers covered and compare. | • Be careful to do what the question asks. For example, do you need to identify, describe, explain or compare? Each requires a different response.<br>• When in doubt, point it out, i.e. give a thorough answer. Dot (bullet) points and labelled diagrams are acceptable as long as you fully address the question.<br>• Tailor your answer specifically to the question being asked. Don't rely on a general answer when given a specific scenario. For example, indicate in which direction the equilibrium shifted, don't just say it shifted.<br>• Look at the marks allocated for an indication of the level of detail required. Generally, 1 mark = 1 key point, 2 marks = 2 key points etc.<br>• Set out your answers clearly, stating the formulas you intend to use because this may earn marks, e.g. $n = \frac{m}{M}$.<br>• Show your working because it is easier for you to check, and you often gain part marks and consequential marks if the marker can see where you made an error.<br>• Read over and use any values, such as relative atomic masses, listed in the data booklet given to you with the exam paper.<br>• Take care to include units in your answers if they are required.<br>• Use an appropriate number of significant figures in your results. It is important to record data to the number of significant figures available from the equipment or observation. Using either a greater or a smaller number of significant figures can be misleading and can lose you marks. |

Every year, examiners comment that many students fail to answer the question: that is, students' answers are 'not relevant' or are 'off the point'. Giving a relevant answer to a question is vital and depends on:

- choosing appropriate content
- using content purposefully to do what is asked (see the key action words list below).

| Key action words | |
|---|---|
| analyse | break down into key points; show the essence of; inquire into; identify components and the relationship between them; draw out and relate implications |
| compare | show points of similarity and difference and their significance |
| contrast | show points of difference |
| describe | give the features/characteristics of |
| design | provide steps for an experiment or procedure |
| evaluate | make a judgement based on criteria; determine the value of |
| explain | provide details to give the reader an understanding of; relate cause and effect; make the relationships between things evident; provide why and/or how |
| justify | demonstrate the correctness of (by giving evidence); give evidence in support of an argument or conclusion |
| suggest | put forward ideas, proposals or recommendations |

## Sample high-level and low-level exam responses

Each year, the examiners write an assessment report on the last exam. Reading these reports is a good way to understand what the examiners expected in the answers. The reports show where students showed strong responses and where improvement could be found.

Looking closely at sample responses will help you identify what is required to gain full marks. Two sample responses are provided below: the first is well done (high-level response), whereas the other has room for improvement (low-level response). The annotations draw your attention to key points to note in each response.

 ISBN 978 0 6557 0027 2

## High-level responses

**Question 1** Carbon disulfide gas ($CS_2$) is used in the manufacture of rayon. $CS_2(g)$ can be made in an endothermic gas-phase equilibrium reaction between sulfur trioxide gas ($SO_3$) and carbon dioxide. Oxygen gas is also produced in the reaction.

**a** Write a balanced chemical equation for the reaction.

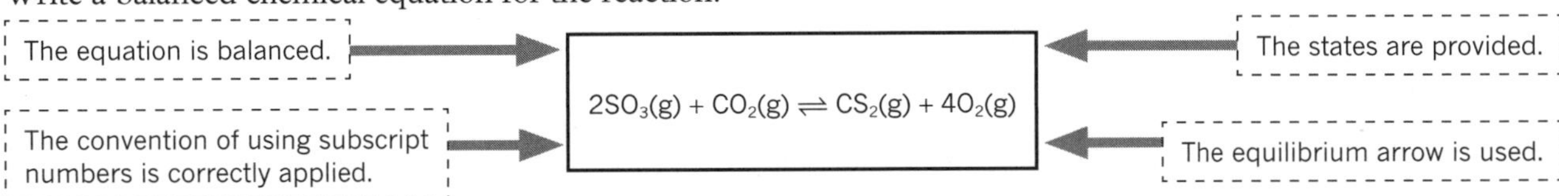


**b** Write an expression for the equilibrium constant of the reaction.

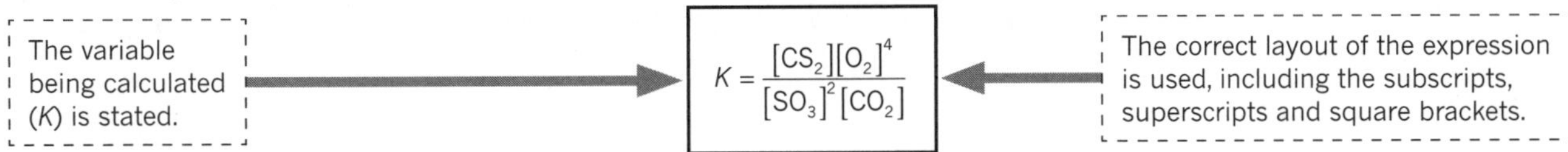


**c** An equilibrium mixture of these gases was made by mixing sulfur trioxide and carbon dioxide. The equilibrium mixture consisted of 0.028 mol of $CS_2$, 0.022 mol of $SO_3$, 0.014 mol of $CO_2$ and an unknown amount of $O_2$ in a 20 L vessel. Calculate the:

**i** amount of $O_2$, in mol, present in the equilibrium mixture

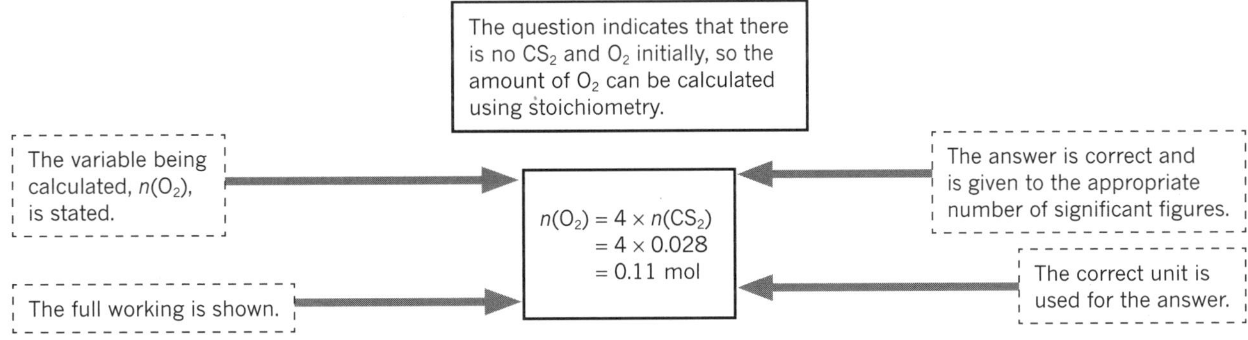


**ii** value of the equilibrium constant at that temperature.

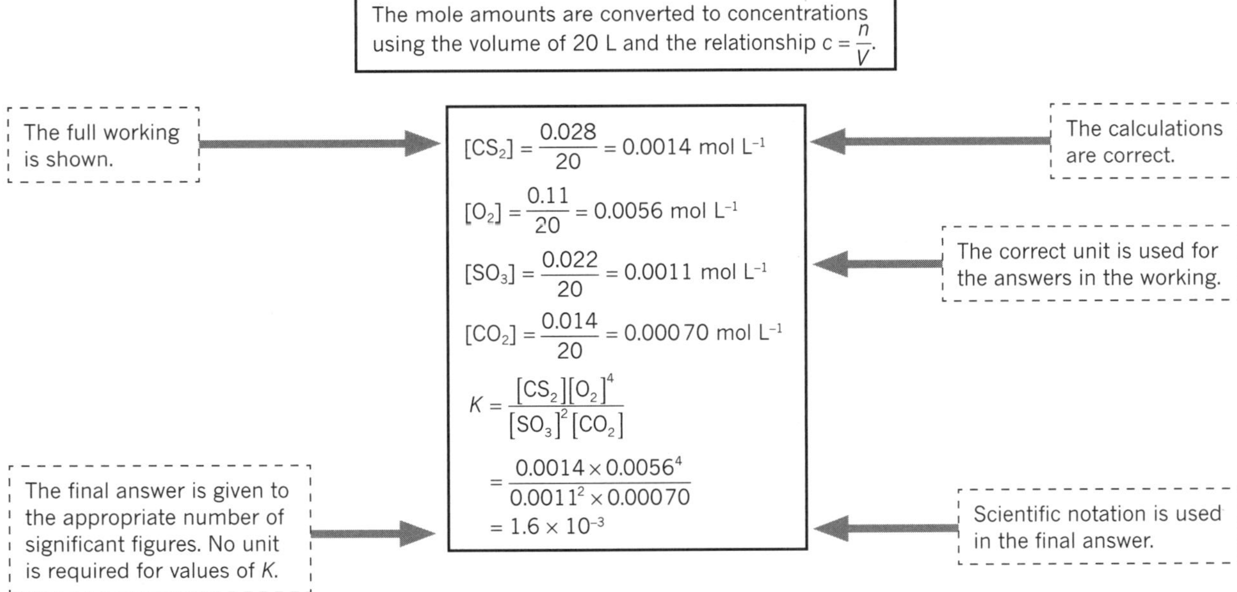

## Low-level responses

**Question 1** Carbon disulfide gas ($CS_2$) is used in the manufacture of rayon. $CS_2(g)$ can be made in an endothermic gas-phase equilibrium reaction between sulfur trioxide gas ($SO_3$) and carbon dioxide. Oxygen gas is also produced in the reaction.

**a** Write a balanced chemical equation for the reaction.

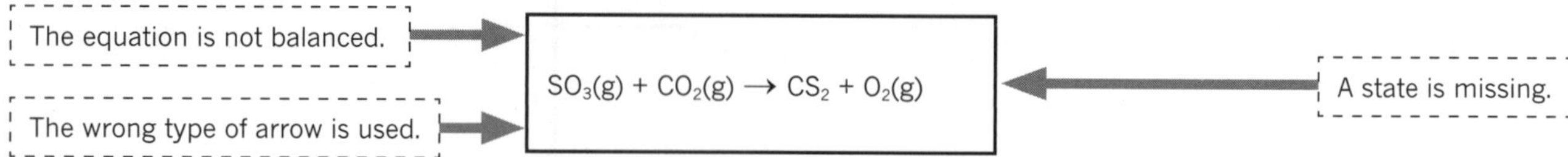


**b** Write an expression for the equilibrium constant of the reaction.

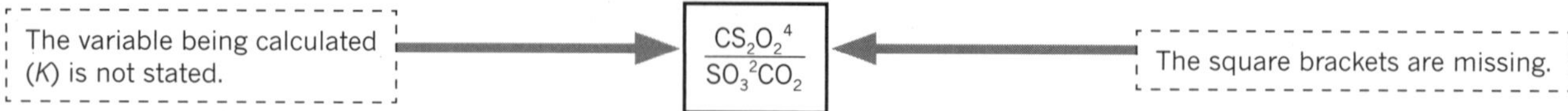


**c** An equilibrium mixture of these gases was made by mixing sulfur trioxide and carbon dioxide. The equilibrium mixture consisted of 0.028 mol of $CS_2$, 0.022 mol of $SO_3$, 0.014 mol of $CO_2$ and an unknown amount of $O_2$ in a 20 L vessel. Calculate the:

**i** amount of $O_2$, in mol, present in the equilibrium mixture

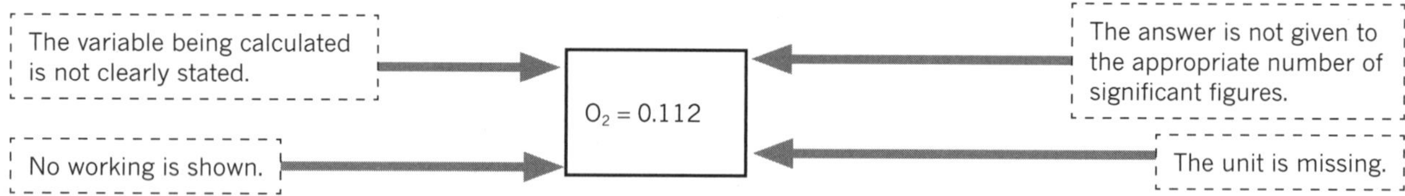


**ii** value of the equilibrium constant at that temperature.

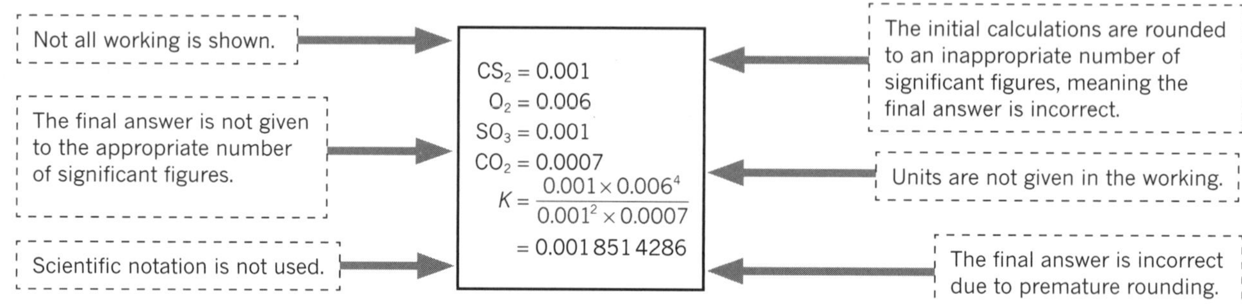


 ISBN 978 0 6557 0027 2

UNIT 3

# How can design and innovation help to optimise chemical processes?

## AREA OF STUDY 1

## What are the current and future options for supplying energy?

### Outcome

On completion of this unit the student should be able to compare fuels quantitatively with reference to combustion products and energy outputs, apply knowledge of the electrochemical series to design, construct and test primary cells and fuel cells, and evaluate the sustainability of electrochemical cells in producing energy for society.

### Key knowledge

#### Carbon-based fuels

- the definition of a fuel, including the distinction between fossil fuels (coal, natural gas, petrol) and biofuels (biogas, bioethanol, biodiesel) with reference to their renewability (ability of a resource to be replaced by natural processes within a relatively short period of time)
- fuel sources for the body measured in kJ $g^{-1}$: carbohydrates, proteins and lipids (fats and oils)
- photosynthesis as the process that converts light energy into chemical energy and as a source of glucose and oxygen for respiration in living things:
  $$6CO_2(g) + 6H_2O(l) \rightarrow C_6H_{12}O_6(aq) + 6O_2(g)$$
- oxidation of glucose as the primary carbohydrate energy source, including the balanced equation for cellular respiration:
  $$C_6H_{12}O_6(aq) + 6O_2(g) \rightarrow 6CO_2(g) + 6H_2O(l)$$
- production of bioethanol by the fermentation of glucose and subsequent distillation to produce a more sustainable transport fuel:
  $$C_6H_{12}O_6(aq) \rightarrow 2C_2H_5OH(aq) + 2CO_2(g)$$
- comparison of exothermic and endothermic reactions, with reference to bond making and bond breaking, including enthalpy changes ($\Delta H$) measured in kJ, molar enthalpy changes measured in kJ $mol^{-1}$ and enthalpy changes for mixtures measured in kJ $g^{-1}$, and their representations in energy profile diagrams
- determination of limiting reactants or reagents in chemical reactions
- combustion (complete and incomplete) reactions of fuels as exothermic reactions: the writing of balanced thermochemical equations, including states, for the complete and incomplete combustion of organic molecules using experimental data and data tables

#### Measuring changes in chemical reactions

- calculations related to the application of stoichiometry to reactions involving the combustion of fuels, including mass–mass, mass–volume and volume–volume stoichiometry, to determine heat energy released, reactant and product amounts and net volume or mass of major greenhouse gases ($CO_2$, $CH_4$ and $H_2O$), limited to standard laboratory conditions (SLC) at 25°C and 100 kPa
- the use of specific heat capacity of water to approximate the quantity of heat energy released during the combustion of a known mass of fuel and food
- the principles of solution calorimetry, including determination of calibration factor and consideration of the effects of heat loss; analysis of temperature–time graphs obtained from solution calorimetry
- energy from fuels and food:
  - calculation of energy transformation efficiency during combustion as a percentage of chemical energy converted to useful energy

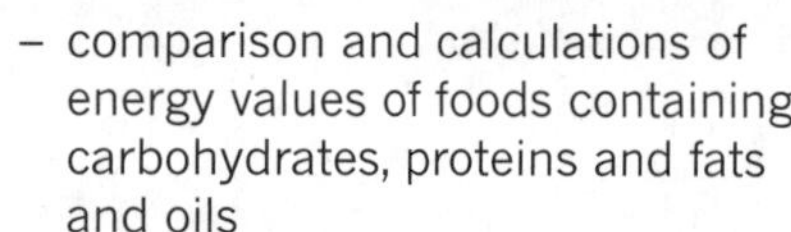

- comparison and calculations of energy values of foods containing carbohydrates, proteins and fats and oils

**Primary galvanic cells and fuel cells as sources of energy**

- redox reactions as simultaneous oxidation and reduction processes, and the use of oxidation numbers to identify the reducing agent, oxidising agent and conjugate redox pairs
- the writing of balanced half-equations (including states) for oxidation and reduction reactions, and the overall redox cell reaction in both acidic and basic conditions
- the common design features and general operating principles of non-rechargeable (primary) galvanic cells converting chemical energy into electrical energy, including electrode polarities and the role of the electrodes (inert and reactive) and electrolyte solutions (details of specific cells not required)
- the use and limitations of the electrochemical series in designing galvanic cells and as a tool for predicting the products of redox reactions, for deducing overall equations from redox half-equations and for determining maximum cell voltage under standard conditions
- the common design features and general operating principles of fuel cells, including the use of porous electrodes for gaseous reactants to increase cell efficiency (details of specific cells not required)
- the application of Faraday's laws and stoichiometry to determine the quantity of galvanic or fuel cell reactant and product, and the current or time required to either use a particular quantity of reactant or produce a particular quantity of product
- contemporary responses to challenges and the role of innovation in the design of fuel cells to meet society's energy needs, with reference to green chemistry principles: design for energy efficiency, and use of renewable feedstocks

VCE Chemistry Study Design extracts © VCAA (2022); reproduced by permission.

# Carbon-based fuels

This topic examines how fuels are used to meet our energy needs and the chemical principles that underpin their use, as well as their individual advantages and disadvantages for particular applications. You will investigate different ways that energy can be produced sustainably while minimising undesirable impacts on the environment. You will also learn about the source of energy in the human body and other living things.

- **To revise Year 11 material, complete Worksheet 1.**

## EXOTHERMIC AND ENDOTHERMIC REACTIONS

Chemical energy is stored in the bonds between atoms, ions and molecules. It results from attractions and repulsions between electrons and protons, and vibration's and rotations of nuclei.

The **chemical energy** of a substance is called its **heat content** or **enthalpy** and is given the symbol *H*. We can measure the change in enthalpy, $\Delta H$, for a chemical reaction:

$$\Delta H = H_{\text{products}} - H_{\text{reactants}}$$

Depending on the enthalpy change that occurs, a reaction is described as **exothermic** or **endothermic** (Table 3.1.1).

**Table 3.1.1** A comparison of exothermic and endothermic reactions

| Exothermic reaction | Endothermic reaction |
|---|---|
| More energy is released than is absorbed. | More energy is absorbed than is released. |
| $\Delta H$ is negative. | $\Delta H$ is positive. |

During a chemical reaction, the bonds of the reactant particles absorb energy and break (Figure 3.1.1 on the following page). The new bonds of the product particles then form, releasing energy. The energy needed for a reaction to occur is called the **activation energy**, as shown in Figure 3.1.2 on the following page.

The combustion of fuels such as wood and petrol are examples of exothermic reactions, which release substantial quantities of energy.

 ISBN 978 0 6557 0027 2

## KEY KNOWLEDGE

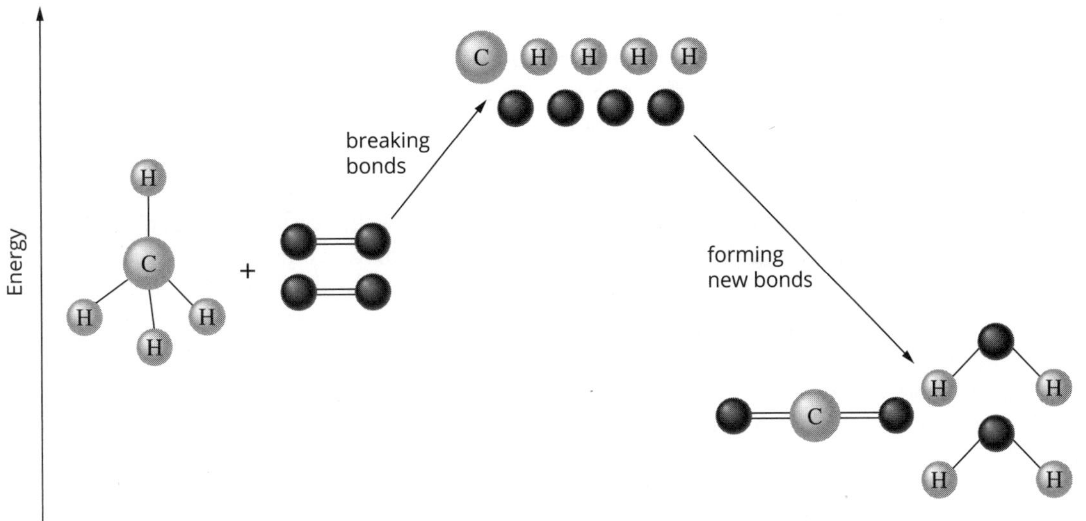


**Figure 3.1.1** Energy changes in the breaking and forming of bonds in the reaction between methane and oxygen

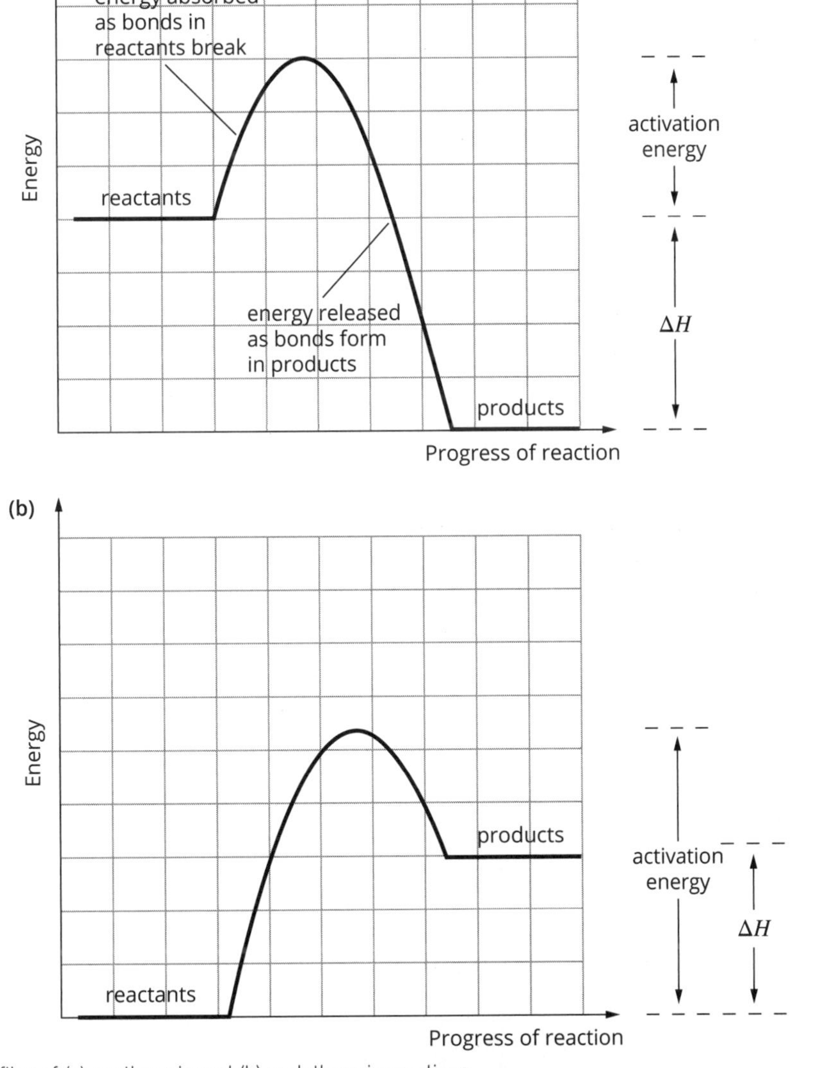


**Figure 3.1.2** Energy profiles of (a) exothermic and (b) endothermic reactions

# KEY KNOWLEDGE

## Thermochemical equations

**Thermochemical equations** are chemical equations that include a $\Delta H$ value, which shows the energy released or absorbed if the number of moles of reactants shown in the equation react completely. For example, when 2 moles of octane ($C_8H_{18}$) are burnt, 10 108 kJ of energy is released. The equation can be written as:

$$2C_8H_{18}(l) + 25O_2(g) \rightarrow 16CO_2(g) + 18H_2O(l) \quad \Delta H = -10\,108 \text{ kJ}$$

If one mole of octane is burnt, only 5054 kJ of energy is released. This indicates that the molar **enthalpy change** for octane is 5054 kJ mol$^{-1}$.

When the equation is reversed, the sign of $\Delta H$ is also reversed:

$$16CO_2(g) + 18H_2O(l) \rightarrow 2C_8H_{18}(l) + 25O_2(g) \quad \Delta H = +10\,108 \text{ kJ}$$

For thermochemical equations, it is important to understand that:

- when an equation is reversed, $\Delta H$ for the reversed reaction has the same magnitude but opposite sign from the original equation
- when the coefficients in an equation are doubled, $\Delta H$ for the new equation is doubled; similarly, if the coefficients in an equation are halved, $\Delta H$ for the new equation is halved
- because a change of state involves a change in enthalpy, the states of the reactants and products are important.

The **joule** is the unit of energy. Other related units are:

- kilojoules: 1 kJ = $10^3$ J
- megajoules: 1 MJ = $10^6$ J.

The energy available from fuels (energy content) can be measured using units such as kJ mol$^{-1}$, kJ g$^{-1}$ and MJ t$^{-1}$. If a fuel contains a mixture of chemicals, rather than a pure substance, then the units are given in kJ g$^{-1}$ or MJ t$^{-1}$.

- **You will now be able to complete Worksheet 2.**

## FUELS

Fuels contain chemical energy that is relatively easily released as a source of useful energy. **Biofuels** are organic fuels that are produced from recently living organisms and include biodiesel, bioethanol, biogas (containing methane) produced from decomposing sewage and solid wastes, combustible wastes (garbage), wood and charcoal. Fossil fuels can also be considered to be organic fuels because they contain carbon, hydrogen and other elements. They are formed by geological action on plant matter. Energy sources can be classified in terms of their renewability.

- **Non-renewable** resources are used at a rate faster than they can be replaced. Examples are fossil fuels, which include coal, natural gas and oil.
- **Renewable** resources are continually replaced by natural processes within a relatively short period of time. Examples are wind, water, tides, biomass and solar energy. Biofuels are renewable but not all biofuels are **sustainable**. Sustainable sources of energy are those that are not expected to be depleted within the lifetime of the human race. The use of these sources causes no long-term damage to the environment.

### Non-renewable fuels

Non-renewable fuels have been formed over millions of years from plant and animal matter (Table 3.1.2). The chemical energy contained in fossil fuels has come from solar energy. The use of fossil fuels has significant environmental disadvantages. For example, using fossil fuels:

- enhances the greenhouse effect because the greenhouse gases $CO_2$ and $H_2O$ are emitted when the fuels are burnt
- produces $SO_2$ gas, which forms acid rain
- produces nitrogen oxides, which increase air pollution.

**Table 3.1.2** A comparison of some non-renewable energy sources

| Fossil fuel | Description | Reasons for use | Energy content and use |
|---|---|---|---|
| coal | • a mixture of organic molecules with molar masses as high as 3000<br>• found as black coal, brown coal and peat, with decreasing percentage carbon content | • cheap to mine<br>• large deposits worldwide | • high energy content<br>• the higher the percentage of carbon, the higher the energy content, the greater the efficiency of energy production and the less the environmental impact<br>• produces more pollutants than other fossil fuels<br>• used in electricity production |
| natural gas | • mostly methane<br>• may be found trapped between rock layers; called shale gas | • more efficient in energy production than coal and other fossil fuels<br>• moderate cost<br>• when burnt, produces less particulate matter in emissions than coal does | • high energy content<br>• used in electricity production, heating, cooking and as a transport fuel |
| petrol | • a component of crude oil | • relatively cheap<br>• widely available<br>• readily transported | • high energy content<br>• used as a transport fuel |


ISBN 978 0 6557 0027 2

# KEY KNOWLEDGE

## Renewable fuels

Either directly or indirectly, living organisms obtain energy from the Sun by the process of **photosynthesis**. Glucose and oxygen are products of photosynthesis:

$$6CO_2(g) + 6H_2O(l) \xrightarrow{\text{sunlight}} \underset{\text{glucose}}{C_6H_{12}O_6(aq)} + 6O_2(g)$$

Plants convert solar energy to chemical energy in glucose, which then undergoes polymerisation to form starch and cellulose. Starch is stored in plants and then hydrolysed to re-form glucose, for the subsequent controlled release of energy for the plant in the process of **respiration**. Animals also use the food they eat to obtain energy by respiration:

$$C_6H_{12}O_6(aq) + 6O_2(g) \rightarrow 6CO_2(g) + 6H_2O(l)$$

Starch, cellulose and other plant material in sugar cane and grains may be converted to biofuels, such as biogas, bioethanol and biodiesel, for transport and other industrial and domestic uses (Table 3.1.3).

Bioethanol is produced by the **fermentation** of glucose, which is an anaerobic process involving enzymes in yeast:

$$C_6H_{12}O_6(aq) \xrightarrow{\text{yeast enzyme}} 2C_2H_5OH(aq) + 2CO_2(g)$$

Much of the water present in the impure ethanol produced by this reaction is removed by **distillation** to produce a more sustainable transport fuel.

Biogas, bioethanol and biodiesel are sometimes described as being 'carbon neutral'. In the process of photosynthesis, carbon dioxide is removed from the air and is converted into plant material that is used to make biofuels. When the fuel derived from the plants undergoes combustion, carbon dioxide is produced that can then be recycled as more plants grow. In reality, energy is needed for the growing and harvesting of these fuels and greenhouse gases are emitted as a consequence. However, there is an overall reduction in the levels of carbon dioxide emitted when these fuels are used instead of fossil fuels.

Another issue that arises with respect to biofuel production is whether it is appropriate to use farmland to produce biofuels instead of food crops.

**Table 3.1.3** A comparison of some renewable energy sources

| Biofuel | How it is produced | Energy content, use and sustainability |
|---|---|---|
| Biogas<br>• mainly a gaseous mixture of $CO_2$ and $CH_4$ | • generated by anaerobic digestion of organic waste matter by microorganisms<br>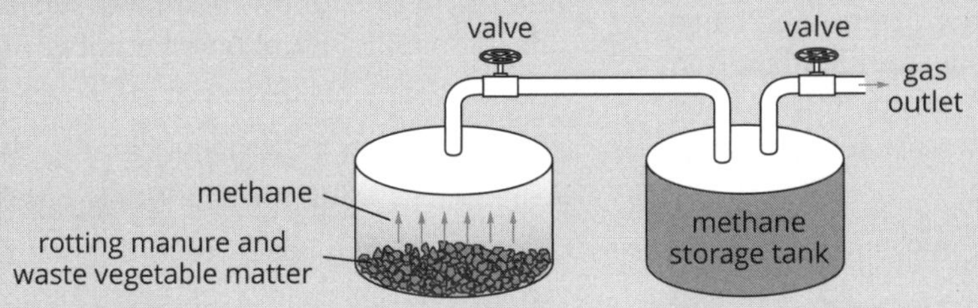<br><br>A biogas digester | • has a low energy content because percentage of $CO_2$ present is low<br>• used for heating and to power homes and farms<br>• is very significant when considering a circular economy in terms of the products and the production of energy (Remember that a circular economy means that maximum value of the resources is achieved and that all waste can be recovered.) |
| Bioethanol | • produced by fermentation of glucose—an anaerobic process involving yeast: $C_6H_{12}O_6(aq) \rightarrow 2C_2H_5OH(l) + 2CO_2(g)$<br>• ethanol produced is distilled to produce a more sustainable transport fuel. Distillation separates the ethanol from the water–ethanol mixture on the basis of their boiling points. | • 12% less greenhouse gas emissions than petrol<br>• has less energy per litre than petrol<br>• used as a transport fuel, but causes greater engine wear than petrol<br>• is a renewable fuel because it is made from biomass |
| Biodiesel<br>• a mixture of esters | • produced by reaction between vegetable oils (esters) and an alcohol, commonly methanol<br>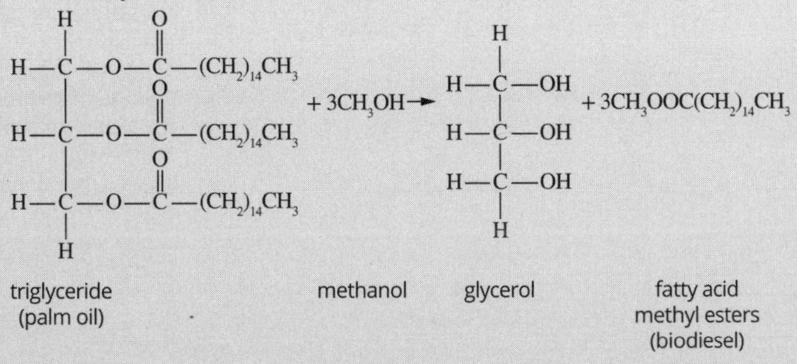<br><br>A chemical equation showing the production of a biodiesel | • 10–20% less gas emissions than petrodiesel<br>• has more energy available per litre than petrodiesel<br>• used as a transport fuel<br>• is biodegradable and non-toxic<br>• made from renewable resources<br>• cleaner refineries |

● **You will now be able to complete Worksheet 3.**

## KEY KNOWLEDGE

### Complete and incomplete combustion

When a hydrocarbon fuel undergoes complete combustion in excess oxygen, the products are $CO_2$ and $H_2O$. However, if the oxygen supply is limited, CO or C can be produced instead of $CO_2$. Carbon monoxide is also produced when methanol and ethanol burn in limited oxygen. For example:

- complete combustion of methanol:
  $2CH_3OH(g) + 3O_2(g) \rightarrow 2CO_2(g) + 4H_2O(l)$
- incomplete combustion of methanol:
  $2CH_3OH(g) + 2O_2(g) \rightarrow 2CO(g) + 4H_2O(l)$

To determine whether the oxygen available for a reaction is limited, you can use the amounts of oxygen and fuel present to calculate the ratio $\frac{n(O_2)}{n(\text{fuel})}$. If this ratio is greater than the stoichiometric ratio for $\frac{n(O_2)}{n(\text{fuel})}$, then oxygen is in excess.

### FUELS FOR THE BODY

Food provides the energy we need for our activities and for the chemical reactions in our bodies. It is composed of three macronutrients: carbohydrates, fats and oils, and proteins. The energy content for each macronutrient is different and is measured in kJ $g^{-1}$ (Table 3.1.4). The energy content is usually determined by simply burning the food, as the energy released in this process is similar to the energy released when food is oxidised in the body.

Table 3.1.4 shows that the energy determined tends to be greater from combustion than from oxidation in the body because of incomplete absorption of food by the body, incomplete oxidation of some nutrients and loss of waste heat.

**Table 3.1.4** Comparison of the energy content of macronutrients in food

| Macronutrient | Energy released when burnt (kJ $g^{-1}$) | Energy released when digested (kJ $g^{-1}$) |
|---|---|---|
| carbohydrates | 17 | 16 |
| fats and oils (lipids) | 39 | 37 |
| proteins | 24 | 17 |

# Measuring changes in chemical reactions

This topic examines how the amounts of energy and gases produced in combustion reactions involving fuels and food can be measured using stoichiometry.

### STOICHIOMETRY INVOLVING REACTIONS OF FUELS

In this course, you will be asked to do different types of stoichiometric calculations. When using stoichiometry to calculate quantities of gases involved in reactions, the conditions are limited to **standard laboratory conditions (SLC)**, which are 25°C and 100 kPa. The molar volume of a gas at SLC is 24.8 L $mol^{-1}$. Stoichiometric calculations can be classified into different categories, including **mass–mass**, **mass–volume** and **volume–volume**, but the steps in the calculations are similar for all. The steps are as follows.

1 Write a balanced equation.
2 Determine the number of moles of each reactant (or **reagent**).
3 Determine the number of moles of the limiting 'known reactant'.
4 Determine the **mole ratio** from the balanced chemical equation and use it to calculate the amount, in mol, of the 'unknown substance' (i.e. the substance with an unknown number of moles):

$$\frac{n(\text{unknown})}{n(\text{known})} = \frac{\text{coefficient of unknown substance}}{\text{coefficient of known substance}}$$

$n(\text{unknown}) = n(\text{known}) \times (\text{mole ratio})$

5 Calculate the quantity required for the unknown substance. Table 3.1.5 provides possible formulas you might use.

**Table 3.1.5** Formulas used for calculating amount in moles

| Formula | Symbol and unit |
|---|---|
| $n = \frac{m}{M}$<br>$n = c \times V$ (for solutions)<br>$n = \frac{pV}{RT}$ (for gases)<br>$n = \frac{V}{V_m}$ (for gases)<br>$n = \frac{N}{N_A}$ | $n$ amount in moles (mol)<br>$m$ mass in grams (g)<br>$M$ molar mass in grams per mole (g $mol^{-1}$)<br>$N_A$ Avogadro's number = $6.02 \times 10^{23}$ particles $mol^{-1}$<br>$N$ number of particles<br>$c$ concentration in moles per litre (mol $L^{-1}$) or molar (M)<br>$V$ volume of gas in litres (L)<br>$p$ pressure in kilopascals (kPa)<br>$T$ temperature in kelvin (K)<br>$R$ general gas constant = 8.31 J $K^{-1}$ $mol^{-1}$<br>$V_m$ molar volume of gas in litres (L $mol^{-1}$) |


ISBN 978 0 6557 0027 2

# KEY KNOWLEDGE

If you are provided with information that allows you to determine the number of moles of both reactants, you need to determine the **limiting reactant** first, so that you can calculate the mole ratio of the limiting reactant and the unknown substance. (A simple way to determine the limiting reactant is to calculate the number of moles of each reactant and divide each by their own coefficient. The reactant for which the smallest number is obtained is the limiting reactant.) Sometimes you are told the limiting reactant and you can then omit this step.

The following worked example shows you how to do a mass–volume stoichiometric calculation. Example 2 shows you how to use stoichiometry to calculate the heat energy released by a reaction.

**Example 1:** Determine the volume of $CO_2$ gas produced at SLC when 4.60 g of ethanol, $CH_3CH_2OH$, reacts with 20.0 g oxygen gas.

| Thinking | Working | |
|---|---|---|
| 1 Write a balanced equation. | $CH_3CH_2OH(g) + 3O_2(g) \rightarrow 2CO_2(g) + 3H_2O(l)$ | |
| 2 Determine the number of moles of each reactant using one of the equations from Table 3.1.5. | $n(CH_3CH_2OH)$:<br>$n = \frac{m}{M}$<br>$= \frac{4.60}{46.0}$<br>$= 0.100$ mol | $n(O_2)$:<br>$n = \frac{m}{M}$<br>$= \frac{20.0}{32}$<br>$= 0.625$ mol |
| 3 Determine the number of moles of the limiting 'known reactant' by:<br>• using the number of moles of each reactant from above and dividing by that reactant's coefficient<br>• comparing the two values—the smaller is the limiting reactant. | For $CH_3CH_2OH$:<br>$\frac{0.100}{1} = 0.100$<br>This is the smaller value, so $CH_3CH_2OH$ is the limiting reactant used. | For $O_2$:<br>$\frac{0.625}{3} = 0.208$ |
| 4 Use the mole ratio from the balanced chemical equation to calculate the amount, in mol, of the 'unknown substance':<br>$\frac{n(\text{unknown substance})}{n(\text{limiting known substance})} = \frac{\text{coefficient (unknown substance)}}{\text{coefficient (known substance)}}$ | $\frac{n(CO_2)}{n(CH_3CH_2OH)} = \frac{2}{1}$<br>$n(CO_2) = \frac{2}{1} \times n(CH_3CH_2OH)$<br>$= \frac{2}{1} \times 0.100 = 0.200$ mol | |
| 5 Calculate the quantity required for the 'unknown substance'. (In this case, calculate the volume of $CO_2$.) | $n(CO_2) = \frac{V(CO_2)}{V_m}$<br>$V(CO_2) = n(CO_2) \times V_m$<br>$= 0.200 \times 24.8$<br>$= 4.96$ L | |

**Example 2:** If the heat of combustion of $CH_3CH_2OH$ is 1360 kJ $mol^{-1}$, calculate the energy released when 4.60 g $CH_3CH_2OH$ is burnt in 20.0 g oxygen gas. Also calculate the energy released per litre of $CO_2$ produced.

| Thinking | Working |
|---|---|
| 1 Given that the heat of combustion is the energy released by combustion of 1 mol of a substance, determine the energy released for the calculated number of moles. | 1360 kJ released for 1 mol ethanol<br>$x$ kJ released for 0.100 mol ethanol burnt (when 4.96 L $CO_2$ is released)<br>$\frac{x}{1360} = \frac{0.100}{1}$<br>$x = 0.100 \times 1360$<br>$= 136$ kJ |
| 2 Calculate the number of kJ per litre of $CO_2$. | Energy released per litre of $CO_2 = \frac{136}{4.96}$ kJ<br>$= 27.4$ kJ |

If the reactants and products are in the gaseous state and at the same temperature and pressure, stoichiometric calculations do not always require the calculation of the number of moles. The same number of moles of different gases occupies the same volume at constant temperature and pressure, so the mole ratio in a balanced chemical equation can also be used as a volume ratio.

# KEY KNOWLEDGE

## Specific heat capacity

The **specific heat capacity** of a substance is the amount of energy needed to raise the temperature of 1 gram of the substance by 1°C.

Heat capacity is a measure of how effectively a substance stores energy. Different substances have different heat capacities; the specific heat capacity of water is 4.18 J $g^{-1}$ $°C^{-1}$. To calculate the estimated energy change that occurs when a mass of a substance undergoes a change in temperature, use the following formula:

heat energy = mass of water × specific heat capacity × temperature change

or

$$q = m \times c \times \Delta T$$

- **You will now be able to complete Worksheet 4 and conduct Practical activities 1 and 2.**

## CALORIMETRY

**Calorimetry** is a technique for measuring the energy change during a chemical reaction. Two instruments are used (Figure 3.1.3):

- a **bomb calorimeter**—for reactions that involve gaseous reactants or products, particularly combustion reactions to determine the energy content of foods
- a **solution calorimeter**—for reactions in solution.

Both types of calorimeters have a large, insulated reservoir of water. The heat released or absorbed by a chemical reaction within the calorimeter causes the temperature of the water to change, and this temperature change is measured.

## Calibration of calorimeters

Calorimeters can be calibrated electrically by releasing a known amount of electrical energy in the calorimeter and measuring the change in temperature, $\Delta T$. The energy required to raise the temperature of a calorimeter and its contents by 1°C is called the **calibration factor** for that particular calorimeter. The calibration factor can be calculated using the following formulas:

$$E = V \times I \times t$$

$$\text{calibration factor} = \frac{E}{\Delta T}$$

where $E$ is the energy input (J), $V$ is the potential difference (V), $I$ is the current (A), $t$ is the time (s) and $\Delta T$ is the temperature change (°C or K).

The energy change for a specific reaction can be calculated by measuring $\Delta T$ when the reaction is performed in the calibrated calorimeter. The energy change can be converted into joules per mole (or per gram or per litre) for that substance and expressed as a $\Delta H$ value or an energy content value. The energy change calculated in a calibrated calorimeter is an accurate measurement.

## Solving problems involving calorimetry

Solving calorimetry problems generally involves three steps (Figure 3.1.4 on the following page). Sometimes one of these steps is unnecessary because of the information given. Worksheet 5 gives you practice in these calculations.

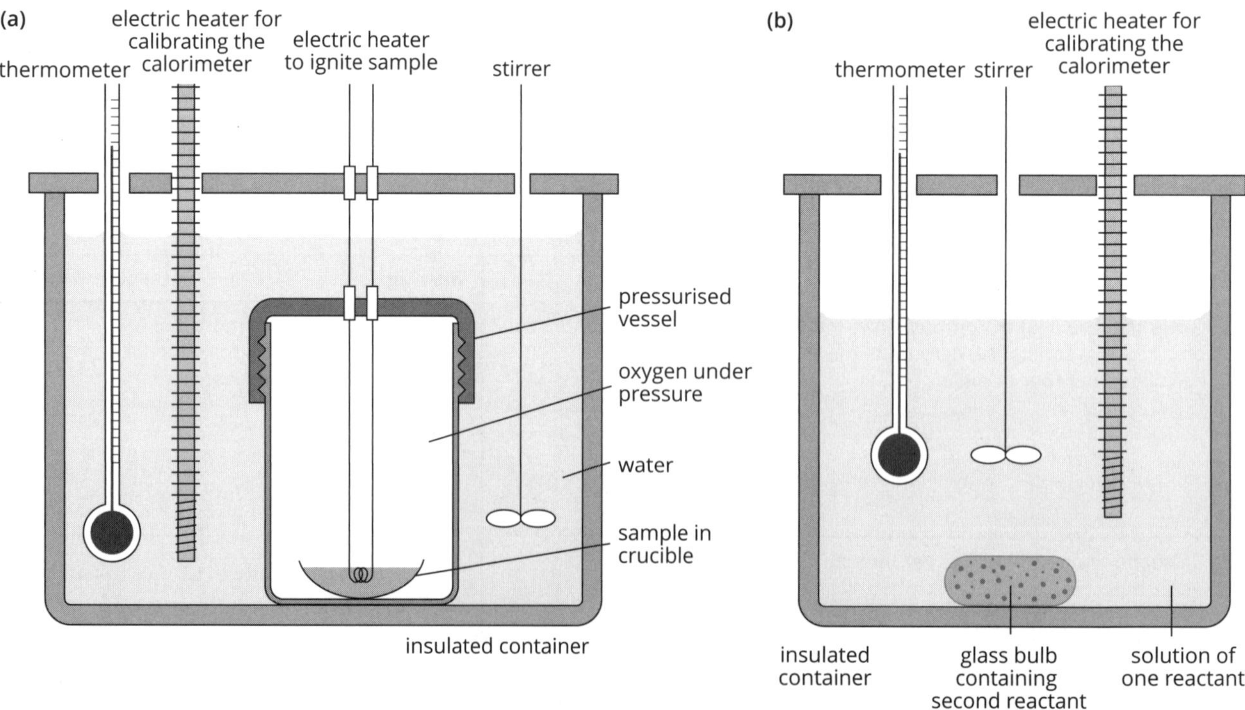


**Figure 3.1.3** (a) A bomb calorimeter and (b) a solution calorimeter are used for measuring energy changes in reactions.

- **You will now be able to conduct Practical activity 3.**


ISBN 978 0 6557 0027 2

# KEY KNOWLEDGE

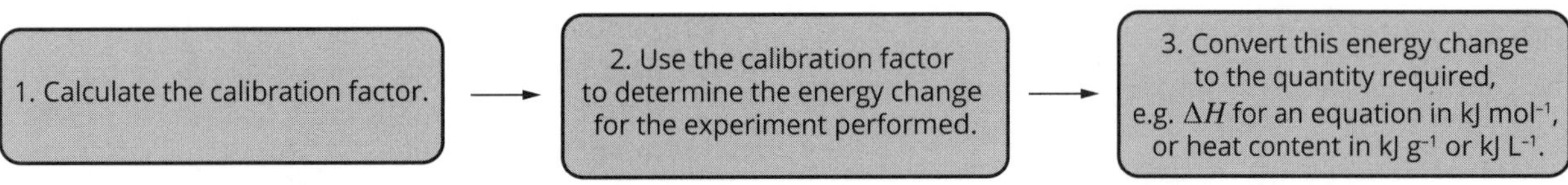


**Figure 3.1.4** The steps involved in calorimetry calculations

## Temperature–time graphs

If a solution calorimeter is not properly insulated, although the measurement results may be repeatable and therefore **precise**, the $\Delta T$ measured will not be **accurate** or close to the true value. Plotting a temperature–time graph and using the extrapolation technique shown in Figure 3.1.5 enables a more accurate $\Delta T$ to be estimated.

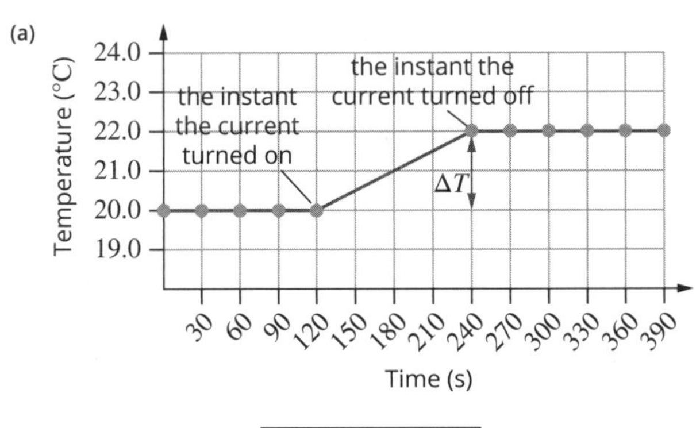


Graph a:

$\Delta T = T_{final} - T_{initial}$

$= 22.0 - 20.0$

$= 2.0°C$

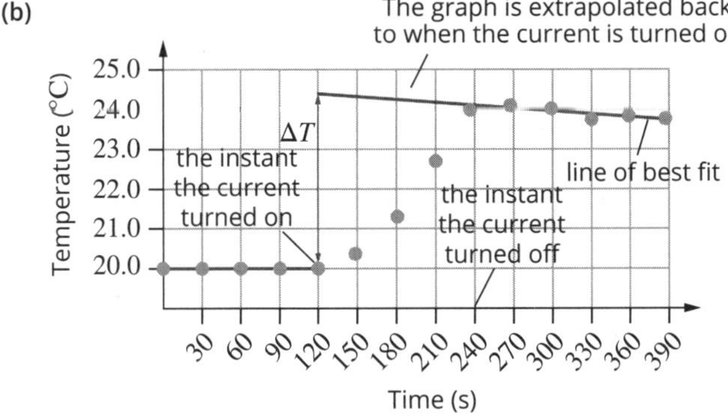


Graph b:

$\Delta T = T_{final} - T_{initial}$

$= 24.4 - 20.0$

$= 4.4°C$

**Figure 3.1.5** Temperature–time graphs for the calibration of two calorimeters. (a) A graph for a well-insulated calorimeter. (b) A graph for a poorly insulated calorimeter with heat loss. Extrapolating the line of best fit for the points after energy was released gives a more accurate value of $\Delta T$.

- **You will now be able to conduct Practical activity 5.**

## ENERGY FROM FUELS AND FOOD

If a bomb calorimeter is not available, you can obtain an estimate of the energy content of food or fuels by using the equipment shown in Figure 3.1.6. The energy transferred to the water can be calculated using the specific heat capacity of water and measuring the volume of water and the change in temperature. The energy is given by:

$$q = m \times c \times \Delta T$$

where $q$ is the energy that is transferred to the water (J), $m$ is the mass of the water (g), which is equal to the volume (mL), $c$ is the specific heat capacity of the water (4.18 J g$^{-1}$ °C$^{-1}$), and $\Delta T$ is the change in temperature (°C or K).

The values obtained by this method will not be the same as the theoretical values for the food or fuel because only some of the heat produced is absorbed by the water, producing a smaller $\Delta T$ and therefore a lower estimate of the energy content.

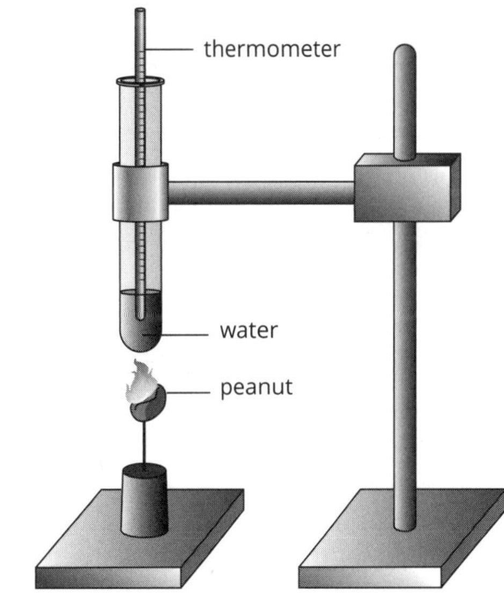


**Figure 3.1.6** Equipment for measuring the energy content of a peanut. The chemical energy released from the combustion of the peanut is transferred to a test tube of water, where it is transformed into heat energy of the water.

# KEY KNOWLEDGE

## Energy transformations

The combustion of a fuel releases the chemical energy in the fuel as heat energy. This heat energy can be transformed into kinetic or electrical energy for use in vehicles and power stations.

**Energy transformation efficiency** is the percentage of energy from a source that is converted into useful energy. It is calculated using the formula:

$$\text{energy transformation efficiency} = \frac{\text{useful energy} \times 100}{\text{chemical energy obtained from the source}}$$

Power stations are 30–40% efficient (Figure 3.1.7). When natural gas, petrol or biofuels are used in the combustion engines of vehicles, the efficiency of energy transformations is about 25%. In petrol combustion engines, the chemical energy in the petrol is converted into thermal or heat energy, which the engine converts into mechanical energy to drive the car. This is converted into kinetic energy as the car moves forward.

As a general rule, the more energy transformations that are involved, the lower the efficiency.

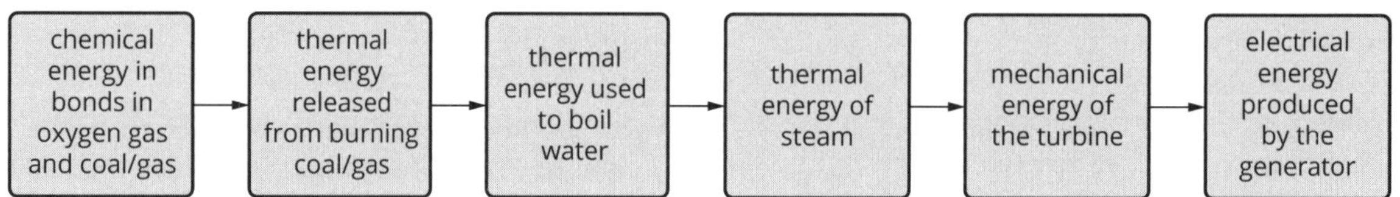


**Figure 3.1.7** Energy transformations in coal and gas-burning power stations. The least efficient transformation is the conversion of thermal energy of steam to mechanical energy of the turbine.

- **You will now be able to complete Worksheet 5 and conduct Practical activity 4.**

# Primary galvanic cells and fuel cells as sources of energy

To supply the energy needs of society, scientists have developed devices that generate electrical energy directly from chemicals, instead of obtaining energy from combustion reactions. This topic examines the operation of electrochemical cells as sources of energy and how they might be used to provide sustainable power.

## OXIDATION AND REDUCTION

**Redox** (reduction–oxidation) reactions involve a transfer of electrons. The term OIL RIG is one way to remember the movement of electrons in the processes of oxidation and reduction.

- In **oxidation**, electrons are lost: Oxidation is Loss (OIL).
- In **reduction**, electrons are gained: Reduction is Gain (RIG).

Consider the burning of magnesium in oxygen:

$$2Mg(s) + O_2(g) \rightarrow 2MgO(s)$$

The electronic configuration of each Mg atom is 2,8,2, which becomes 2,8 when the $Mg^{2+}$ ion is formed. The configuration of each O atom changes from 2,6 to 2,8 when the $O^{2-}$ ion is formed.

- Each Mg atom has lost electrons and been oxidised. This can be represented as a half-equation:

$$Mg \rightarrow Mg^{2+} + 2e^-$$

- Each O atom has gained these electrons and been reduced. The half-equation is:

$$O_2 + 4e^- \rightarrow 2O^{2-}$$

Magnesium is called the **reducing agent** (reductant) and oxygen is called the **oxidising agent** (oxidant). During the reaction, some of the chemical energy in the reactants (magnesium solid and oxygen gas) is converted into heat energy and light energy, as the solid magnesium oxide product forms.

Some of the concepts involved in redox reactions are shown in Figure 3.1.8.

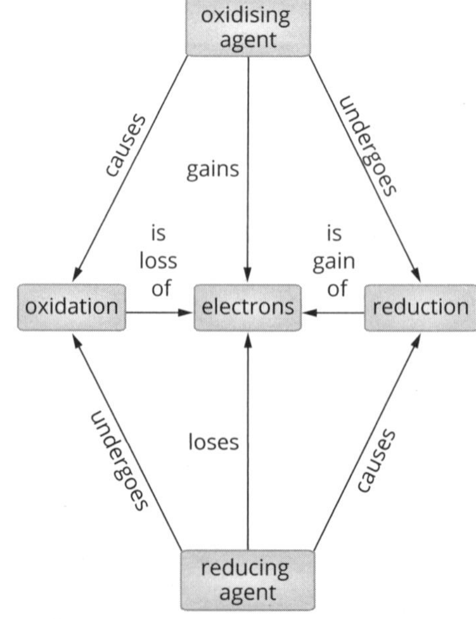


**Figure 3.1.8** Concepts associated with redox reactions

 ISBN 978 0 6557 0027 2

# KEY KNOWLEDGE

## Oxidation numbers

Oxidation can also be defined as occurring when there is an increase in **oxidation number** and reduction as occurring when there is a decrease in oxidation number.

Oxidation numbers:

- are used to decide if a reaction is a redox reaction
- have no physical meaning; they are defined to enable reactions involving molecules as well as those involving ions to be described as redox reactions
- are determined by using the rules in Table 3.1.6.

**Table 3.1.6** Rules for determining oxidation numbers

| Species | Oxidation number | Examples |
|---|---|---|
| element | 0 | $Cl_2$, Mg, C, $O_2$, $H_2$ |
| ion | • charge on the ion | $Na^+ = +1$; $Cl^- = -1$ |
| oxygen in compound | • −2 in its compounds<br>• exception is $H_2O_2$, where it is −1 | O = −2 in $H_2O$, $CO_2$, $Na_2O$ |
| hydrogen in compound | • +1 in compounds with non-metals<br>• −1 in compounds with metals | H = +1 in HCl, $H_2S$, $CH_4$<br>H = −1 in NaH |
| molecular ion and molecule | • sum of oxidation numbers equals charge on molecular ion (zero in the case of neutral molecules)<br>• the most electronegative element has the negative oxidation number | For $MnO_4^-$, oxygen is defined as −2. Because there are 4 oxygen atoms, in order to have an overall charge on the $MnO_4^-$ ion of −1, Mn must have an oxidation number of +7.<br>Let oxidation number of Mn be $x$:<br>$x + 4(-2) = -1$<br>So $x = +7$ |

## Balancing redox equations

Half-equations can be written for oxidation and reduction reactions. The number of atoms of each element and the charge on both sides of the equation must be balanced.

Once you have balanced the half-equations, add them together, ensuring there are equal numbers of moles of electrons on each side. In the overall redox equation, cancel electrons and any other species as required. Consider the following examples.

*Example 1:* In the reaction between $Fe^{3+}(aq)$ and Zn(s), Zn(s) has an oxidation number of 0 and undergoes oxidation because its oxidation number increases to +2 when the reaction occurs. Therefore, Zn is the reducing agent. $Fe^{3+}$ is the oxidising agent because its oxidation number decreases from +3 to 0 when it undergoes reduction.

- Reduction: $Fe^{3+}(aq) + 3e^- \rightarrow Fe(s)$
- Oxidation: $Zn(s) \rightarrow Zn^{2+}(aq) + 2e^-$
- Overall equation (multiply the half-equations by integers and add so electrons cancel):

$$2Fe^{3+}(aq) + 3Zn(s) \rightarrow 2Fe(s) + 3Zn^{2+}(aq)$$

The $Zn^{2+}$/Zn and $Fe^{3+}$/Fe pairs in the reaction above are called **conjugate redox pairs**. A conjugate redox pair is made up of an oxidising agent and the reducing agent that is formed when it gains electrons.

*Example 2:* $Fe^{2+}$ and $Cr_2O_7^{2-}$ react in acidic solution. $Fe^{2+}$ is the reducing agent and $Cr_2O_7^{2-}$ is the oxidising agent.

The half-equation for $Cr_2O_7^{2-}$ reacting to form $Cr^{3+}$ is balanced for reactions in acidic conditions by following the rules in Table 3.1.7.

**Table 3.1.7** Rules for balancing redox half-equations in acidic solutions

| Rule | Example: reduction of $Cr_2O_7^{2-}(aq)$ to $Cr^{3+}(aq)$ |
|---|---|
| 1 Balance all elements except hydrogen and oxygen. | $Cr_2O_7^{2-} \rightarrow 2Cr^{3+}$ |
| 2 Balance oxygen atoms by adding $H_2O$ molecules. | $Cr_2O_7^{2-} \rightarrow 2Cr^{3+} + 7H_2O$ |
| 3 Balance hydrogen atoms by adding $H^+$ ions. | $Cr_2O_7^{2-} + 14H^+ \rightarrow 2Cr^{3+}(aq) + 7H_2O$ |
| 4 Balance charge by adding electrons, and add states. | $Cr_2O_7^{2-}(aq) + 14H^+(aq) + 6e^- \rightarrow 2Cr^{3+}(aq) + 7H_2O(l)$ |

To balance half-equations in alkaline solutions, follow the rules in Table 3.1.8.

**Table 3.1.8** Rules for balancing redox half-equations in alkaline solutions

| Rule | Example: reduction of $Cr_2O_7^{2-}(aq)$ to $Cr^{3+}(aq)$ |
|---|---|
| 1 Look at $n(H^+)$ and add the same number of moles of $OH^-$ ions to *both* sides. | $Cr_2O_7^{2-}(aq) + 14H^+(aq) + 6e^- + 14OH^-(aq) \rightarrow 2Cr^{3+}(aq) + 7H_2O(l) + 14OH^-(aq)$ |
| 2 Because $H^+$ ions readily react with $OH^-$ ions to form the equivalent moles of $H_2O$ molecules, adjust each side. | $Cr_2O_7^{2-}(aq) + 14H_2O(l) + 6e^- \rightarrow 2Cr^{3+}(aq) + 7H_2O(l) + 14OH^-(aq)$ |
| 3 Cancel $H_2O$ molecules so they are on only one side of the equation. | $Cr_2O_7^{2-}(aq) + 7H_2O(l) + 6e^- \rightarrow 2Cr^{3+}(aq) + 14OH^-(aq)$ |

● **You will now be able to complete Worksheet 6.**

# KEY KNOWLEDGE

## GALVANIC CELLS

**Galvanic cells** are electrochemical cells that convert chemical energy to electrical energy. They involve **spontaneous** redox reactions (a redox reaction that occurs naturally). The two half-reactions occur in separate **half-cells**. This prevents direct contact between the oxidising agent and the reducing agent, so electrons can only be transferred by travelling through an external circuit, from the negative to the positive electrode. The flow of electrons provides electrical energy.

### The Daniell cell

The general features of galvanic cells are illustrated by the Daniell cell shown in Figure 3.1.9.

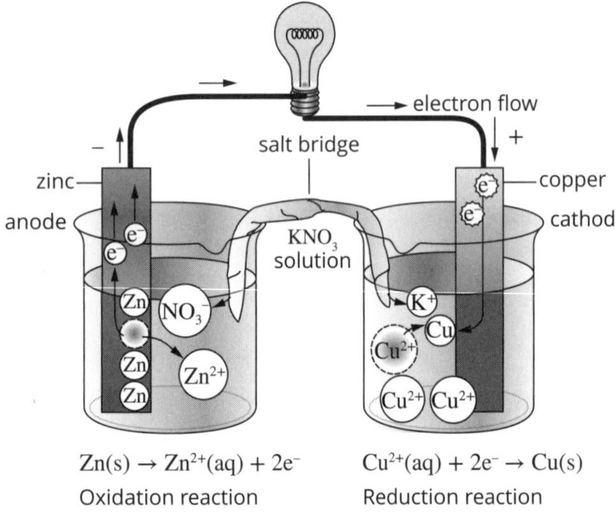

**Figure 3.1.9** The general features of a galvanic cell

The zinc metal electrode is oxidised to zinc ions:

$$Zn(s) \rightarrow Zn^{2+}(aq) + 2e^-$$

The electrons produced move through the wire (the external circuit) towards the positive electrode, where they reduce copper ions:

$$Cu^{2+}(aq) + 2e^- \rightarrow Cu(s)$$

The number of electrons produced is equal to the number consumed at the other half-cell. The overall equation is the sum of the two half-equations:

$$Cu^{2+}(aq) + Zn(s) \rightarrow Cu(s) + Zn^{2+}(aq)$$

Zn is oxidised and acts as the reducing agent. $Cu^{2+}$ is reduced and acts as the oxidising agent.

The **salt bridge** (often a piece of filter paper soaked in a solution of a soluble ionic compound) balances the charges formed and consumed in each half-cell. At the negative electrode, the positive $Zn^{2+}$ ions being formed are balanced by negative ions moving from the salt bridge. At the positive electrode, the $Cu^{2+}$ ions being consumed are balanced by positive ions from the salt bridge.

By definition, the electrode where oxidation occurs is the **anode** and the electrode where reduction occurs is the **cathode**. (You can remember this because because oxidation and anode both start with vowels; reduction and cathode both start with consonants.) In galvanic cells, the anode is negative and the cathode is positive. In the salt bridge, cations move towards the cathode and anions move towards the anode.

Each half-cell contains a conjugate redox pair. In the cell in Figure 3.1.9 the pairs can be written as $Zn^{2+}(aq)/Zn(s)$ and $Cu^{2+}(aq)/Cu(s)$.

If the reducing agent of a conjugate pair is not a metal, an inert electrode is used in the half-cell. This electrode can be either a graphite carbon rod (C) or a piece of platinum (Pt). An inert electrode allows the flow of charge through the solutions but does not take part in the reactions. Such electrodes are required when both the reducing and oxidising agents are ions in solution, such as in the half-cell $Fe^{3+}(aq)/Fe^{2+}(aq)$, or when the conjugate pair involves a gas and ions, such as in $H^+(aq)/H_2(g)$.

## THE ELECTROCHEMICAL SERIES

The **electrochemical series** (Appendix 2) shows the relative strengths of oxidising agents and reducing agents. Half-reactions in the series are written as reduction reactions, with the strongest oxidising agent at the top of the table on the left-hand side and the strongest reducing agent at the bottom of the table on the right-hand side of each equation.

The **standard hydrogen half-cell** is used as a reference (Figure 3.1.10) and assigned a **standard electrode potential (*E*°)** of 0 V. This half-cell operates at the standard conditions of 100 kPa, 1 M and 25°C. The half-equation for the cell is:

$$2H^+(aq) + 2e^- \rightleftharpoons H_2(g)$$

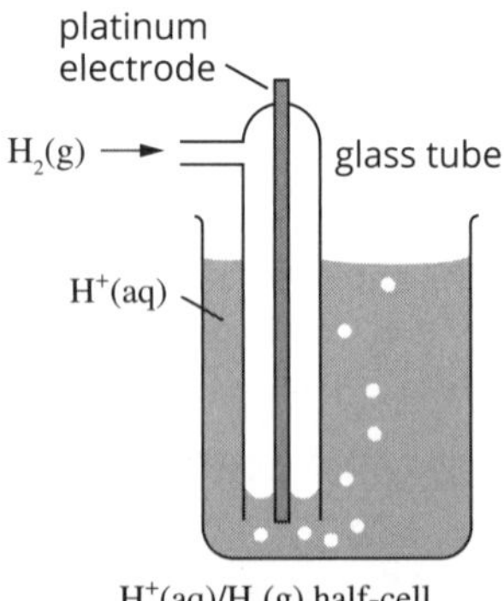

**Figure 3.1.10** The standard hydrogen half-cell

 ISBN 978 0 6557 0027 2

# KEY KNOWLEDGE

The electrochemical series is constructed by measuring the voltages of different half-cells when they are connected to the standard hydrogen electrode. Equations for half-reactions that occur in half-cells that are positive with respect to the standard hydrogen electrode are placed above the $H^+(aq)/H_2(g)$ equation in the series and those that are negative are placed below it.

The maximum cell voltage for any combination of two half-cells can be calculated using the $E^\circ$ values from the electrochemical series.

$E^\circ_{cell}$ = $E^\circ$(of half-equation containing oxidising agent) − $E^\circ$(of half-equation containing reducing agent)

Measurements of $E^\circ$ values are repeatable because almost the same measurement results are obtained when the experiments are performed by the same person under the same conditions. Measurements are reproducible if a person setting up similar equipment obtains values that closely resemble those in the published electrochemical series.

● **You will now be able to complete Worksheet 7.**

## Using the electrochemical series

A spontaneous reaction occurs when the strongest oxidising agent and strongest reducing agent react, which means the $E^\circ$ of the oxidising agent is more positive (higher in the table) than the $E^\circ$ of the reducing agent. If several possible oxidising agents and reducing agents are present, the pair most likely to react is the pair furthest apart in the series.

**Example:** A cell is made from $Ag^+(aq)/Ag(s)$ and $Zn^{2+}(aq)/Zn(s)$ half-cells. Use the electrochemical series to predict the electrode reactions, write the overall cell equation and determine the maximum cell voltage under standard conditions.

| Thinking | Working |
|---|---|
| 1 Write the relevant half-equations. | $Ag^+(aq) + e^- \rightarrow Ag(s)$<br>$Zn^{2+}(aq) + 2e^- \rightarrow Zn(s)$ |
| 2 Of the chemicals in the cell, the chemical highest on the left of the electrochemical series (the strongest oxidising agent) will react with the chemical lowest on the right of the electrochemical series (the strongest reducing agent).<br>Write the anode reaction and the cathode reaction. | $Ag^+$ is the strongest oxidising agent.<br>Zn is the strongest reducing agent.<br>So $Ag^+$ and Zn will react.<br>Anode reaction:<br>$Zn(s) \rightarrow Zn^{2+}(aq) + 2e^-$<br>Cathode reaction:<br>$Ag^+(aq) + e^- \rightarrow Ag(s)$ |
| 3 Write the overall cell equation, including states. | $2Ag^+(aq) + Zn(s) \rightarrow 2Ag(s) + Zn^{2+}(aq)$ |
| 4 Calculate the cell voltage. | $E^\circ_{cell} = E^\circ_{at\ cathode} - E^\circ_{at\ anode}$<br>$= +0.80 - (-0.76)$<br>$= 1.56$ V |

## Limitations of predictions using the electrochemical series

There are two significant limitations with the use of the electrochemical series.

- The electrochemical series applies to reactions under standard conditions. For non-standard conditions, the series cannot be used to reliably predict reactions.
- The electrochemical series predicts the likelihood of a reaction, but not the rate of reaction. A reaction may not be observed when chemicals are directly mixed together if the reaction occurs slowly.

## Primary and secondary cells

Commercial galvanic cells can be classified as:

- **primary cells**, which cannot be recharged because the products of the cell reaction during electricity production do not all remain in contact with the electrodes. For example, the alkaline cells used in torches are primary cells.
- **secondary cells**, which can be recharged. This type of cell will be examined in Unit 3 Area of Study 2.

● **You will now be able to complete Worksheet 8 and conduct Practical activity 6.**

# FUEL CELLS

In **fuel cells** (Figure 3.1.11 on the following page), a spontaneous reaction occurs that directly converts chemical energy to electrical energy for as long as reactants are continuously provided. Because chemical energy is being directly transformed into electrical energy, fuel cells are significantly more efficient (as high as 80%) than other forms of energy production. Unlike the cells described earlier, chemical energy is not stored within the cell. Fuel cells using hydrogen gas as a fuel can use either an alkaline or an acid electrolyte, and produce about 1 V. Water is the only chemical product and heat energy as well as electrical energy is released.

The nature of the electrodes is crucial to the operation and efficiency of the fuel cell. Electrodes enable contact between the fuel (a reducing agent) and the electrolyte, and between the oxidising agent and the electrolyte. Electrodes must:

- be conductors
- have catalytic properties to speed up the rate of the electrode reactions; different catalysts are used in the construction of the anode (often Pt) and the cathode (Ni powder or nanomaterials)
- be porous so that they keep the gaseous fuel and oxidising agent apart but allow each to be in contact with the electrolyte.

The electrolyte enables movement of ions to balance charge and complete the circuit; cations move towards the cathode and anions towards the anode. Various electrolytes are used in fuel cells, including solutions of KOH, $H_3PO_4$ and soluble metal carbonates; permeable polymers and ceramics are also in use.

# KEY KNOWLEDGE

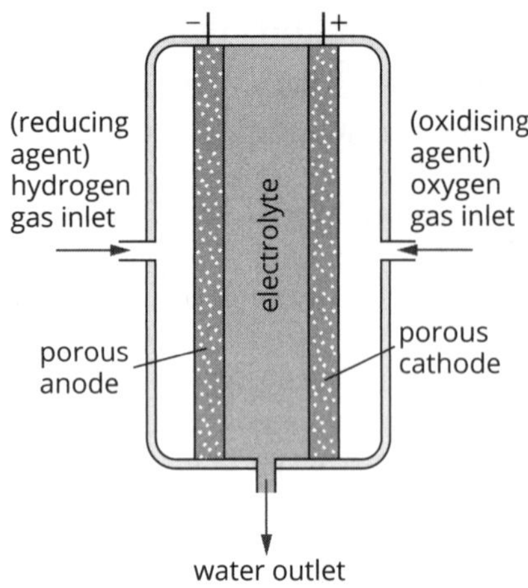


For a cell using an acidic electrolyte:
Anode reaction: $H_2(g) \rightarrow 2H^+(aq) + 2e^-$
Cathode reaction: $O_2(g) + 4H^+(aq) + 4e^- \rightarrow 2H_2O(l)$
Overall reaction: $2H_2(g) + O_2(g) \rightarrow 2H_2O(l)$

**Figure 3.1.11** A simple hydrogen–oxygen fuel cell

## INNOVATION TO OVERCOME DESIGN CHALLENGES

To reduce greenhouse gas emissions and prevent further climate change by global warming, scientists and engineers are attempting to apply the following green chemistry principles for processes and technologies that produce energy:

- design for energy efficiency
- use of renewable feedstocks.

Hydrogen–oxygen fuel cells offer the prospect of a clean, efficient-energy resource. Vehicles using a hydrogen fuel cell to power an electric motor use 40–60% of the fuel's energy and are much more energy efficient than the 25% efficiency of petrol combustion engines. However, there are difficulties associated with the production of 'green' hydrogen fuel.

Most of the hydrogen generated at present is produced using hydrocarbon fuels, with the release of $CO_2$ and water vapour. A 'cleaner' process would use solar or wind power to generate electricity, which can then be used to electrolyse a renewable raw material (water) and produce hydrogen (Table 3.1.9). Using a sustainable process to produce hydrogen for use in highly efficient fuel cells could greatly reduce the global warming issues associated with power production and transport energy sources.

**Table 3.1.9** Production of hydrogen

| Renewable energy sources | Non-renewable energy sources |
|---|---|
| • collection of methane gas from landfill sites for use in steam reforming to produce $H_2$ gas<br>• splitting water using electricity from solar cells and wind turbines | • using fossil fuels in a steam reforming process to produce $H_2$ gas: $CH_4(g) + H_2O(g) \rightleftharpoons CO(g) + 3H_2(g)$<br>• using the carbon monoxide generated to generate further hydrogen: $CO(g) + H_2O(g) \rightleftharpoons CO_2(g) + H_2(g)$ |

● **You will now be able to complete Worksheet 9 and conduct Practical activity 7.**

## FARADAY'S LAWS

**Faraday's laws** can be stated as follows.

- The quantity of electricity passing through a cell is directly proportional to the mass of the material evolved, deposited or dissolved at an electrode.
- To produce one mole of a product, one, two, three or another whole number of moles of electrons must be consumed.

The **electric charge**, *Q*, passing through a cell is measured in coulombs, *C*. It can be calculated by measuring the current, *I*, in amps, which flows for a time, *t*, in seconds:

$$Q = I \times t$$

The charge on one mole of electrons is 96 500 C. The value of 96 500 C is called the **faraday**, *F*, where 1 F = 96 500 C $mol^{-1}$.

The charge passing through a cell can be determined from the formula:

$$Q = n(e^-) \times F = n(e^-) \times 96\,500$$

where $n(e^-)$ is the number of moles of electrons.

Faraday's laws can be used to determine the mass of product formed at an electrode during reaction in a galvanic or fuel cell, as in the following worked example.

 ISBN 978 0 6557 0027 2

## KEY KNOWLEDGE

**Example:** In a Daniell cell, copper is deposited onto a copper electrode at the cathode. The cell produces a current of 10.5 A for 25 minutes. What mass of copper will be deposited?

| Thinking | Working |
|---|---|
| 1 Write the half-equation for the reaction at the electrode. | $Cu^{2+}(aq) + 2e^- \rightarrow Cu(s)$ |
| 2 Use the formula $Q = I \times t$ to calculate the charge, $Q$, that flows through the cell. | $Q = 10.5 \times 25.0 \times 60$<br>$= 15\,750$ C |
| 3 Use the formula $Q = n(e^-) \times F$ to calculate the number of mole of electrons flowing through the cell. | $Q = n(e^-) \times 96\,500$<br>$n(e^-) = \frac{Q}{96500}$<br>$= \frac{15\,750}{96\,500}$<br>$= 0.163$ mol |
| 4 Use the half-equation and the mole ratio to determine $n$(product) at the cathode. | $\frac{n(Cu)}{n(e^-)} = \frac{1}{2}$<br>$n(Cu) = \frac{1}{2} \times 0.163$<br>$= 0.0816$ mol |
| 5 Calculate the mass of product at the cathode. | $m(Cu) = n \times M$<br>$= 0.0816 \times 63.5$<br>$= 5.18$ g |

- **You will now be able to complete Worksheet 10.**

# WORKSHEET 1

Classification and identification

## Knowledge review—mole, stoichiometry and redox

1 Identify the correct term from the list below for each definition and complete the following table. This will help you check your knowledge and understanding of the key ideas involved in electronic configuration and electron transfer in preparation for your study of redox reactions and electrochemical cells. Terms may be used more than once or not at all.

| octet rule | anion | cation | corrosion | oxidation | reduction | valence electrons |
|---|---|---|---|---|---|---|
| electrolyte | ion | OIL RIG | reducing agent | oxidising agent | redox reaction | electron |

| Definition | Term |
|---|---|
| a negatively charged subatomic particle that moves around the nucleus of an atom | |
| a solution that conducts electricity | |
| a negatively charged ion | |
| the collective name for outer shell electrons | |
| the name of the rule describing the transfer or sharing of electrons to achieve 8 electrons in the outer shell | |
| a reaction in which a metal reacts with water and oxygen gas | |
| a reaction in which electrons are transferred | |
| a reaction in which a substance loses one or more electrons | |
| a reaction in which a substance gains one or more electrons | |
| a useful way to remember the definitions of oxidation and reduction | |
| a substance that causes another substance to undergo an oxidation reaction and is itself reduced | |
| a substance that causes another substance to undergo a reduction reaction and is itself oxidised | |

2 Indicate which of the following reactions are redox reactions and, where appropriate, state which reactant is the reducing agent. Write N/A if the reaction is not a redox reaction.

| Reaction | Redox reaction? | Reactant that is the reducing agent |
|---|---|---|
| $Cu^{2+}(aq) + Mg(s) \longrightarrow Cu(s) + Mg^{2+}(aq)$ | | |
| $MgCl_2(aq) + 2AgNO_3(aq) \longrightarrow Mg(NO_3)_2(aq) + 2AgCl(s)$ | | |
| $Zn(s) + 2Fe^{3+}(aq) \longrightarrow Zn^{2+}(aq) + 2Fe^{2+}(aq)$ | | |

3 Mass–mass stoichiometry: Determine the mass of $CO_2$ gas produced when 4.60 g of $CH_3CH_2OH$ reacts with excess oxygen gas.

| Thinking | Working |
|---|---|
| 1 Write a balanced equation. | |
| 2 Determine the number of moles of the known reactant, then use the mole ratio from the equation to determine the number of moles of the unknown. | |
| 3 Calculate the quantity required. | |


ISBN 978 0 6557 0027 2

4 Mass–volume stoichiometry: Determine the volume, in litres, of $CO_2$ gas at SLC (25°C and 100 kPa) produced when 4.60 g of $CH_3CH_2OH$ reacts with excess oxygen gas.

| Thinking | Working |
|---|---|
| 1 Steps 1–3 are the same as in Question 3, to calculate the number of moles of unknown substance ($CO_2$). | |
| 2 Calculate the quantity required. | |

5 Volume–volume stoichiometry: When 300 L of propane is burnt in excess oxygen gas, calculate the volumes of $CO_2$ gas and water vapour produced at constant temperature and pressure.

Remember that mole ratios are also equal to gas volume ratios when temperature and pressure are constant.

| Thinking | Working |
|---|---|
| 1 Write a balanced equation. | |
| 2 Determine the volume of $CO_2$. | |
| 3 Determine the volume of $H_2O$. | |

# WORKSHEET 2

Classification and identification

## Combustion of fuels—energy profile diagrams and calculations

Use the terms in the box to complete questions in this worksheet. Terms may be used more than once or not at all.

| | | | | |
|---|---|---|---|---|
| chemical energy | enthalpy | activation energy | energy source | exothermic |
| endothermic | $H$ | $\Delta H$ | $H_{products} - H_{reactants}$ | negative |
| positive | activation energy | joules | $\Delta T$ | $\Delta H = -890$ kJ |
| energy absorbed as bonds break | energy released as bonds form | $CH_4(g) + 2O_2(g)$ | $CO_2(g) + 2H_2O(l)$ | $H_{products} - H_{reactants}$ |

**1** The combustion of fuels releases a significant quantity of heat that can be used as a/an ________________ ________________. ________________ ________________ is the energy stored in the bonds between atoms and between molecules arising from attractions and repulsions between subatomic particles. It is called ________________ and is represented by the symbol ______. All combustion reactions are ________________ reactions. In such reactions, the enthalpy change, symbol ______, has a ________________ sign. An expression for $\Delta H$ can be written as $\Delta H$ = ________________.

**2** Energy profile diagrams can be used to represent the energy changes in chemical reactions. Using the terms and symbols from the box above, draw and fully label an energy profile diagram for the reaction:

$CH_4(g) + 2O_2(g) \rightarrow CO_2(g) + 2H_2O(l) \quad \Delta H = -890$ kJ

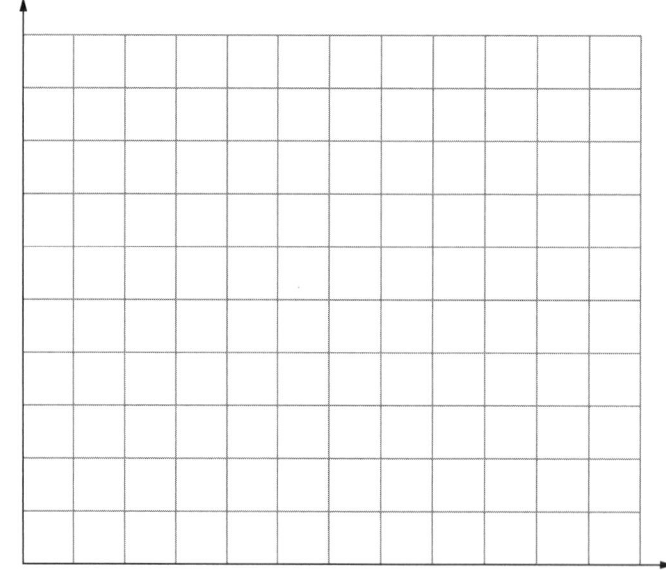

**3** Thermochemical equations are balanced chemical equations that include $\Delta H$ for the reaction. Consider the thermochemical equation for octane:

$C_8H_{18}(g) + \frac{25}{2}O_2(g) \rightarrow 8CO_2(g) + 9H_2O(l) \qquad \Delta H = -5464$ kJ

**a** Determine the value and sign of $\Delta H$ for the following equations:

**i** $2C_8H_{18}(g) + 25O_2(g) \rightarrow 16CO_2(g) + 18H_2O(l) \qquad \Delta H =$ ____________ kJ

**ii** $4CO_2(g) + \frac{9}{2}H_2O(l) \rightarrow \frac{1}{2}C_8H_{18}(g) + \frac{25}{4}O_2(g) \qquad \Delta H =$ ____________ kJ

**b** Select the correct alternatives from the italicised terms to complete this sentence.

The reaction represented by equation i is an *exothermic/endothermic* reaction, whereas the reaction represented by equation ii is an *exothermic/endothermic* reaction.

 ISBN 978 0 6557 0027 2

**WORKSHEET 3**

**Simulation**

# Motor fuels—today and tomorrow

This worksheet explores some of the factors that influence the choice of a fuel. These factors include energy content, mass of carbon dioxide produced, renewability and environmental impact.

1 Using the data in the table below, calculate the energy released and the mass of carbon dioxide produced by complete combustion of 1.00 L each of petrol (octane) and bioethanol. The mass of carbon dioxide produced is one measure of the environmental impact of the fuels.

| Assume petrol contains only octane, $C_8H_{18}$ | Combustion equations: |
|---|---|
| Density of octane: 0.918 g $mL^{-1}$ | Octane: $2C_8H_{18}(l) + 25O_2(g) \rightarrow 16CO_2(g) + 18H_2O(l)$ |
| Density of ethanol: 0.79 g $mL^{-1}$ | Bioethanol: $CH_3CH_2OH(l) + 3O_2(g) \rightarrow 2CO_2(g) + 3H_2O(l)$ |
| Heat of combustion of octane: 5460 kJ $mol^{-1}$ | |
| Heat of combustion of bioethanol: 1360 kJ $mol^{-1}$ | |

**Petrol (octane)**

**a** Energy, in MJ, released by complete combustion of 1.00 L of fuel:

**b** Mass of carbon dioxide, in grams, produced by complete combustion of 1.00 L of fuel:

**Bioethanol**

**c** Energy, in MJ, released by complete combustion of 1.00 L of fuel:

**d** Mass of carbon dioxide, in grams, produced by complete combustion of 1.00 L of fuel:

2 Summarise the environmental effects of the production of these fuels.

| Fuel type | Method of production | Environmental issues (with respect to production and use) |
|---|---|---|
| octane | | |
| bioethanol | | |

3 Briefly summarise the main advantages and disadvantages that you have identified for the use of the two fuels in motor vehicles.

| Fuel | Advantages | Disadvantages |
|---|---|---|
| octane | | |
| bioethanol | | |

4 A particular motor vehicle running on petrol (octane) converts 35% of the energy released by the burning fuel into mechanical energy. If the combustion of petrol releases $3.2 \times 10^4$ kJ $L^{-1}$ of chemical energy, calculate the useful energy output of the fuel in the car.

5 Use the terms in the box to complete the following questions about biofuels. Terms may be used more than once or not at all.

| | | | | | | |
|---|---|---|---|---|---|---|
| 6 | anaerobic | aerobic | $O_2$ | yeast | plant | $CO_2$ |
| methane | hydrogen | biodiesel | microorganisms | fossil fuels | alcohol | fermentation |
| photosynthesis | biogas | glucose | $H_2$ | efficiency | ethanol | |

**a** Complete the summary statements.

As the world's population grows and the demand for energy increases, governments are examining ways of improving the _______________ of energy generation, as well as reducing the use of _______________ _______________. Alternative fuels that may replace fossil fuels in the future are biogas, bioethanol and biodiesel. Biofuels derived from renewable sources such as _______________ matter can be blended or used pure. In villages and on farms, collection of agricultural waste and the use of a biogas digester enable _______________ to be generated under _______________ conditions and in the presence of _______________. Biogas is mainly _______________ and carbon dioxide. Fermentation of sugars, such as _______________, in sugar cane or sugar beet involves the use of _______________ and produces bioethanol. Bioethanol can be produced from a wide variety of _______________ matter and is made by a reaction between a vegetable oil, or a triglyceride, and a/an _______________, usually methanol.


ISBN 978 0 6557 0027 2

b Complete the diagram, writing equations where appropriate.

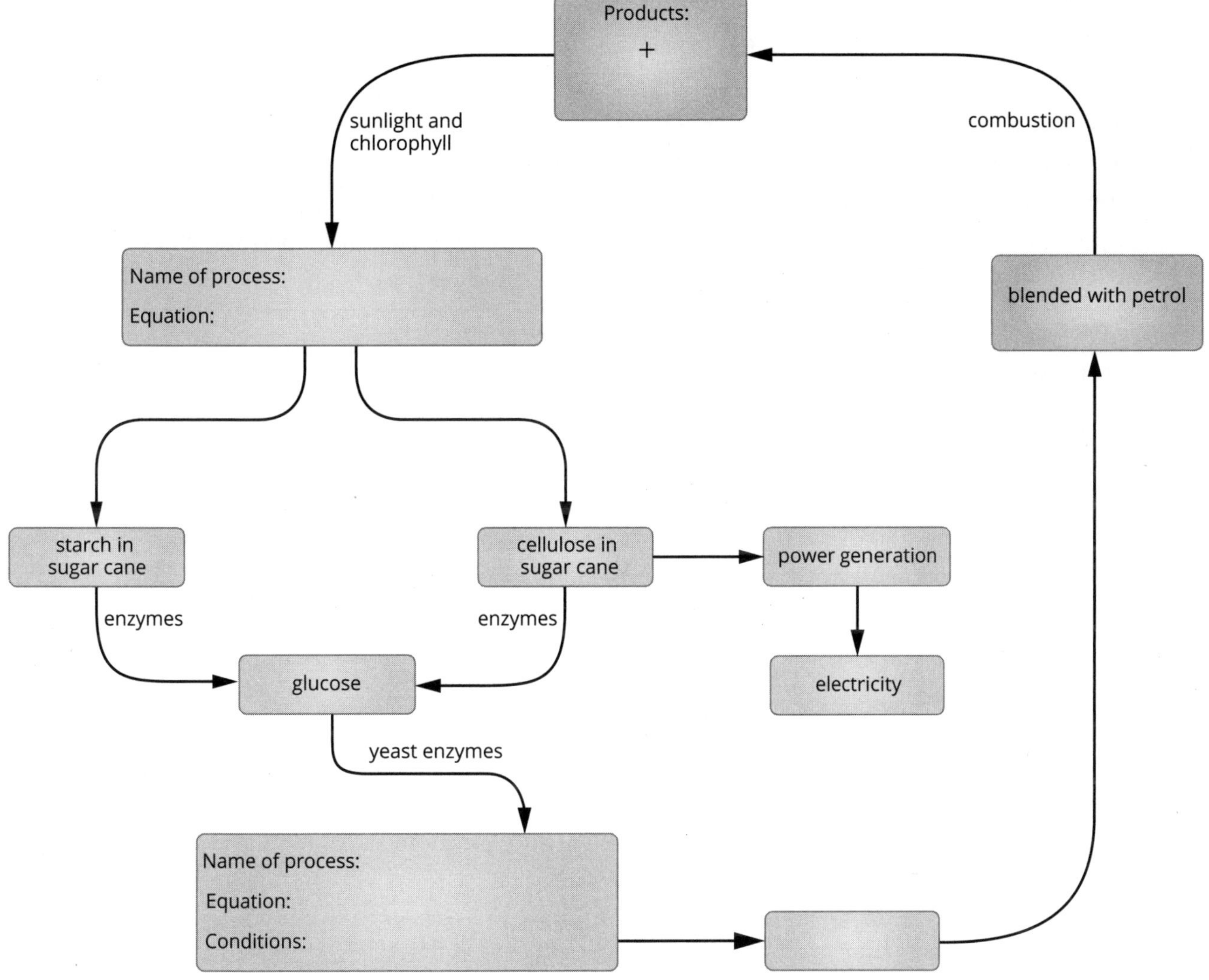


ISBN 978 0 6557 0027 2

# Energy changes and stoichiometry—limiting reactants

**1** Butane gas, $C_4H_{10}$, can be used as a fuel to heat water. If 1.00 L of water is heated from 20.0°C to 100.0°C, calculate the:

**a** energy absorbed by the water from the combustion of butane

**b** mass of butane consumed to supply the energy in part **a**.

Use the following information:

- Specific heat capacity of water = 4.18 J $g^{-1}$ °$C^{-1}$
- Molar mass of butane = 58.48 g $mol^{-1}$
- Heat of combustion of butane = 2880 kJ $mol^{-1}$
- Density of water = 1.00 g $mL^{-1}$

| Thinking | Working |
|---|---|
| Determine $\Delta T$ for the experiment. | $\Delta T$ = |
| **a** The formula required to convert volume to mass is: $\text{density} = \frac{\text{mass}}{\text{volume}}$ The formula to determine the energy released by the fuel is: Energy = specific heat capacity × mass × $\Delta T$ | |
| **b** Use the heat of combustion of the fuel to calculate the mass of fuel consumed to produce the energy calculated in part **a**. | |

**2** **a** Determine, the volume, in litres, of the greenhouse gas $CO_2$ produced at SLC (25°C and 100 kPa) when 9.20 g of $CH_3CH_2OH$ reacts with 45.0 g of oxygen gas.

| Thinking | Working |
|---|---|
| 1 Write a balanced equation. | |
| 2 Determine the limiting known reactant and the number of moles. | |
| 3 Using the answer to step 2, determine the ratio of moles of unknown reactant to moles of limiting known reactant. | |
| 4 Calculate the number of moles of $CO_2$ produced in the reaction. Now convert this to the volume of $CO_2$ at SLC that is produced. | |

**b** For the reaction in part **a**, given the molar enthalpy of combustion of ethanol is –1360 kJ $mol^{-1}$, calculate the amount of energy produced per litre of $CO_2$ released, measured at SLC, in kJ $L^{-1}$.

| Thinking | Working |
|---|---|
| 1 Determine the energy released for calculated mol of $CO_2$ in part **a**. | |
| 2 Calculate the number of kJ $L^{-1}$ of $CO_2$. | |


ISBN 978 0 6557 0027 2

# WORKSHEET 5

Case study

## Calorimetry—energy from a pizza

The nutritional guide on a packaged pizza states that in a 47.0 g slice of a pizza, there is 4.1 g fat, 17.3 g carbohydrate and 5.0 g protein.

**1** **a** Complete the following statements, which describe steps in the electrical calibration of a calorimeter, using the terms, symbols and formulas from the box. (Terms may be used more than once or not at all.) The order of the steps is incorrect.

| calorimeter | $\frac{E}{\Delta T}$ | $\Delta T$ | measured | electric heater | $I$ |
|---|---|---|---|---|---|
| temperature | time | $VIt$ | $E \times \Delta T$ | calibration | $V$ |

**A** Record the initial and final _______________ of the water.

**B** Calculate the change in temperature, _______________.

**C** Calculate the thermal energy in joules released in the calorimeter during electrical calibration, $E$ = _______________.

**D** Fill the _______________ with a _______________ volume of water.

**E** To calculate the energy during calibration, record the current, _______________ (in amps), and the voltage, _______________ (in volts), during this time (in seconds).

**F** Use a/an _______________ _______________ to heat the closed calorimeter for a _______________ period of _______________.

**G** Calculate the _______________ factor, which is equal to _______________.

**b** Order the statements A–G correctly to describe the process of electrical calibration: _______________

**2** Use terms from the box below to complete the statements describing the steps needed to calculate the energy content, in kJ $g^{-1}$, of the pizza, using the calorimeter. (Some terms may not be used.)

| water | pizza | temperature | energy released | burn | combustion |
|---|---|---|---|---|---|
| calibration | $E \times$ mass of pizza | calorimeter | maximum temperature | $\frac{\text{calibration factor}}{\Delta T}$ | $\Delta T$ |
| energy | kJ $g^{-1}$ | kJ $mol^{-1}$ | calibration factor $\times \Delta T$ | | |

**a** Use the same volume of _______________ in the experiment as for the _______________.

**b** Weigh a sample of the _______________ and place it in the calorimeter.

**c** Measure the _______________ of the _______________ in the calorimeter.

**d** _______________ the pizza in the bomb calorimeter.

**e** Measure the _______________ _______________ of the water after combustion.

**f** Calculate _______________________.

**g** Calculate the _______________ _______________ for this measured mass, using the formula: $E$ = _______________

**h** Convert this energy into the energy content of the pizza in units of _______________, using the formula: Energy content = _______________/mass of pizza

3 Use the following experimental data, obtained using a bomb calorimeter, to calculate the energy released by a slice of the pizza and the energy content of the pizza in kJ $g^{-1}$.

| Voltage | 4.95 V | Time of calibration | 236 s |
|---|---|---|---|
| Current | 8.81 A | Temperature before combustion | 19.54°C |
| Temperature before calibration | 18.44°C | Temperature after combustion | 79.04°C |
| Temperature after calibration | 19.54°C | Mass of pizza slice | 47.0 g |

4 Complete the following questions in the table below.

a Calculate the percentage of fat, carbohydrate and protein in the pizza.

b Using the data in the first row of the table, calculate the energy released by each of these nutrients in the slice of pizza.

| | Fat (4.1 g) | Carbohydrate (17.3 g) | Protein (5.0 g) |
|---|---|---|---|
| Energy released when digested (kJ $g^{-1}$) | 37 | 16 | 17 |
| **a** Percentage contained in the pizza | | | |
| **b** Energy released | | | |

c Use this data to calculate the total energy content of a 47.0 g slice of pizza, in kJ.

d Compare your answer to part **c** with your answer from Question 3, which was obtained using a calibrated bomb calorimeter. Why do these energy content values for the pizza differ?

 ISBN 978 0 6557 0027 2

# WORKSHEET 6



## Redox reactions and oxidation numbers

1 Balance the following redox half-equations and overall equation in acidic solution.

a Reduction half-equation: $MnO_4^-(aq) \rightarrow Mn^{2+}(aq)$ ______

b Oxidation half-equation: $Fe^{2+}(aq) \rightarrow Fe^{3+}(aq)$ ______

c Overall redox equation:

______

d Select the correct words in italics to make the following statement true.

In the above reaction, $MnO_4^-(aq)$ is the *oxidising agent/reducing agent* and is *oxidised/reduced* by $Fe^{2+}(aq)$.

2 Consider the following overall redox equation and complete the question.

$2S_2O_4^{2-}(aq) + H_2O(l) \rightarrow S_2O_3^{2-}(aq) + 2HSO_3^-(aq)$

a Reduction half-equation: ______

b Oxidation half-equation: ______

c Explain how oxidation numbers can be used to show that $S_2O_3^{2-}(aq)$ is the product of a reduction half-reaction.

______

______

3 Determine the oxidation state of Cr in each of the following ions and compounds.

$Cr_2O_7^{2-}(aq)$: ______ $Cr^{3+}(aq)$: ______ $K_2CrO_4(s)$: ______

4 Consider the overall redox equation: $2HCHO(aq) + O_2(g) \rightarrow 2HCOOH(aq)$ and complete the question.

a Reduction half-equation: ______

b Oxidation half-equation: ______

c Use oxidation numbers to describe what happens to the HCHO(aq) and $O_2(g)$ in this reaction.

______

______

5 For the following reaction, balance the half-equations *under alkaline conditions* and name the process for each half-equation. Deduce the balanced overall equation.

______ half-equation: $Cr_2O_7^{2-}(aq) \rightarrow Cr^{3+}(aq)$

______ half-equation: $C_2H_6O(aq) \rightarrow C_2H_4O_2(aq)$

Overall redox equation: ______

6 Use oxidation numbers to explain which species has been oxidised and which species has been reduced, where appropriate. Using this reasoning, confirm which reactions are redox reactions.

a $PbO(s) + 2HCl(aq) \rightarrow PbCl_2(s) + H_2O(l)$

______

______

______

b $PbO_2(s) + 4HCl(aq) \rightarrow PbCl_2(s) + 2H_2O(l) + Cl_2(g)$

______

______

______

c $Fe_2O_3(s) + 3CO(g) \rightarrow 2Fe(l) + 3CO_2(g)$

d $FeO(l) + SiO_2(s) \rightarrow FeSiO_3(l)$

 ISBN 978 0 6557 0027 2

# WORKSHEET 7

Simulation

## Investigating galvanic cells—developing an electrochemical series

The following terms may be needed to complete this worksheet.

| | | | | | |
|---|---|---|---|---|---|
| oxidation | reduction | electron flow | direction of positive ion flow | | $Zn^{2+}$ solution |
| inert | $Cu^{2+}$ solution | reducing agent | anode | cathode | opposite |
| $KNO_3$ solution | $AgNO_3$ solution | salt bridge | copper | electrode | oxidising agent |
| same | positive | negative | polarity | zinc | |

A student carried out a series of experiments to compare the relative strengths of a number of oxidising agents and reducing agents by constructing cells made up from the four half-cells in the table below. (Platinum electrodes were used where required.)

| Experimental half-cells | |
|---|---|
| **Half-cell** | **Half-reaction** |
| $I_2(aq)/I^-(aq)$ | $I_2(aq) + 2e^- \rightleftharpoons 2I^-(aq)$ |
| $Zn^{2+}(aq)/Zn(s)$ | $Zn^{2+}(aq) + 2e^- \rightleftharpoons Zn(s)$ |
| $Sn^{4+}(aq)/Sn^{2+}(aq)$ | $Sn^{4+}(aq) + 2e^- \rightleftharpoons Sn^{2+}(aq)$ |
| $Cu^{2+}(aq)/Cu(s)$ | $Cu^{2+}(aq) + 2e^- \rightleftharpoons Cu(s)$ |

The aim of the experiment was to construct an electrochemical series and determine the strongest oxidising agent and strongest reducing agent in the series. The student's measurement results are shown in the following table.

| Electrochemical series results | | | |
|---|---|---|---|
| **Cell number** | **Electrochemical cell** | **Voltage (V)** | **Half-cell with positive electrode** |
| 1 | $I_2(aq)/I^-(aq)$ and $Zn^{2+}(aq)/Zn(s)$ | 1.30 | $I_2(aq)/I^-(aq)$ |
| 2 | $I_2(aq)/I^-(aq)$ and $Sn^{4+}(aq)/Sn^{2+}$ | 0.39 | $I_2(aq)/I^-(aq)$ |
| 3 | $I_2(aq)/I^-(aq)$ and $Cu^{2+}(aq)/Cu(s)$ | 0.20 | $I_2(aq)/I^-(aq)$ |
| 4 | $Zn^{2+}(aq)/Zn(s)$ and $Sn^{4+}(aq)/Sn^{2+}(aq)$ | 0.91 | $Sn^{4+}(aq)/Sn^{2+}(aq)$ |
| 5 | $Zn^{2+}(aq)/Zn(s)$ and $Cu^{2+}(aq)/Cu(s)$ | 1.10 | $Cu^{2+}(aq)/Cu(s)$ |
| 6 | $Sn^{4+}(aq)/Sn^{2+}(aq)$ and $Cu^{2+}(aq)/Cu(s)$ | 0.19 | $Cu^{2+}(aq)/Cu(s)$ |

**1** Identify the chemicals and equipment needed for the student to carry out the experiment by completing the list below.

wires, voltmeter, ______________, ______________

0.1 M $I_2/I^-$ solution

0.1 M ______________ solution

0.1 M ______________ solution

______________ electrode

0.1 M ______________ solution

0.1 M ______________ solution

______________ electrode

______________ electrode

**2** Describe the safety precautions needed for this experiment.

______________________________________________

______________________________________________

______________________________________________

______________________________________________

3 Identify the missing terms in the method for this experiment.

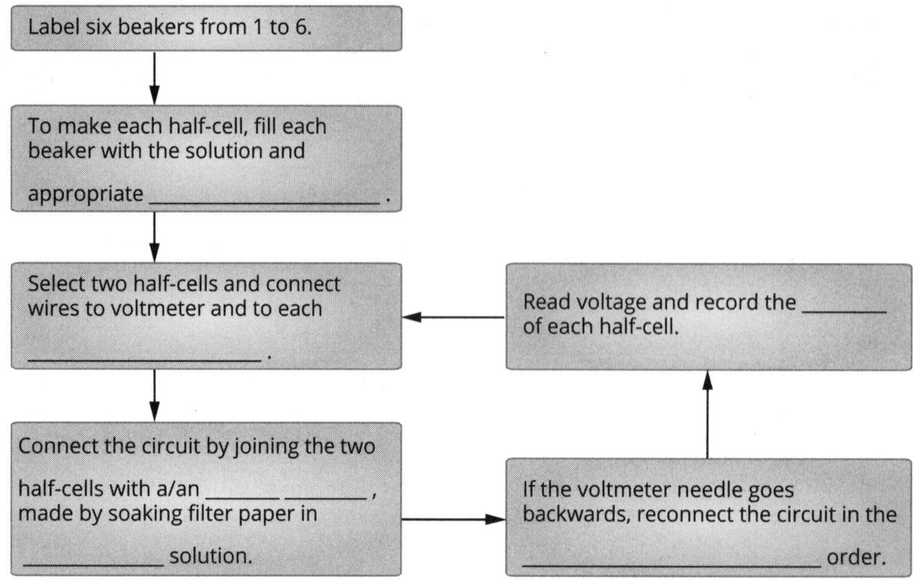


4 a What is the independent variable? ______________________________

b What is the dependent variable and how is it measured?

______________________________

5 Draw and fully label a diagram for cells 1 and 6 on the templates provided.

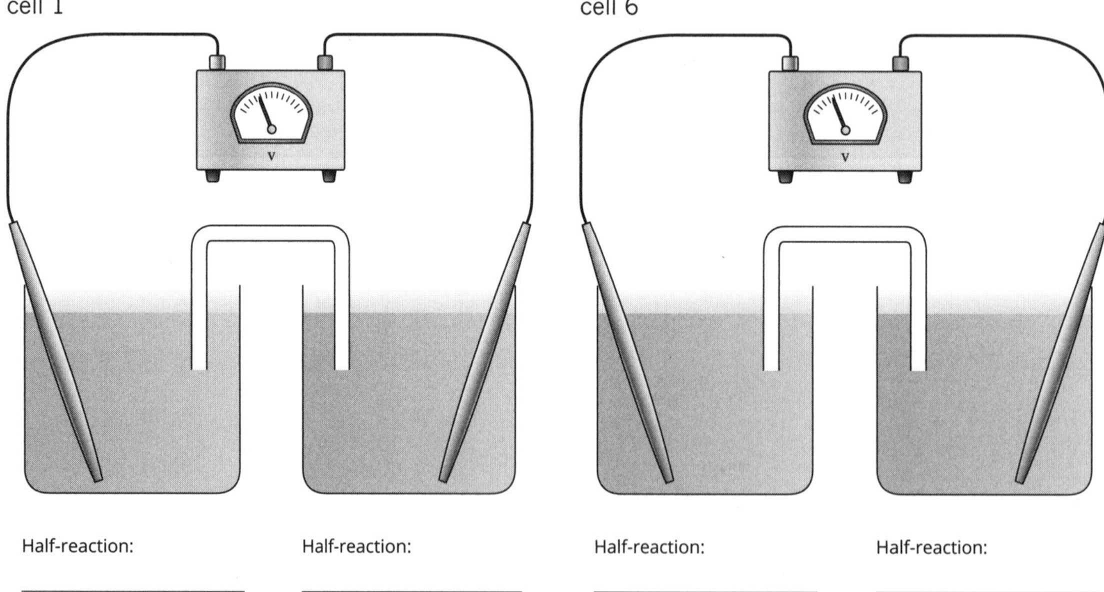


Half-reaction: ______________ Half-reaction: ______________ Half-reaction: ______________ Half-reaction: ______________

 ISBN 978 0 6557 0027 2

6 Use the results to complete the following table.

| Redox reactions results | | | |
|---|---|---|---|
| Cell number | Anode equation | Cathode equation | Overall equation |
| 1 | | | |
| 2 | | | |
| 3 | | | |
| 4 | | | |
| 5 | | | |
| 6 | | | |

7 Construct a simple electrochemical series for the half-reactions using the experimental results.

| strongest oxidising agent | $\rightleftharpoons$ | |
|---|---|---|
| | $\rightleftharpoons$ | |
| | $\rightleftharpoons$ | |
| | $\rightleftharpoons$ | strongest reducing agent |

8 Suggest why this electrochemical series may not exactly match the one printed in a textbook.

9 Develop a conclusion for this experiment, naming the strongest oxidising agent and reducing agent.

ISBN 978 0 6557 0027 2

# WORKSHEET 8

Modelling

## The electrochemical series—predicting reactions

1 The electrochemical series is used to compare the relative strengths of oxidising agents and reducing agents.

**a** Draw and label a diagram of the standard reference half-cell used for comparing half-cells. Show the construction of the electrode and the composition of the electrolyte. Use the following labels: platinum electrode (inert electrode), $H_2(g)$, glass tube, $H^+(aq)$.

**b** The electrochemical series shows the relative strengths of oxidising agents and reducing agents at standard conditions. Explain what is meant by standard conditions.

**c** Explain the function of an inert electrode.

2 Consider the three conjugate redox pairs $Br_2(aq)/Br^-(aq)$, $Zn^{2+}(aq)/Zn(s)$ and $Sn^{4+}(aq)/Sn^{2+}(aq)$.

**a** Use the electrochemical series in Appendix 2 to list the oxidising agents in order of increasing strength.

**b** Use the electrochemical series to list the reducing agents in order of increasing strength.

**c** Suppose inert electrodes, zinc metal and 1 M solutions of $Br_2$, $Br^-$, $Zn^{2+}$, $Sn^{2+}$ and $Sn^{4+}$ are available. Set up a cell using two of these redox pairs to give the highest voltage. Draw the cell and label the anode, cathode, positive and negative electrodes, direction of electron flow, salt bridge and direction of movement of positive ions in the salt bridge. Write the half-cell equations under the appropriate half-cells.

3 **a** Use the electrochemical series to predict the reaction that occurs in each of the following galvanic cells, writing the appropriate reactions in the table. The $E°$ for the $MnO_4^-/Mn^{2+}$ half-cell is 1.49 V.

| Galvanic cell | Anode reaction | Cathode reaction | Overall cell reaction |
|---|---|---|---|
| Half-cell 1: Cu metal electrode, $CuSO_4(aq)$<br>Half-cell 2: Mg metal electrode, $MgSO_4(aq)$ | | | |
| Half-cell 1: $Cl_2(g)$, $Cl^-(aq)$, inert electrode<br>Half-cell 2: $MnO_4^-(aq)$, $Mn^{2+}(aq)$ , $H^+(aq)$, inert electrode | | | |

**b** Suppose that all the contents of the half-cells for one of the cells were mixed together in one container. Give two reasons why the predicted reaction might not be observed.


ISBN 978 0 6557 0027 2

# WORKSHEET 9

Case study

# Fuel cells—applications and efficiency

**1** The alcohol concentration in a motorist's breath can be determined by using a breathalyser based on a fuel cell. In the cell, oxygen gas from the air is reduced to water, and ethanol, $CH_3CH_2OH$, in the breath is oxidised to ethanoic acid, $CH_3COOH$. The electrolyte is a solution of phosphoric acid, $H_3PO_4$, and a platinum–silver–plastic interface is used for each electrode. The voltage produced by the cell is proportional to the ethanol concentration and, when calibrated, light-emitting diodes forming part of the instrument show green for zero ethanol in the breath, amber for a small amount of ethanol and red when above the limit.

**a** Write half-cell equations for the reaction at each electrode in acidic solution.

Anode: ______________________

Cathode: ______________________

**b** Write the overall equation in the fuel cell. ______________________

**c** Correctly label the diagram for this fuel cell, including the anode, cathode and electrolyte.

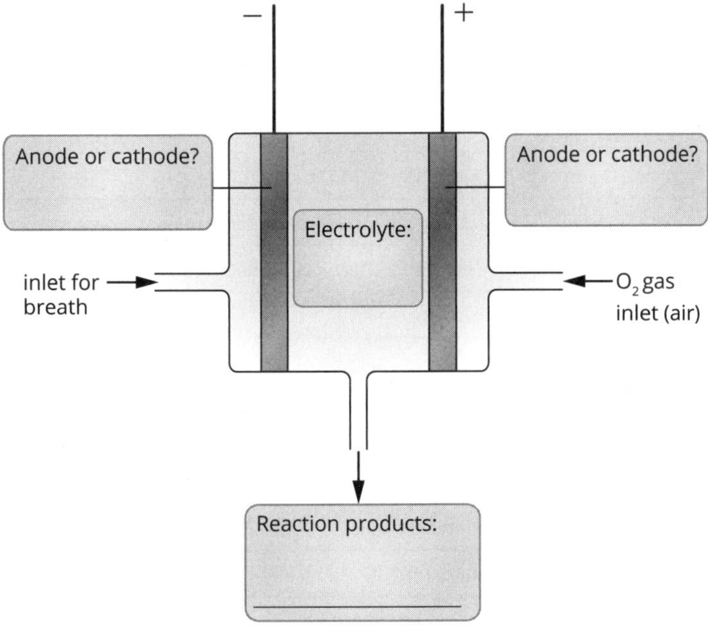


**2** The nature of the electrodes in a fuel cell determines the efficiency of the cell. List three functions of the electrodes in a fuel cell and explain how they increase the efficiency of fuel cells.

______________________

______________________

**3** When considering the use of fuel cells in motor vehicles, it is interesting to compare the efficiency of the engine when using hydrogen in a fuel cell to power an electric engine and petrol in a traditional combustion engine. Complete the table below.

| Engine type | Efficiency of energy transformations (%) | Energy transformations involved |
|---|---|---|
| electric engine driven by a hydrogen fuel cell | | |
| petrol combustion engine | | |

**4** Using hydrogen gas as a fuel in fuel cells offers the potential of a clean, efficient, renewable energy source. Discuss the use of hydrogen as a future fuel to produce renewable energy in terms of energy efficiency, atmospheric emissions and the use of renewable feedstocks.

______________________

______________________

______________________

ISBN 978 0 6557 0027 2

# WORKSHEET 10



## Literacy review—electrochemistry and thermochemistry

**1** To review your understanding of thermochemistry (the branch of chemistry concerned with the quantities of heat evolved or absorbed during chemical reactions) and electrochemistry, select the correct term for each definition from the list below. Some terms may be used several times and others not at all.

| | | | | |
|---|---|---|---|---|
| enthalpy | galvanic cell | fuel cell | oxidation | reduction |
| anode | cathode | salt bridge | specific heat capacity | spontaneous |
| $\Delta H$ | exothermic | endothermic | renewable | non-renewable |
| glucose | fossil fuel | biofuel | activation energy | |

| Definition | Term or expression |
|---|---|
| a fuel that can be produced from crops or other organic material | |
| another name for chemical energy or heat content of a substance | |
| an energy source that can be replaced at an equal or faster rate than it can be consumed | |
| the electrode where reduction occurs | |
| a reaction that releases energy to the surroundings | |
| the energy needed to break the bonds between atoms in the reactants | |
| a device used to complete a circuit by balancing the ions in each half-cell | |
| an electrochemical cell that spontaneously converts chemical to electrical energy | |
| a chemical reaction that occurs of its own accord | |
| a process that involves a loss of electrons | |
| the property of a substance, which is a measure of the energy required to raise the temperature of 1 gram of the substance by 1°C | |
| the electrode where oxidation occurs | |
| a process that involves a decrease in oxidation number | |
| the symbol for the energy content of a chemical reaction, usually measured in kJ | |
| a type of cell that continually converts chemical energy to electrical energy as reactants are supplied to it | |
| the main energy source for cells in the body | |

**2** Complete the concept map summarising the key ideas about galvanic cells by using terms from the box. (Some terms may be used more than once; others may not be used at all.)

| | | | | |
|---|---|---|---|---|
| oxidation | reduction | cathode | spontaneous reaction | |
| chemical | electrical | oxidising agent | reducing agent | anode |

 ISBN 978 0 6557 0027 2

## WORKSHEET 10

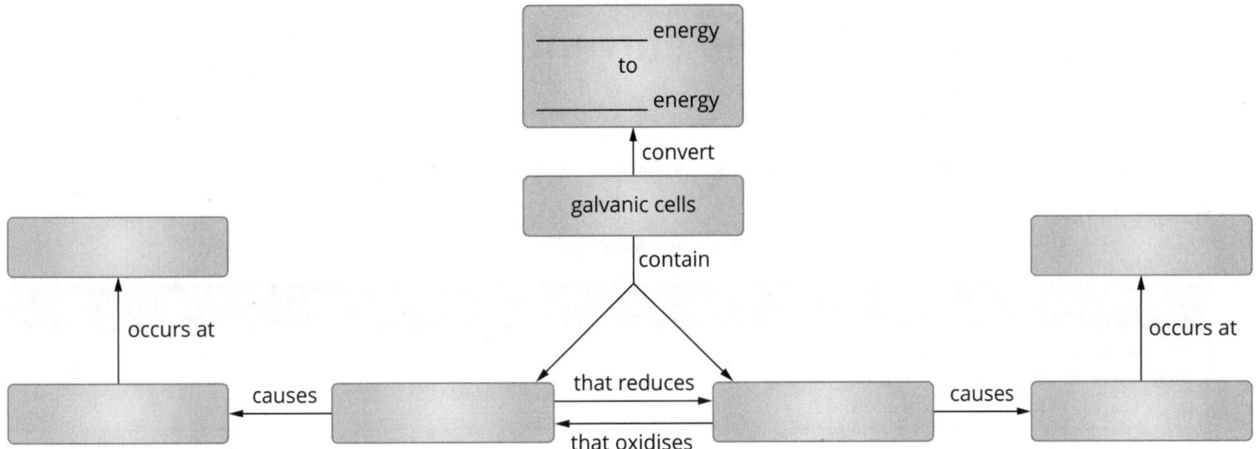


**3** Give the meaning of the term that was not used in the concept map and use an everyday example to illustrate the term.

Term: ____________________

Meaning: ____________________________________________________________

Example: ____________________

**4** Match the terms in the left-hand list (1–11) with the corresponding text from the right-hand side list (A–K).

| | | | |
|---|---|---|---|
| 1 | renewable fuel | A | a reaction with negative change in enthalpy |
| 2 | energy transformation efficiency | B | burning in excess oxygen |
| 3 | photosynthesis | C | a unit of energy |
| 4 | fermentation | D | a reaction in which plants convert solar energy to chemical energy |
| 5 | thermochemical equation | E | a reaction with positive change in enthalpy |
| 6 | joule | F | an anaerobic process involving enzymes in yeast |
| 7 | activation energy | G | a measure of the useful energy obtained from an energy conversion |
| 8 | complete combustion | H | a fuel that is continually replaced by natural processes within a relatively short period of time |
| 9 | exothermic reaction | I | a reaction equation that includes the energy change involved |
| 10 | endothermic reaction | J | the energy required for a reaction to occur |
| 11 | cellular respiration | K | $C_6H_{12}O_6(aq) + 6O_2(g) \rightarrow 6CO_2(g) + 6H_2O(l)$ |

**5** As a consequence of Faraday's laws, there are two equations that relate charge, time, current, moles of electrons and faraday. Write these equations and state the units for each quantity.

# WORKSHEET 11

## Reflection—What are the current and future options for supplying energy?

The following table summarises the key knowledge covered in this area of study.

**1** Reflect on how well you understand the concepts listed. Rate your learning by shading the circle that corresponds to your current level of understanding for each one.

| Key knowledge | Not confident ◄ | | | ► Very confident |
|---|---|---|---|---|
| The definition of a fuel; distinction between fossil fuels and biofuels; fuel sources for the body (carbohydrates, proteins and lipids) | ○ | ○ | ○ | ○ |
| Photosynthesis; oxidation of glucose; production of bioethanol | ○ | ○ | ○ | ○ |
| Exothermic and endothermic reactions; energy profile diagrams; combustion reactions; writing balanced thermochemical equations | ○ | ○ | ○ | ○ |
| Stoichiometric calculations for reactions involving the combustion of fuels; determination of limiting reactants | ○ | ○ | ○ | ○ |
| The use of specific heat capacity of water; principles of solution calorimetry | ○ | ○ | ○ | ○ |
| Calculations of energy transformation efficiency; energy values of foods containing carbohydrates, proteins and fats and oils | ○ | ○ | ○ | ○ |
| Redox reactions; oxidation numbers; balancing redox half-equations | ○ | ○ | ○ | ○ |
| Common design features and operating principles of primary galvanic cells | ○ | ○ | ○ | ○ |
| The use and limitations of the electrochemical series in designing galvanic cells | ○ | ○ | ○ | ○ |
| Common design features and operating principles of fuel cells | ○ | ○ | ○ | ○ |
| Applications of Faraday's laws and stoichiometry | ○ | ○ | ○ | ○ |
| Contemporary responses to challenges and innovation in the design of fuel cells for meeting society's energy needs | ○ | ○ | ○ | ○ |

**2** Consider the points you have shaded from 'Not confident' to 'Very confident'. List specific ideas you can identify that were challenging.

______________________________

______________________________

______________________________

**3** Write down two different strategies that you will apply to help further your understanding of these ideas.

______________________________

______________________________

______________________________

 ISBN 978 0 6557 0027 2

# PRACTICAL ACTIVITY 1

**Controlled experiment**

# Synthesis and energy content of biodiesel

## SUGGESTED DURATION

- 60 minutes

## INTRODUCTION

The biodiesel produced in this experiment may be suitable for use in a diesel engine. It is a methyl ester synthesised from a vegetable oil and methanol, according to the reaction in Figure 3.1.12. The length of the hydrocarbon chains depends on the vegetable oil used.

$$\begin{array}{l} H_2C-O-C(=O)-(CH_2)_n-CH_3 \\ HC-O-C(=O)-(CH_2)_n-CH_3 \\ H_2C-O-C(=O)-(CH_2)_n-CH_3 \end{array} + 3CH_3OH \xrightarrow{KOH} \begin{array}{l} H_2C-OH \\ HC-OH \\ H_2C-OH \end{array} + 3H_3C-O-C(=O)-(CH_2)_n-CH_3$$

triglyceride | methanol | glycerol | methyl ester (biodiesel)

**Figure 3.1.12** The production of a biodiesel

The energy released by burning the biodiesel is used to heat water. Knowing that 4.18 J of energy is required to heat 1 g of water by 1°C, you can calculate the quantity of energy released by combustion of the biodiesel.

## MATERIALS

- distilled water
- potassium hydroxide pellets, KOH(s)
- methanol, $CH_3OH(l)$
- canola oil (or similar vegetable oil)
- saturated sodium chloride solution, NaCl(aq)
- empty aluminium soft-drink can with an opening that has been enlarged to hold a thermometer and allow stirring
- crucible or aluminium tea light case
- paper towel, to form wick
- 125 mL conical flask with stopper
- stirring rod
- 10 mL measuring cylinder
- 50 mL measuring cylinder
- 100 mL measuring cylinder
- 2 × 100 mL beakers
- separating funnel (or dropping pipette)
- stopwatch
- thermometer, –10 to 110°C
- test tubes with stoppers
- retort stand and clamp
- tripod and gauze mat
- heatproof mat
- safety glasses
- gloves

ISBN 978 0 6557 0027 2

## PRACTICAL ACTIVITY 1

### AIM

To prepare a sample of biodiesel from vegetable oil, and to compare the energy content of this biodiesel with the energy content of petrol

| PRE-LAB SAFETY INFORMATION | | |
|---|---|---|
| **Material or procedure** | **Hazard** | **Control** |
| potassium hydroxide pellets, KOH(s) | • corrosive | Wear safety glasses, lab coat and gloves. |
| methanol, $CH_3OH(l)$ | • highly flammable, toxic and an irritant to the skin | Keep away from naked flames or sources of ignition and wear safety glasses, lab coat and gloves. |
| biodiesel | • flammable, toxic and an irritant to the skin | Wear safety glasses, lab coat and gloves. |
| Please indicate that you have understood the information in the safety table.<br>Name (print):<br>I understand the safety information (signature): | | |

### METHOD

## PART A • SYNTHESIS OF BIODIESEL

1 ▪ Place approximately 0.4 g of KOH pellets into a 125 mL conical flask.
2 ▪ Add 10 mL of methanol to the flask. Insert the stopper and swirl until the KOH pellets dissolve.
3 ▪ Add 50 mL of canola oil to the flask. Insert the stopper and shake the mixture vigorously for 10 minutes.
4 ▪ Pour the mixture into a separating funnel, and allow the mixture to separate overnight (or for at least 30 minutes) into two layers. The upper layer will be the biodiesel, and the lower layer will be mostly glycerol. Record your observations. (If a separating funnel is not available, the mixture can be left in the flask and the glycerol removed with a dropping pipette.)
5 ▪ Drain the glycerol (bottom layer) into a 100 mL waste beaker.
6 ▪ Add 5 mL of saturated NaCl solution to the separating funnel. Shake the mixture gently two or three times, and allow the mixture to separate for at least 30 minutes. Record your observations.
7 ▪ Drain the salt water (bottom layer) into the same waste beaker used for the glycerol (or use a dropping pipette to remove as much of the bottom layer as possible).
8 ▪ Determine and record the mass of a clean 100 mL beaker in Table 1. Drain the biodiesel (top layer) into this beaker. Determine and record the mass of the biodiesel produced.

## PART B • ENERGY CONTENT OF BIODIESEL

1 ▪ Add approximately 2 mL of biodiesel to a new, empty tea light case or crucible. The mass should be around 1.5 g.
2 ▪ Determine and record the mass of the tea light case and biodiesel in Table 3.
3 ▪ Form a small wick from a piece of paper towel and place it into the tea light case.
4 ▪ Set up a tripod and gauze mat on a retort stand in a well-ventilated area or in a fume hood.
5 ▪ Place the tea light case on the gauze mat.
6 ▪ Add 100 mL of water to the aluminium can. Clamp the aluminium can so that it sits about 10 cm above the gauze mat.
7 ▪ Using another clamp, insert a –10 to 110°C thermometer so that it is immersed in the water (but not touching the bottom of the can).
8 ▪ Record the temperature of the water.
9 ▪ Light the wick with a match or butane lighter.
10 ▪ Stir the water gently with a stirring rod, and note the temperature of the water every 30 seconds in Table 2.
11 ▪ Once the flame is extinguished, continue to record the temperature every 30 seconds. Keep measuring the temperature for at least 90 seconds after the maximum temperature is reached. Record the maximum temperature reached in Table 3, together with three or more readings after the maximum. Record the maximum $\Delta T$ in Table 3.


ISBN 978 0 6557 0027 2

## PRACTICAL ACTIVITY 1

### RESULTS

## PART A • SYNTHESIS OF BIODIESEL

Record your observations and measurement results for Part A.

**Table 1** Measurement results for Part A

| Item | Mass (g) |
|---|---|
| mass of clean 100 mL beaker | |
| mass of clean 100 mL beaker and biodiesel | |
| mass of biodiesel synthesised | |

## PART B • ENERGY CONTENT OF BIODIESEL

Record your observations and measurement results for Part B and plot a graph of your results.

**Table 2** Water temperature (°C) at time intervals after biodiesel lit

| Time (s) | Temp. (°C) | Time (s) | Temp. (°C) | Time (s) | Temp. (°C) |
|---|---|---|---|---|---|
| 0 | | 210 | | 420 | |
| 30 | | 240 | | 450 | |
| 60 | | 270 | | 480 | |
| 90 | | 300 | | 510 | |
| 120 | | 330 | | 540 | |
| 150 | | 360 | | 570 | |
| 180 | | 390 | | 600 | |

**Table 3** Summary of results for Part B

| Fuel | Mass of aluminium tea light case and fuel initially (g) | Mass of aluminium tea light case and fuel finally (g) | Mass of fuel burnt (g) | Initial temperature (°C) | Final temperature (°C) | $\Delta T$ (°C) |
|---|---|---|---|---|---|---|
| biodiesel | | | | | | |

### DISCUSSION

**1** Write an unbalanced equation for the combustion of your biodiesel, showing all the reactants and products.

**2** **a** What do you notice adhering to the base and sides of the aluminium can?

**b** What is this material likely to be?

**c** From your equation in Question 1 and your observations in Part A, what can you conclude about the nature of the combustion of the biodiesel in this experiment?

**d** How is this likely to affect the calculated heat of combustion of your biofuel? Explain your answer.

**e** What improvements can you suggest for this experimental design?

**3** For your biofuel, calculate the following quantities.

**a** Calculate the energy absorbed by the water when the fuel was burnt, in kJ.

**b** Determine the energy released by the fuel when it was burnt, in kJ.

**c** Calculate the heat of combustion for your biofuel in units of kJ $g^{-1}$.

**4** Compare the value obtained for your biofuel in units of kJ $g^{-1}$ with that of petrol, which has a heat of combustion of approximately 48 kJ $g^{-1}$. Explain why your experimental measurement of the heat content of biodiesel is likely to be less than the true value. On the basis of energy content, would the biofuel be a suitable alternative fuel to petrol?

**5** Discuss the purity of the biodiesel you made, suggesting the contaminants it might contain.

**6** Explain the meaning of the statement: 'Biodiesel is a carbon-neutral fuel'. Give reasons why this may not be always true.

## CONCLUSION

 ISBN 978 0 6557 0027 2

# PRACTICAL ACTIVITY 2

Controlled experiment

## Energy content of different alcohols

### SUGGESTED DURATION

- 50 minutes

### INTRODUCTION

Alcohols are useful fuels. Both methanol and ethanol have been used as substitutes for petrol. In this experiment, the energy released by burning alcohol is used to heat water. Knowing that 4.18 J of energy is required to heat 1 g (1 mL) of water by 1°C, you can calculate the quantity of energy released by combustion of the alcohol.

4.18 J $g^{-1}$ $°C^{-1}$ is the specific heat capacity of water.

### AIM

To measure and compare the thermal energy released during the combustion of various alcohols

### MATERIALS

- spirit burners containing one of the following alcohols: ethanol, $CH_3CH_2OH(l)$, propan-1-ol, $CH_3CH_2CH_2OH(l)$, or butan-1-ol, $CH_3CH_2CH_2CH_2OH(l)$
- 250 mL measuring cylinder
- steel can
- thermometer, –10 to 110°C
- retort stand and clamp
- bench mat
- electronic balance
- safety glasses

**PRE-LAB SAFETY INFORMATION**

| Material or procedure | Hazard | Control |
|---|---|---|
| ethanol, $CH_3CH_2OH(l)$ | • toxic and highly flammable | Avoid skin contact. Do not breathe vapour. |
| propan-1-ol, $CH_3CH_2CH_2OH(l)$ | • toxic and highly flammable | Avoid skin contact. Do not breathe vapour. |
| butan-1-ol, $CH_3CH_2CH_2CH_2OH(l)$ | • toxic and highly flammable | Avoid skin contact. Do not breathe vapour. |
| filling spirit burners | • flammable liquids | Do not fill near open flame. To minimise the risk of fire, spirit burners should be filled with alcohol prior to the commencement of the laboratory session. |

Please indicate that you have understood the information in the safety table.

Name (print): ______________________

I understand the safety information (signature): ______________________

### METHOD

1. Using a retort stand and clamp, set up the steel can 3–4 cm above the wick of the spirit burner.
2. Pour 200 mL of water into the can and record its temperature.
3. Weigh the spirit burner and alcohol. Record the mass and the name of the alcohol.
4. Light the burner and heat the water. Stir continuously.
5. Once the water temperature has increased by about 20°C, extinguish the burner and record the highest temperature reached by the water.
6. Record the mass of the burner and hence deduce the mass of alcohol consumed. Record this mass.
7. Repeat the experiment using another alcohol.

## PRACTICAL ACTIVITY 2

### RESULTS

Record your measurement results for the energy content of alcohols in the following table.

| Alcohol | Mass of spirit burner and alcohol initially (g) | Mass of spirit burner and alcohol finally (g) | Mass of alcohol burnt (g) | Initial temp. of water (°C) | Final temp. of water (°C) | $\Delta T$ of 200 mL water (°C) |
|---|---|---|---|---|---|---|
| ethanol | | | | | | |
| propan-1-ol | | | | | | |
| butan-1-ol | | | | | | |

### DISCUSSION

1 Write balanced equations for the complete combustion of ethanol, propan-1-ol and butan-1-ol.

2 For each alcohol, calculate the following quantities.

a Calculate the energy absorbed by the water when the alcohol was burnt.

| Ethanol | Propan-1-ol | Butan-1-ol |
|---|---|---|
| | | |

b Determine the energy released by the experimental mass of each alcohol burnt.

| Ethanol | Propan-1-ol | Butan-1-ol |
|---|---|---|
| | | |

c Calculate the energy content for each alcohol in units of kJ $g^{-1}$.

| Ethanol | Propan-1-ol | Butan-1-ol |
|---|---|---|
| | | |

d Calculate the heat of combustion for each alcohol in kJ $mol^{-1}$.

| Ethanol | Propan-1-ol | Butan-1-ol |
|---|---|---|
| | | |

 ISBN 978 0 6557 0027 2

## PRACTICAL ACTIVITY 2

3 Compare the values you obtained for each alcohol in units of kJ $g^{-1}$.

a On the basis of energy content calculated from your experiment, which alcohol would be the most suitable alternative fuel to petrol? (Petrol has a heat of combustion of approximately 48 kJ $g^{-1}$.)

b Why is a high energy content per unit mass an important consideration in selecting a fuel for vehicles, particularly for air transport?

4 The heats of combustion of ethanol, propan-1-ol and butan-1-ol are 1364, 2021 and 2670 kJ $mol^{-1}$.

a How accurate are your calculated results when compared to these true values?

b Comment on the main sources of error in this experiment and suggest how the experiment could be improved.

5 a Graph the published heats of combustion for each alcohol as stated in Question 4 (vertical axis) against the number of carbon atoms in each alcohol (horizontal axis).

b Estimate the value of the heat of combustion for pentan-1-ol using these published values.

c Check reference tables of chemical data to see how close your graphical estimate for pentan-1-ol is to the true value.

6 Discuss the validity of your results in terms of the experimental design and the results you obtained.

## CONCLUSION

# PRACTICAL ACTIVITY 3

Controlled experiment

## Energy content of a snack food

### SUGGESTED DURATION

- 40 minutes

### INTRODUCTION

The energy that a particular food supplies to the body is often almost the same as the energy released when that food is burnt. In this experiment, the energy content of a Cheezel is estimated by using the heat evolved when the Cheezel burns to heat water. You can determine how much heat is absorbed by water using the heat capacity of water. One gram of water absorbs 4.18 J for a rise in temperature of 1°C.

### AIM

To measure the thermal energy released during the combustion of a Cheezel

As directed by your teacher, complete the pre-lab safety information by referring to the safety data sheets (SDSs) or your teacher's risk assessment for the activity.

| PRE-LAB SAFETY INFORMATION | | |
|---|---|---|
| **Material used** | **Hazard** | **Control** |
| | | |

Please indicate that you have understood the information in the safety table.

Name (print):

I understand the safety information (signature):

### MATERIALS

- Cheezel®
- 50 mL measuring cylinder
- retort stand and clamp
- steel/aluminium can
- bench mat
- thermometer, −10 to 110°C, or temperature probe and data collection system
- needle probe
- micro or semi-micro Bunsen burner
- electronic balance
- safety glasses

### METHOD

1. Attach a Cheezel to the needle probe. Determine the accurate mass of the needle probe with the Cheezel attached. (Try to position the needle so that it passes through both sides of the Cheezel.)
2. Record the energy content reported on the packet in Table 1.
3. Use a retort stand and clamp a can so that its base is about 30 cm above a bench mat.
4. Add 50 mL of water to the can; measure and record its temperature.
5. Clamp the needle probe and Cheezel so that the Cheezel is 5–10 cm below the base of the can. Make sure the Cheezel is centred below the can.
6. Use a micro burner to ignite the Cheezel. Use the thermometer to stir the water gently. (Do not leave the thermometer in contact with the can.) Heat the water with the burning Cheezel until it has stopped burning (there will be residual Cheezel on the needle probe).
7. Record the highest temperature reached by the water in the can.
8. Allow what remains of the Cheezel to cool. Weigh it together with the needle probe to prevent the burnt Cheezel from breaking up. Record its mass.

### RESULTS

| Table 1 Burnt Cheezel results | |
|---|---|
| Energy content of food | |
| Initial mass of food and needle probe (g) | |
| Final mass of burnt food and needle probe (g) | |
| Initial temperature of water (°C) | |
| Final temperature of water after heating (°C) | |

 ISBN 978 0 6557 0027 2

## PRACTICAL ACTIVITY 3

### DISCUSSION

1 Calculate the change in temperature of the water. ___________

2 Calculate the thermal energy absorbed by the water in the can when the Cheezel was burnt.

3 Use your answer to Question 2 to calculate the energy released per mass of Cheezel burnt, in units of $kJ\,g^{-1}$.

4 Compare the result you obtained for the energy content of the Cheezel with the specified true value on the packet.

Account for any difference.

5 How could you calibrate your equipment to obtain a more accurate result? Suggest improvements and modifications to this experiment to improve the reproducibility of the data.

6 Discuss the errors associated with this practical activity.

### CONCLUSION

# PRACTICAL ACTIVITY 4

Controlled experiment

# Fermentation

## SUGGESTED DURATION

- 30 minutes to set up and up to 50 minutes over 5 days for completion of fermentation

## INTRODUCTION

In this activity, you will design an experiment to produce ethanol by fermenting sugar. A recent use of the age-old process of the fermentation of sugar is in the production of bioethanol fuel for motor vehicles. In this investigation, ethanol is made from sucrose. Fermentation takes about 5 days. During this time, the reaction is monitored by measuring the mass of the fermentation vessel. The process is complete when constant mass is reached.

## AIM

- To design an experiment to produce ethanol by fermenting sugar and to verify the product gas is $CO_2$ by testing with limewater
- To monitor the reaction by measuring the mass of the fermentation vessel and determine when the process is complete (Fermentation takes about 5 days.)

## SUGGESTED APPROACH

1. Using research from secondary sources for information, write a method of how you will perform this experiment. Include a labelled diagram of the experimental set-up. List the materials and submit your method and materials list to your teacher for approval.
2. Complete a risk assessment for your chosen method in the table provided. Show it to your teacher.
3. Set up your experiment so you can monitor the mass changes.

| PRE-LAB SAFETY INFORMATION | | |
|---|---|---|
| **Material used** | **Hazard** | **Control** |
| limewater, $Ca(OH)_2$ solution | | |
| | | |
| | | |
| Please indicate that you have understood the information in the safety table.<br>Name (print):<br>I understand the safety information (signature): | | |

## STUDENT MATERIALS


ISBN 978 0 6557 0027 2

## PRACTICAL ACTIVITY 4

## STUDENT METHOD

**Variables**

Independent variable: ______________________________

Dependent variable: ______________________________

Controlled variables: ______________________________

## DISCUSSION

1 Construct a results table for recording the measurements and observations and draw a labelled diagram of the experimental set-up.

2 State the gas that was produced during fermentation and indicate how you knew this.

______________________________

3 Write a balanced equation for the fermentation process involving glucose, which can be used to produce bioethanol.

______________________________

4 Briefly discuss the advantages and disadvantages associated with the use of ethanol as a biofuel.

______________________________

______________________________

______________________________

5 Identify any errors in the experimental method and the effects they may have had. When evaluating your method and results, suggest improvements and modifications to improve the reproducibility of the results.

______________________________

______________________________

______________________________

## CONCLUSION

______________________________

______________________________

# PRACTICAL ACTIVITY 5

Controlled experiment

# Calibration of a calorimeter

## SUGGESTED DURATION

- 50 minutes

## INTRODUCTION

A calorimeter is often calibrated by passing an electric current through a heating coil in the instrument. The temperature of the calorimeter and its contents increases in proportion to the amount of energy supplied by the heating coil. The calibration factor (joules of energy required per degree Celsius of temperature increase) is determined by measuring the temperature increase that results from a measured energy input. The value of the calibration factor depends upon the characteristics of the particular calorimeter.

## MATERIALS

- 100 mL measuring cylinder
- calorimeter (with a code number for identification)
- DC power supply
- 5 wire leads
- thermometer, −10 to 50°C
- stopwatch
- ammeter
- voltmeter

## AIM

To calibrate a calorimeter by measuring the increase in temperature that results from a measured input of electrical energy

**PRE-LAB SAFETY INFORMATION**

| Material used | Hazard | Control |
|---|---|---|
| electrical equipment | • electric shock or damage to instruments from incorrect connection of circuit | Use electrical equipment carefully. Check all connections. |

Please indicate that you have understood the information in the safety table.

Name (print): ______________________

I understand the safety information (signature): ______________________

## METHOD

1. Record the code number of your calorimeter in the space below the Results tables. You may need to use this calorimeter for future experiments after you have calibrated it.
2. Pour 100 mL of water into the calorimeter. Stir the water and record its temperature every 30 seconds in Table 1 for several minutes. (You could enter data directly into a computer spreadsheet program for later processing.)
3. Once the temperature is steady, apply a measured voltage of approximately 6 V for exactly 3 minutes using the circuit shown in Figure 3.1.13. Stir continuously and record the temperature every 30 seconds.
4. Record the potential difference (voltage) and current while the water is heating. After 3 minutes, turn off the power. Continue to stir the water in the calorimeter and record its temperature every 30 seconds for a further 3 minutes.
5. Discard the water in the calorimeter and repeat steps 1–4 using fresh water. Record your results in Table 2 .

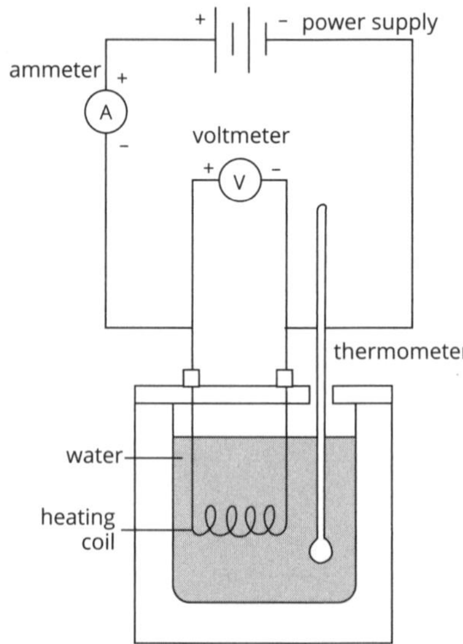

**Figure 3.1.13** The experimental circuit set-up

 ISBN 978 0 6557 0027 2

## RESULTS

**Table 1** Results for calibration 1

| Time (min) | Temperature (°C) | Time (min) | Temperature (°C) | Time (min) | Temperature (°C) |
|---|---|---|---|---|---|
| | | | | | |
| | | | | | |
| | | | | | |
| | | | | | |
| | | | | | |

**Table 2** Results for calibration 2

| Time (min) | Temperature (°C) | Time (min) | Temperature (°C) | Time (min) | Temperature (°C) |
|---|---|---|---|---|---|
| | | | | | |
| | | | | | |
| | | | | | |
| | | | | | |
| | | | | | |

Code number of your calorimeter: ___________

Potential difference during heating for calibration 1: ___________ V

Current during heating for calibration 1: ___________ A

Time of heating for calibration 1: _______________ s

Potential difference during heating for calibration 2: ___________ V

Current during heating for calibration 2: ___________ A

Time of heating for calibration 2: _______________ s

**Variables**

Independent variable: ___________________________________

Dependent variable: ___________________________________

Controlled variables: ___________________________________

## DISCUSSION

1 Graph the two sets of data you collected, plotting temperature against time.

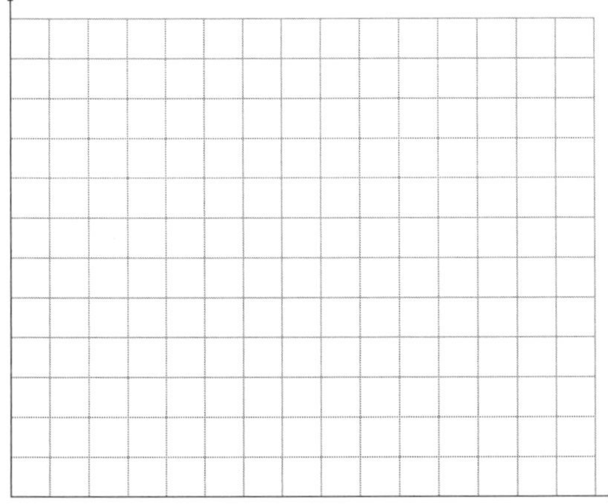

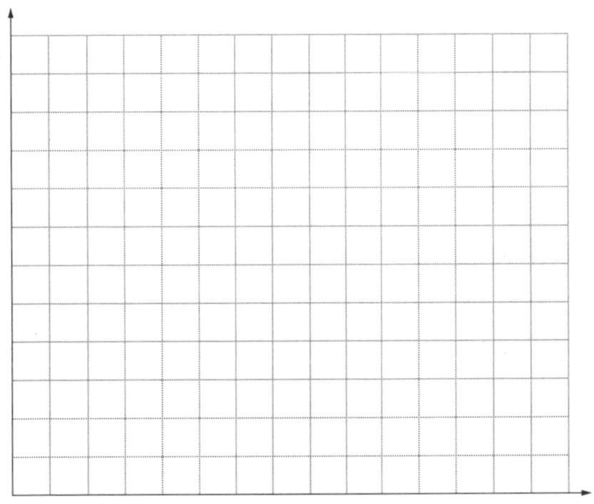

2 For each set of data, determine the temperature increase over the 3-minute heating period.

Temperature increase for first calibration: __________°C

Temperature increase for second calibration: __________°C

3 For each set of data, calculate the thermal energy supplied by the heating coil in the calorimeter.

4 Calculate the calibration factor for each set of data.

Mean measurement value of the two calibration factors: __________ J °C$^{-1}$

5 What energy transformation occurs when the calorimeter and its contents are being heated?

______________________________

6 **a** Sketch the temperature against time graph that you would expect to obtain for a:

**i** well-insulated calorimeter. **ii** poorly insulated calorimeter.

**b** Comment on the insulation of your own calorimeter.

______________________________

7 What would be the effect on the calibration factor obtained from this experiment of unknowingly using 50 mL of water instead of 100 mL of water to calibrate the calorimeter?

______________________________

8 In future experiments, what volume of solution should you place in the calorimeter to measure the enthalpy changes of chemical reactions?

______________________________

9 Discuss the systematic and random errors associated with this experiment and their effect on the accuracy and precision of your results. Comment on the validity of this experiment.

______________________________

______________________________

______________________________

**CONCLUSION**

______________________________

______________________________

______________________________


ISBN 978 0 6557 0027 2

# PRACTICAL ACTIVITY 6

Controlled experiment

## Half-cells and the electrochemical series

### SUGGESTED DURATION

- 50 minutes

### INTRODUCTION

During the operation of a galvanic cell, the relative abilities of the chemical species in the half-cells to donate and accept electrons determine which electrode is positive and which is negative. The half-cell containing the stronger reducing agent will undergo an oxidation reaction; the electrode in this half-cell will be negative. The other half-cell will contain the stronger oxidising agent and undergo a reduction reaction.

### AIM

To construct a number of different electrochemical cells and use them to determine the relative positions of various pairs of oxidising agents and reducing agents in the electrochemical series

As directed by your teacher, complete the pre-lab safety information by referring to the safety data sheets (SDSs) or your teacher's risk assessment for the activity.

### MATERIALS

- 50 mL of 0.1 M copper(II) nitrate solution, $Cu(NO_3)_2(aq)$
- 50 mL of an equal mixture of 0.2 M iron(III) nitrate solution, $Fe(NO_3)_3(aq)$, and 0.2 M iron(II) sulfate solution, $FeSO_4(aq)$
- 50 mL of an equal mixture of 0.05 M iodine, $I_2(aq)$ and 0.1 M potassium iodide solution, KI(aq)
- 50 mL of 0.1 M zinc nitrate solution, $Zn(NO_3)_2(aq)$
- 50 mL of 0.1 M potassium nitrate solution, $KNO_3(aq)$
- copper electrode
- zinc electrode
- 2 carbon rods
- 5 × 100 mL beakers
- voltmeter
- 2 wire leads with alligator clips
- 6 strips of 2 cm × 15 cm chromatography paper or cartridge paper
- safety glasses

**PRE-LAB SAFETY INFORMATION**

| Material used | Hazard | Control |
|---|---|---|
| 0.1 M copper(II) nitrate solution, $Cu(NO_3)_2(aq)$ | | |
| 0.2 M iron(III) nitrate solution, $Fe(NO_3)_3(aq)$ | | |
| 0.2 M iron(II) sulfate solution, $FeSO_4(aq)$ | | |
| 0.1 M zinc nitrate solution, $Zn(NO_3)_2(aq)$ | | |

Please indicate that you have understood the information in the safety table.

Name (print): ______________________

I understand the safety information (signature): ______________________

**Variables**

Independent variable: ______________________

Dependent variable: ______________________

Controlled variables: ______________________

### METHOD

1. Construct the following half-cells:
   a. $Cu^{2+}/Cu$: a 100 mL beaker containing a copper electrode and a 50 mL copper(II) nitrate solution
   b. $Fe^{3+}/Fe^{2+}$: a 100 mL beaker containing a carbon rod and a 50 mL iron(III) nitrate/iron(II) sulfate solution (25 mL of each)
   c. $I_2/I^-$: a 100 mL beaker containing a carbon rod and a 50 mL iodine/potassium iodide solution
   d. $Zn^{2+}/Zn$: a 100 mL beaker containing a zinc electrode and a 50 mL zinc nitrate solution.
2. Make a salt bridge by soaking a strip of filter paper in a beaker of potassium nitrate solution. Join the $Cu^{2+}/Cu$ and $Zn^{2+}/Zn$ half-cells using the salt bridge. Attach two wires to the voltmeter and clip the other end of one wire to the copper electrode. Momentarily touch the loose end of the second wire to the zinc electrode. If the needle of the voltmeter is deflected onto the scale, clip this loose end of the wire to the zinc electrode and proceed with the next step. If the needle of the voltmeter is deflected below zero, swap the wires at the terminals of the voltmeter before continuing.

3 ▪ Record the voltage in Table 1 and identify the positive and negative electrodes. (The positive electrode is connected to the positive terminal of the voltmeter.)
4 ▪ Disconnect the leads. Remove and discard the salt bridge.
5 ▪ Repeat steps 2–4 for each of the other five combinations of half-cells.

## RESULTS

**Table 1 Voltage and electrode polarity**

| Cell | Voltage (V) | Positive electrode |
|---|---|---|
| $Cu^{2+}/Cu$ and $Zn^{2+}/Zn$ | | |
| $Fe^{3+}/Fe^{2+}$ and $Zn^{2+}/Zn$ | | |
| $Fe^{3+}/Fe^{2+}$ and $I_2/I^-$ | | |
| $Cu^{2+}/Cu$ and $I_2/I^-$ | | |
| $Cu^{2+}/Cu$ and $Fe^{3+}/Fe^{2+}$ | | |
| $I_2/I^-$ and $Zn^{2+}/Zn$ | | |

## DISCUSSION

1 Draw a diagram of the cells listed. Label the contents of the half-cells, the positive and negative electrodes, the anode and cathode and the direction of electron flow through the external circuit.

$Cu^{2+}/Cu$ and $I_2/I^-$ cell

$Fe^{3+}/Fe^{2+}$ and $Zn^{2+}/Zn$ cell

2 Write half-equations for the reactions occurring in each half-cell and then write an overall equation for the reaction in each cell.

**a** $Cu^{2+}/Cu$ and $Zn^{2+}/Zn$

Anode reaction: ____________________

Cathode reaction: ____________________

Overall reaction: ____________________

**b** $Fe^{3+}/Fe^{2+}$ and $Zn^{2+}/Zn$

Anode reaction: ____________________

Cathode reaction: ____________________

Overall reaction: ____________________

**c** $Fe^{3+}/Fe^{2+}$ and $I_2/I^-$

Anode reaction: ____________________

Cathode reaction: ____________________

Overall reaction: ____________________

**d** $Cu^{2+}/Cu$ and $I_2/I^-$

Anode reaction: ____________________

Cathode reaction: ____________________

Overall reaction: ____________________


ISBN 978 0 6557 0027 2

**e** $Cu^{2+}/Cu$ and $Fe^{3+}/Fe^{2+}$

Anode reaction: ______

Cathode reaction: ______

Overall reaction: ______

**f** $I_2/I^-$ and $Zn^{2+}/Zn$

Anode reaction: ______

Cathode reaction: ______

Overall reaction: ______

**3** Examine the results you have obtained and write the four half-cell reactions as reduction reactions, listing them in order of increasing reducing agent strength. (Your list should show the strongest oxidising agent in the upper left corner and the strongest reducing agent in the lower right corner.) This is an electrochemical series.

**4** Compare the order of your half-equations in Question 3 with that of those equations listed in the electrochemical series in Appendix 2. Are they the same? Why might differences arise?

**5** What is the function of the salt bridge in an electrochemical cell and why was it necessary to use a new salt bridge for each electrochemical cell?

**6** Consider a $Pb^{2+}/Pb$ half-cell. If lead is a stronger reducing agent than copper, but $Pb^{2+}(aq)$ is a stronger oxidising agent than $Zn^{2+}(aq)$, where in your electrochemical series would you place the following equation?

$Pb^{2+}(aq) + 2e^- \rightarrow Pb(s)$

**7** Briefly evaluate the method and how you could test the repeatability and reproducibility of the data (explaining the difference). Indicate the accuracy of your measurement results in your evaluation. State any random errors and systematic errors.

## CONCLUSION

List $Cu^{2+}$, $Fe^{3+}$, $I_2$ and $Zn^{2+}$ in order of their increasing strength as oxidising agents.

# PRACTICAL ACTIVITY 7

Controlled experiment

# Fuel cells

## SUGGESTED DURATION

- 50 minutes

## INTRODUCTION

Fuels cells provide an efficient, clean energy source that can be used as a means of energy production for cities or for motor vehicles.

## AIM

To investigate the chemistry of two fuel cells: the hydrogen–oxygen fuel cell and the aluminium–air fuel cell

| PRE-LAB SAFETY INFORMATION | | |
|---|---|---|
| **Material used** | **Hazard** | **Control** |
| 1 M potassium hydroxide solution, KOH(aq) | • harmful if swallowed; causes severe burns | Avoid breathing spray from this solution and avoid contact. Wear gloves, safety glasses and a laboratory coat. Do not leave the potassium hydroxide solution in the soft-drink can. |
| electrical equipment | • electric shock or damage to instruments from incorrect connection of circuit | Use electrical equipment carefully. Check all connections. |
| Please indicate that you have understood the information in the safety table. | | |
| Name (print): | | |
| I understand the safety information (signature): | | |

## MATERIALS

- 500 mL of 1 M potassium hydroxide solution, KOH(aq)
- 375 mL aluminium soft-drink can
- 100 mL measuring cylinder
- 2 semi-micro test tubes
- DC power supply
- fuel cell assembly (prepared in advance)
  - plastic cup
  - silicone sealant
  - 2 carbon rods
- retort stand and clamp
- voltmeter
- air pump (e.g. fish tank pump), tubing and tubing clamp (optional)
- 2 wire leads with alligator clips
- spatula
- safety glasses
- safety gloves

## METHOD

## PART A • HYDROGEN–OXYGEN FUEL CELL

1 ▪ The apparatus required for this part of the practical activity should be prepared in advance. It consists of two carbon rods inserted through the base of a plastic cup and sealed in place with silicone sealant (see Figure 3.1.14).

2 ▪ Use a stand or other suitable equipment to clamp the cup in an upright position. Wearing gloves, fill the cup with 1 M potassium hydroxide solution so that the carbon rods are just covered.

3 ▪ Fill a semi-micro test tube with potassium hydroxide solution. Carefully invert the test tube while holding it over the fuel cell apparatus. Surface tension should hold the potassium hydroxide solution in the inverted test tube. Lower the test tube, placing it over one of the graphite electrodes. Repeat this procedure with another semi-micro test tube and the other electrode.

4 ▪ Connect the bottom ends of the graphite electrodes to the terminals of a DC power supply. Adjust the output of the power supply to 6 V and turn the supply on. A gas will begin to form at each of the electrodes. Continue until one of the test tubes is three-quarters full of gas. Turn off and disconnect the power supply.

5 ▪ Connect the electrodes to a voltmeter and record the voltage of the cell in in Table 1.

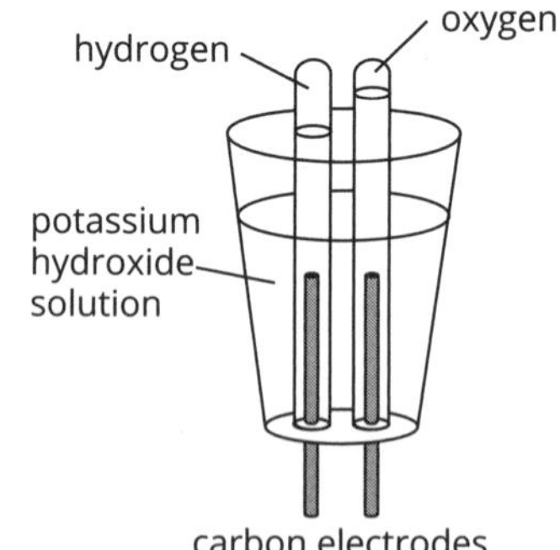


**Figure 3.1.14** The construction of a hydrogen–oxygen fuel cell


ISBN 978 0 6557 0027 2

## PART B • ALUMINIUM–AIR FUEL CELL

1 ▪ Rinse an aluminium soft-drink can with tap water to remove any soft-drink residue. Using a spatula, scratch the interior of the can to expose fresh aluminium. Pour 300 mL of 1 M potassium hydroxide solution into the can.
2 ▪ Using a retort stand, clamp a carbon rod into position so that it is immersed in the potassium hydroxide electrolyte without touching the aluminium can.
3 ▪ Bubble air slowly into the solution using an air pump. Be careful not to splash the electrolyte because it is corrosive. (This step can be omitted if an air pump is not available because the electrolyte will already contain some dissolved oxygen.)
4 ▪ Record the voltage of the cell and the polarity of the electrodes in Table 1.
5 ▪ Discard the contents of the can immediately after use. Do not allow to stand overnight as the KOH will react with the aluminium can.

### RESULTS

| Table 1 Results | |
|---|---|
| hydrogen–oxygen fuel cell voltage | |
| aluminium–air fuel cell voltage | |
| polarity of the carbon rod in the aluminium–air fuel cell | |

### DISCUSSION

## PART A • HYDROGEN–OXYGEN FUEL CELL

1 Write half-equations for the reactions occurring at each electrode when the cell is connected to the power supply. Remember that an alkaline electrolyte has been used.

Anode reaction: ______________________

Cathode reaction: ______________________

2 Explain the difference in the volume of gas produced at each electrode when current flows into the cell from the DC power supply.

______________________

______________________

______________________

3 Write half-equations for the reactions that occur at each electrode when the cell is discharging. Remember that an alkaline electrolyte has been used.

Anode reaction: ______________________

Cathode reaction: ______________________

4 What are the major differences between this hydrogen–oxygen fuel cell that you constructed and the fuel cells being developed for commercial use?

______________________

______________________

______________________

## PART B • ALUMINIUM–AIR FUEL CELL

1 On the basis of the polarity of the electrodes, identify the anode and the cathode of the cell.

______________________

______________________

2 Write equations for the electrode reactions that occur as the cell discharges.

3 What advantages can you see in using the aluminium–air fuel cell over the lead–acid cell to power a car?

4 Briefly discuss any limitations encountered in this experiment and possible modifications.

5 Explain how you could determine the precision of your results.

## CONCLUSION

 ISBN 978 0 6557 0027 2

# EXAM QUESTIONS

## Multiple-choice questions

### Question 1 VCE Chemistry 2018 (A) 10

Bioethanol, $C_2H_5OH$, is produced by the fermentation of glucose, $C_6H_{12}O_6$, according to the following equation.

$C_6H_{12}O_6(aq) \rightarrow 2C_2H_5OH(aq) + 2CO_2(g)$

The mass of $C_2H_5OH$ obtained when 5.68 g of carbon dioxide, $CO_2$, is produced is

**A.** 0.168 g

**B.** 0.337 g

**C.** 2.97 g

**D.** 5.94 g

### Question 2 VCE Chemistry 2019 (A) 23

Which one of the following statements about enthalpy change is correct?

**A.** The sign of the enthalpy change for an endothermic reaction is negative.

**B.** The sign of the enthalpy change for the condensation of a gas to a liquid is negative.

**C.** The enthalpy change is the difference between the activation energy and the energy of the reactants.

**D.** The enthalpy change is the difference between the activation energy and the energy of the products.

### Question 3 VCE Chemistry 2019 (A) 9

A reaction has the energy profile diagram shown below.

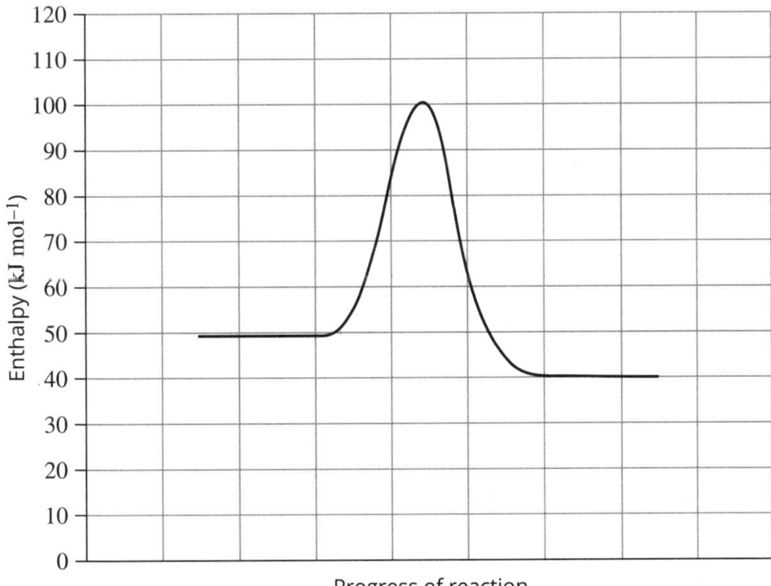


Which of the following represents the energy profile of the reverse reaction?

| | Final product energy (kJ mol$^{-1}$) | $\Delta H$ (kJ mol$^{-1}$) |
|---|---|---|
| **A.** | 40 | +10 |
| **B.** | 50 | +10 |
| **C.** | 50 | −10 |
| **D.** | 40 | −10 |

# EXAM QUESTIONS

**Question 4** VCE Chemistry 2018 (A) 14

An equation for the complete combustion of methanol is

$2CH_3OH(l) + 3O_2(g) \rightarrow 2CO_2(g) + 4H_2O(g)$

$\Delta H$ for this equation would be

**A.** +726 kJ $mol^{-1}$

**B.** –726 kJ $mol^{-1}$

**C.** +1452 kJ $mol^{-1}$

**D.** –1452 kJ $mol^{-1}$

**Question 5** VCE Chemistry 2019 (A) 26

The calibration factor of a bomb calorimeter was determined by connecting the calorimeter to a power supply. The calibration was done using 100 mL of water, 6.5 V and a current of 3.6 A for 4.0 minutes. The temperature of the water increased by 0.48°C during the calibration.

4.20 g of sucrose underwent complete combustion in the bomb calorimeter. The temperature of the 100 mL of water increased from 19.6°C to 25.8°C.

$M(C_{12}H_{22}O_{11}) = 342$ g $mol^{-1}$

The experimental heat of combustion of pure sucrose, in joules per gram, is

**A.** $5.9 \times 10^6$

**B.** $7.3 \times 10^4$

**C.** $1.7 \times 10^4$

**D.** $1.2 \times 10^4$

**Question 6** VCE Chemistry 2018 (A) 15

The following table contains the percentage composition by mass of the nutritional value of some common foods.

| Food | % Carbohydrates | % Fats and oils | % Protein |
|---|---|---|---|
| fish | 0 | 8 | 29 |
| bread | 50 | 4 | 8 |
| cheese | 1 | 34 | 25 |
| milk | 5 | 4 | 3 |

Which one of the following servings has the highest energy content?

**A.** 100 g of fish

**B.** 80 g of bread

**C.** 40 g of cheese

**D.** 258 g (250 mL) of milk

**Question 7** VCE Chemistry 2021 (A) 19

A food chemist conducted an experiment in a bomb calorimeter to determine the energy content, in joules per gram, of a muesli bar. A 3.95 g sample of the muesli bar was combusted in the calorimeter and the temperature of the water rose by 16.7°C.

The calibration factor of the calorimeter was previously determined to be 4780 J $°C^{-1}$.

The energy content of the muesli bar is

**A.** $3.51 \times 10^5$ J $g^{-1}$

**B.** $2.02 \times 10^4$ J $g^{-1}$

**C.** $1.13 \times 10^3$ J $g^{-1}$

**D.** $7.25 \times 10$ J $g^{-1}$


ISBN 978 0 6557 0027 2

# EXAM QUESTIONS

**Question 8** VCE Chemistry 2018 (A) 11

A galvanic cell is set up as shown in the diagram below.

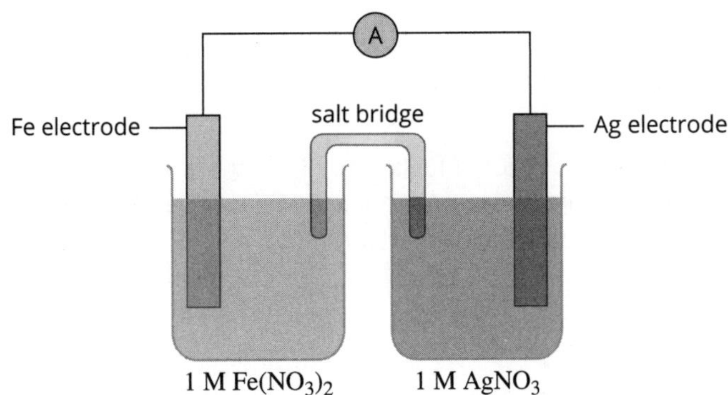


When this cell is operating

**A.** a gas forms at the Ag electrode.

**B.** the mass of the Ag electrode increases.

**C.** $Ag^+$ ions move towards the Fe electrode.

**D.** electrons move from the Ag electrode to the Fe electrode.

**Question 9** VCE Chemistry 2018 (A) 12

The overall reaction for an acidic fuel cell is shown below.

$2H_2 + O_2 \rightarrow 2H_2O$

Porous electrodes are often used in acidic fuel cells because they

**A.** are highly reactive.

**B.** are cheap to produce and readily available.

**C.** are more efficient than solid electrodes at moving charges and reactants.

**D.** provide a surface for the hydrogen and oxygen to directly react together.

**Question 10** VCE Chemistry 2020 (A) 30

Consider the following half-equation.

$ClO_2(g) + e^- \rightleftharpoons ClO_2^-(aq)$

It is also known that:

- $ClO_2(g)$ will oxidise HI(aq), but not HCl(aq)
- $Fe^{3+}(aq)$ will oxidise HI(aq), but not $NaClO_2(aq)$.

Based on this information, $Fe^{2+}(aq)$ can be oxidised by

**A.** $Cl_2(g)$ and $I_2(aq)$.

**B.** $Cl_2(g)$, but not $ClO_2(g)$.

**C.** $ClO_2(g)$ and $Cl_2(g)$, but not $I_2(aq)$.

**D.** $Cl_2(g)$, $ClO_2(g)$ and $I_2(aq)$.

# EXAM QUESTIONS

## Short-answer questions

**Question 1** (7 marks) VCE Chemistry 2019 (B) 4

Internal combustion engines are used in large numbers of motor vehicles. Historically, internal combustion engines have used fuels obtained from crude oil as a source of power. As concerns for the environment have grown, efforts have been made to obtain fuel for combustion engines from other sources.

**a.** One way of reducing the environmental effects of fossil fuels is to blend them with biofuels. A common method is to blend petrol with ethanol in varying ratios. A fuel can be obtained by blending 1 mole of octane, $C_8H_{18}$, and 1 mole of ethanol, $C_2H_5OH$.

The chemical equation for the complete combustion of this fuel mixture is

$C_8H_{18}(l) + C_2H_5OH(l) + 15\frac{1}{2}O_2(g) \rightarrow 10CO_2(g) + 12H_2O(g)$

Calculate the energy released, in kilojoules, when 80 g of this fuel mixture undergoes complete combustion. Show your working. 3 marks

---

---

---

**b.** Some car manufacturers are exploring the use of an acidic ethanol fuel cell to power vehicles. In this fuel cell, the ethanol at one electrode reacts with water that has been produced at the other electrode.

A membrane is used to transport ions between the electrodes. A diagram of an acidic ethanol fuel cell is shown below.

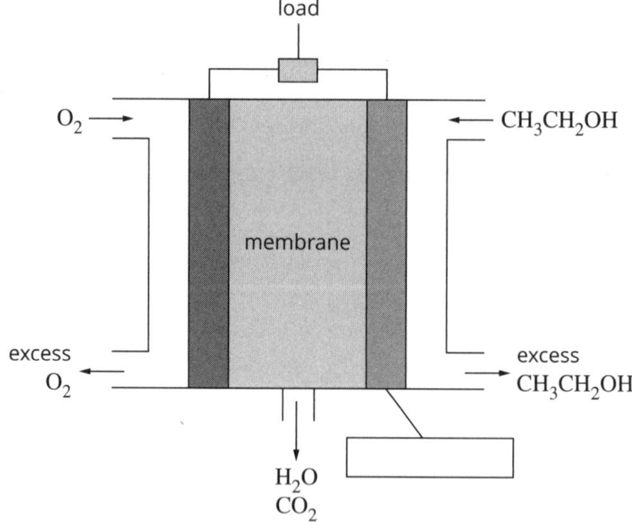


i. Identify the electrode as either the cathode or the anode in the box provided in the diagram above. 1 mark

ii. Write the half-equation for the reaction occurring at the anode. 1 mark

---

iii. The combustion of ethanol and the combustion of octane release about the same amount of energy per mole of carbon dioxide produced. Identify two advantages of powering a vehicle using an ethanol fuel cell instead of an internal combustion engine powered by octane. 2 marks

---

---

---

 ISBN 978 0 6557 0027 2

# EXAM QUESTIONS

**Question 2** (8 marks) VCE Chemistry 2020 (B) 6

Methane gas, $CH_4$, can be captured from the breakdown of waste in landfills. $CH_4$ is also a primary component of natural gas. $CH_4$ can be used to produce energy through combustion.

**a.** Write the equation for the incomplete combustion of $CH_4$ to produce carbon monoxide, CO. 1 mark

**b.** If 20.0 g of $CH_4$ is kept in a 5.0 L sealed container at 25°C, what would be the pressure in the container? 2 marks

**c.** A Bunsen burner is used to heat a beaker containing 350.0 g of water. Complete combustion of 0.485 g of $CH_4$ raises the temperature of the water from 20°C to 32.3°C.

Calculate the percentage of the Bunsen burner's energy that is lost to the environment. 3 marks

**d.** Compare the environmental impact of $CH_4$ obtained from landfill to the environmental impact of $CH_4$ obtained from natural gas. 2 marks

**Question 3** (9 marks) VCE Chemistry 2021 (B) 1

Digesters use bacteria to convert organic waste into biogas, which contains mainly methane, $CH_4$. Biogas can be used as a source of energy.

**a.** Both biogas and coal seam gas contain $CH_4$ as their main component.

Why is biogas considered a renewable energy source but coal seam gas is not? 1 mark

**b.** A digester processed 1 kg of organic waste to produce 496.0 L of biogas at standard laboratory conditions (SLC). The biogas contained 60.0% $CH_4$.

**i.** Write the thermochemical equation for the complete combustion of $CH_4$ at SLC. 2 marks

**ii.** Calculate the amount of energy that could be produced by $CH_4$ from 1 kg of organic waste. 3 marks

**c.** Biogas was combusted to release $1.63 \times 10^3$ kJ of energy. This energy was used to heat 100 kg of water in a tank. The initial temperature of the water was 25.0°C.

**i.** What is the maximum temperature that the water in the tank could reach? 2 marks

**ii.** State why this temperature may not be reached. 1 mark

**Question 4** (9 marks) VCE Chemistry 2021 (B) 4

**a.** What is a fuel cell? 2 marks

The diagram below shows part of an ethanol fuel cell, which produces carbon dioxide and uses an acidic electrolyte.

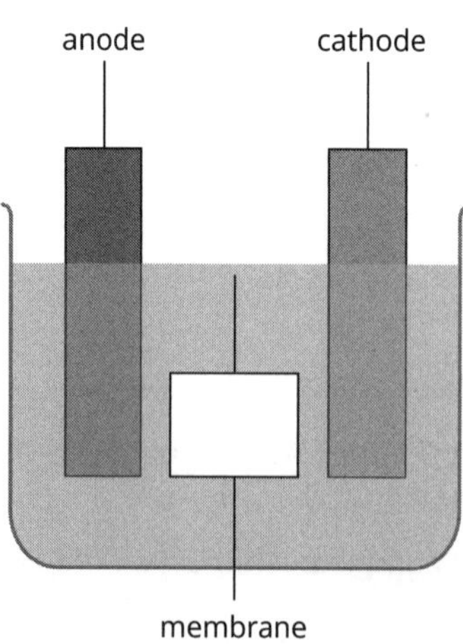


 ISBN 978 0 6557 0027 2

**b. i.** Name the species that crosses the membrane to enable fuel cell operation. 1 mark

______________________________

**ii.** In the box provided on the diagram above, indicate the direction of flow of the species named in part b(i). 1 mark

**c.** Write the equation for the reaction that occurs at the anode of an ethanol fuel cell, which produces carbon dioxide and uses an acidic electrolyte. 1 mark

______________________________

**d.** If an ethanol fuel cell was operating at 25°C and at 100% efficiency, how much electrical energy could be produced from 1.0 g of ethanol? 1 mark

______________________________

______________________________

**e.** Identify **two** aspects of electrode design that can improve the efficiency of a fuel cell. 2 marks

______________________________

______________________________

______________________________

**f.** State how the environmental impact of using an ethanol fuel cell operating at 100% efficiency can be minimised. 1 mark

______________________________

______________________________

**Question 5** (10 marks) VCE Chemistry 2018 (B) 6

Redox reactions occur in the human body as well as in electrochemical cells.

**a.** Nicotinamide adenine dinucleotide (NAD) is a vital coenzyme for energy production in the human body. It exists in two forms: an oxidised form, $NAD^+$, and a reduced form, NADH. NAD is involved in the conversion of ethanol, $CH_3CH_2OH$, to ethanal, $CH_3CHO$, in the human body.

The overall equation for this redox reaction is

$CH_3CH_2OH + NAD^+ \rightarrow CH_3CHO + NADH + H^+$

**i.** Write the two half-equations for this redox reaction. States are not required. 2 marks

Oxidation half-equation ______________________________

Reduction half-equation ______________________________

**ii.** Identify the reducing agent in this redox reaction. 1 mark

______________________________

ISBN 978 0 6557 0027 2

# EXAM QUESTIONS

**b.** The Daniell cell, a type of galvanic cell, was first constructed in the mid-1800s and this type of cell is still in use today. A diagram of the Daniell cell is shown below.

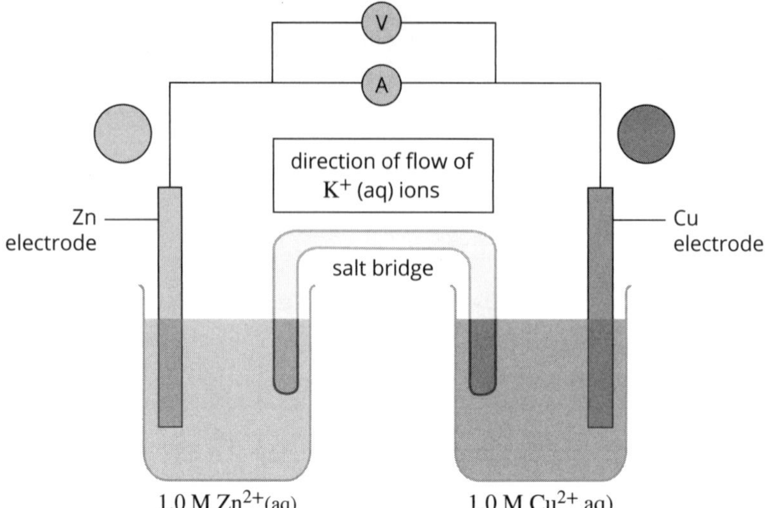


**i.** Label the polarity of the electrodes by placing a positive (+) or negative (–) sign in each of the circles next to the electrodes on the diagram above. 1 mark

**ii.** Use the electrochemical series to determine the theoretical voltage of this cell. 1 mark

______________________________________________

**iii.** The electrolyte in the salt bridge is a potassium nitrate solution, $KNO_3$(aq).

In the box above the salt bridge, use an arrow to indicate the direction of flow of $K^+$(aq) ions. 1 mark

**iv.** List two visible changes that are likely to be observed when the Daniell cell has been operating for some time. 2 marks

______________________________________________

______________________________________________

______________________________________________

**c.** What design features of the Daniell cell structure would allow it to produce electrical energy? 2 marks

______________________________________________

______________________________________________

______________________________________________

 ISBN 978 0 6557 0027 2

UNIT 3

# How can design and innovation help to optimise chemical processes?

AREA OF STUDY 2

## How can the rate and yield of chemical reactions be optimised?

### Outcome 2

On completion of this unit the student should be able to experimentally analyse chemical systems to predict how the rate and extent of chemical reactions can be optimised, explain how electrolysis is involved in the production of chemicals, and evaluate the sustainability of electrolytic processes in producing useful materials for society.

### Key knowledge

#### Rates of chemical reactions

- factors affecting the frequency and success of reactant particle collisions and the rate of a chemical reaction in open and closed systems, including temperature, surface area, concentration, gas pressures, presence of a catalyst, activation energy and orientation
- the role of catalysts in increasing the rate of specific reactions, with reference to alternative reaction pathways of lower activation energies and represented using energy profile diagrams

#### Extent of chemical reactions

- the distinction between reversible and irreversible reactions, and between rate and extent of a reaction
- the dynamic nature of homogenous equilibria involving aqueous solutions or gases, and their representation by balanced chemical or thermochemical equations (including states) and by concentration–time graphs
- the change in position of equilibrium that can occur when changes in temperature or species or volume (concentration or pressure) are applied to a system at equilibrium, and the representation of these changes using concentration–time graphs
- the application of Le Châtelier's principle to identify factors that favour the yield of a chemical reaction
- calculations involving equilibrium expressions (including units) for a closed homogeneous equilibrium system and the dependence of the equilibrium constant ($K$) value on the system temperature and the equation used to represent the reaction
- the reaction quotient ($Q$) as a quantitative measure of the extent of a chemical reaction: that is, the relative amounts of products and reactants present during a reaction at a given point in time
- responses to the conflict between optimal rate and temperature considerations in producing equilibrium reaction products, with reference to the green chemistry principles of catalysis and designing for energy efficiency

#### Production of chemicals using electrolysis

- the use and limitations of the electrochemical series to explain or predict the products of the electrolysis of particular chemicals, given their state (molten liquid or in aqueous solution) and the electrode materials used, including the writing of balanced equations (with states) for the reactions occurring at the anode and cathode and the overall redox reaction for the cell

- the common design features and general operating principles of commercial electrolytic cells (including, where practicable, the removal of products as they form), and the selection of suitable electrode materials, the electrolyte (including its state) and any chemical additives that result in a desired electrolysis product (details of specific cells not required)
- the common design features and general operating principles of rechargeable (secondary) cells, with reference to discharging as a galvanic cell and recharging as an electrolytic cell, including the conditions required for the cell reactions to be reversed and the electrode polarities in each mode (details of specific cells not required)
- the role of innovation in designing cells to meet society's energy needs in terms of producing 'green' hydrogen (including equations in acidic conditions) using the following methods:
  - polymer electrolyte membrane electrolysis powered by either photovoltaic (solar) or wind energy
  - artificial photosynthesis using a water oxidation and proton reduction catalyst system
- the application of Faraday's laws and stoichiometry to determine the quantity of electrolytic reactant and product, and the current or time required to either use a particular quantity of reactant or produce a particular quantity of product

VCE Chemistry Study Design extracts © VCAA (2022); reproduced by permission.

# Rates of chemical reactions

In this topic, you will explore the factors that affect the rate of reactions involved in producing important materials for society. You will investigate the role of catalysts and learn to predict the effect of different changes to the rate of a reaction.

The rate of a reaction can be determined by measuring the change in concentration of the reactants or products with time. The unit of reaction rate is $M\ s^{-1}$.

- **You will now be able to complete Worksheet 12.**

## FACTORS AFFECTING THE RATE OF REACTION

When reaction conditions change, the rate of the reaction may also change. Factors that affect the rate of reaction can be explained by **collision theory**. Reactions occur when particles of the reactants collide, so reaction rate depends on the frequency of collisions (number of collisions per second). However, not all collisions between reactant particles result in a successful reaction. The likelihood of a reaction occurring depends on the:

- energy of particles during collision, which must be equal to or greater than the activation energy
- orientation or arrangement of the particles when they collide.

The activation energy is the minimum energy required for a reaction to occur. At a given temperature, a reaction with a lower activation energy will have a higher reaction rate than if it had a higher activation energy. Collision theory can be used to predict the effect of changes that occur when there is an increase in pressure, concentration, surface area or temperature, or when a catalyst is used (Table 3.2.1 on the following page).

## CATALYSTS

A **catalyst** is a substance that increases the rate of a chemical reaction without itself being consumed. The effect of a catalyst is seen in the energy profile diagram in Table 3.2.1. Catalysts:

- need only be present in small quantities
- do not appear as reactants or products in the equation for the reaction
- are often used in powdered form to increase their surface area, which increases the rate of reaction
- are present in biological systems; **enzymes** (biological catalysts) are the most important examples
- are used extensively in industry to make reactions economically viable
- do not alter the amount of product formed by the reaction, just the reaction rate
- increase the rate of the forward and back reactions equally, for reversible reactions (see Reversible reactions on page 66).


ISBN 978 0 6557 0027 2

## KEY KNOWLEDGE

**Table 3.2.1** Factors affecting the rate of reaction

| Factor | Effect on rate | |
|---|---|---|
| increasing concentration or pressure | Increasing the concentration of reactant particles causes collisions to occur more frequently. This results in faster formation of products. For reactions involving gases, an increase in pressure caused by a decrease in volume has a similar effect. | 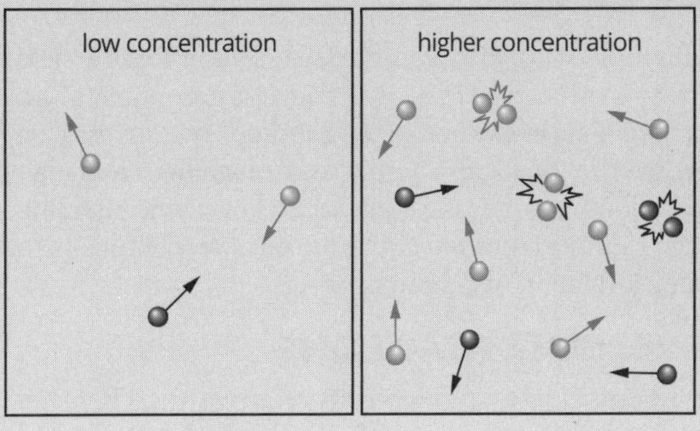 low concentration; higher concentration |
| increasing surface area of solids | Increasing the surface area increases the frequency with which the reactant particles collide. The rate of product formation also increases. | 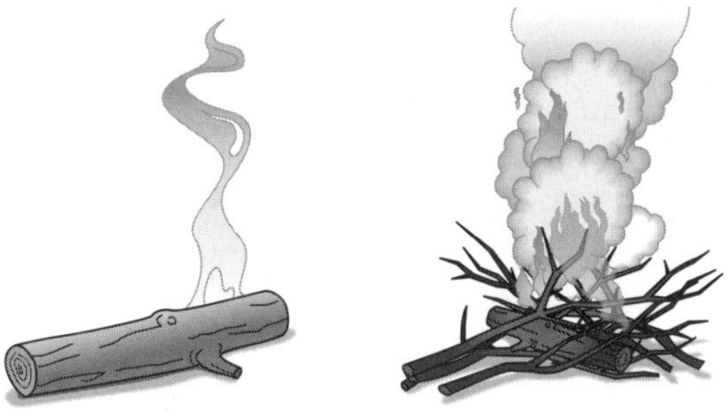 Log of wood (1 piece) (smoking only); Many thin pieces of wood (flames and smoke) |
| increasing temperature | At any specified temperature, particles have a range of velocities and kinetic energies. Increasing the temperature causes more reactant particles to have more kinetic energy and so collide more frequently. More importantly, a higher proportion of particles will have kinetic energies greater than the activation energy ($E_a$). The result is faster product formation. | 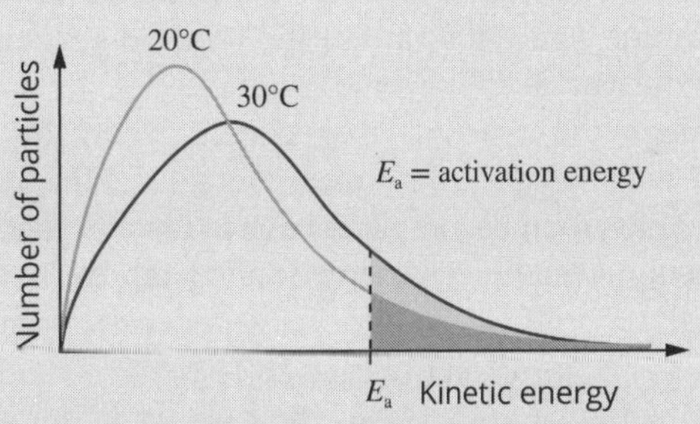 20°C; 30°C; $E_a$ = activation energy; Number of particles; $E_a$ Kinetic energy<br>The distribution of energies of particles at a particular temperature. As temperature increases, more particles have higher energies. |
| using a catalyst | A catalyst provides an alternative pathway for a reaction to occur, with a lower activation energy. A greater proportion of reactant particles will have energies greater than the activation energy, causing the rate of a reaction to increase. | 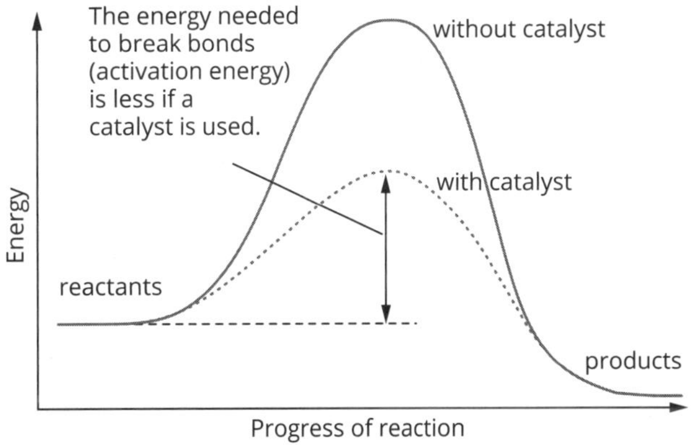 The energy needed to break bonds (activation energy) is less if a catalyst is used.; without catalyst; with catalyst; reactants; products; Energy; Progress of reaction<br>Energy profile diagram showing effect of catalyst |

- **You will now be able to complete Worksheet 13 and conduct Practical activity 8.**

## KEY KNOWLEDGE

# Extent of chemical reactions

In this topic, you will explore the factors that affect the yield of a type of chemical reaction known as an equilibrium reaction. The behaviour of equilibrium systems will be discussed in qualitative terms and quantified using an expression known as the equilibrium law. Understanding equilibrium helps us to control the reaction conditions to favour formation of desired products.

### REVERSIBLE REACTIONS

In chemistry, the term **system** refers to the chemical reaction being studied and the **surroundings** are everything else. The types of systems include:

- **open systems** that exchange matter and energy with the surroundings
- **closed systems** that exchange only energy with the surroundings.

Systems can be defined as **homogeneous**, where all reactants and products are in the same state or phase, or **heterogeneous**, where different states are present (these are not studied in this course).

Chemical reactions that can occur in one direction only are **irreversible** or non-reversible. Reactions that can occur in both directions are **reversible**.

Reversible reactions in a closed system eventually reach a point where the rate of the forward reaction is equal to the rate of the reverse reaction.

Consider the reaction that occurs when $N_2$ gas is mixed with $H_2$ gas in a sealed container. $NH_3$ gas is formed, which then decomposes back to form $N_2$ and $H_2$. The reaction can be represented using double-headed arrows as:

$$N_2(g) + 3H_2(g) \rightleftharpoons 2NH_3(g)$$

When the rates of the forward and reverse reactions are equal, the reaction is described as being at **equilibrium** (Figure 3.2.1).

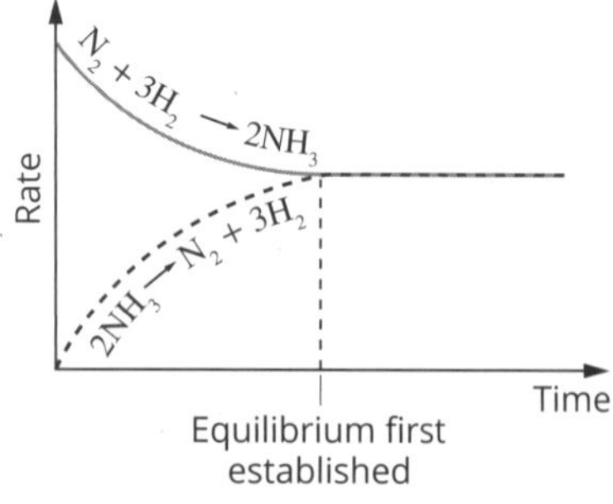


**Figure 3.2.1** A graph for $N_2(g) + 3H_2(g) \rightleftharpoons 2NH_3(g)$, showing the changes in the rates of the forward and reverse reactions with time as a reaction reaches equilibrium

At equilibrium, the concentrations of reactants and products are constant. The reaction appears to have 'stopped' before all the reactants have been converted into products, as shown by the **concentration–time graph** in Figure 3.2.2.

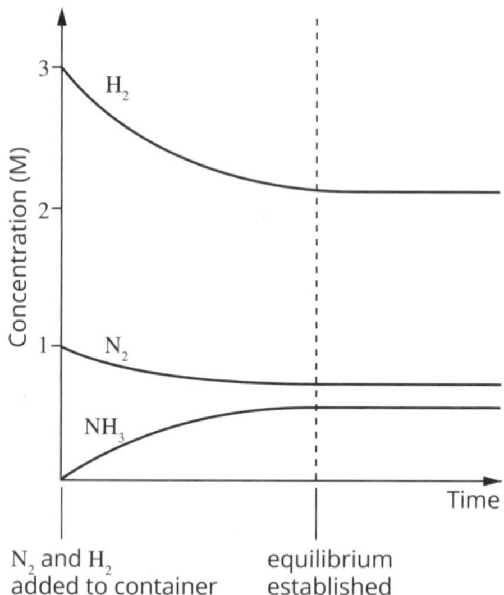


**Figure 3.2.2** A concentration–time graph showing the changes in concentrations of $N_2$, $H_2$ and $NH_3$ as a mixture of nitrogen and hydrogen gases that react at a constant temperature and reach equilibrium

Note that the concentration–time graph indicates that $[H_2]$ decreases at three times the rate at which $[N_2]$ decreases. This is because $H_2$ reacts with $N_2$ in a 3:1 mole ratio. Also, as shown by the coefficients of the equation, $[NH_3]$ increases at twice the rate at which $[N_2]$ decreases.

A reaction does not actually 'stop' at equilibrium. Reactants are continuously forming products and products are continuously changing back into reactants at the same rate; such systems are described as being in **dynamic equilibrium**.

### Reversible reactions and energy profile diagrams

When particles collide and reactions occur, the energy associated with collisions breaks the bonds of reactant particles and the new bonds of the product particles are formed. The energy required to break the bonds of the reactants is the activation energy of the reaction. An energy profile diagram shows that once the products form, the reverse process can occur, in which product particles collide and absorb energy to break their bonds and re-form the reactants.

In Figure 3.2.3 on the following page, the forward reaction is exothermic and the reverse reaction is endothermic. The activation energy for the endothermic reaction is the sum of $\Delta H$ for the reaction and the activation energy for the exothermic reaction.

 ISBN 978 0 6557 0027 2

# KEY KNOWLEDGE

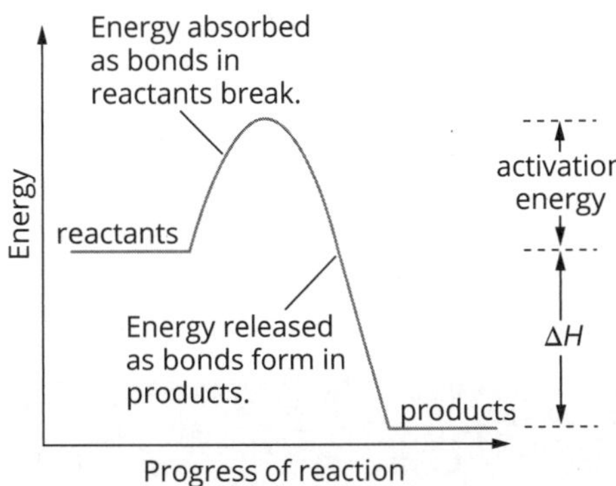


**Figure 3.2.3** An energy profile diagram for an exothermic reaction

## THE EQUILIBRIUM LAW—*K* AND *Q*

The point at which equilibrium is reached is not the same for all reactions. Some reactions go almost to completion, so that when equilibrium is reached almost all the reactants have been converted into products. Other reactions yield almost no products and most of the reactants are present at any time. The **position of equilibrium** indicates the extent to which the reaction has proceeded. The relative concentrations of reactants and products at equilibrium can be described using the **equilibrium law**.

The equilibrium law for a general reaction:

$$aA + bB \rightleftharpoons cC + dD$$

is given by:

$$K = \frac{[C]^c[D]^d}{[A]^a[B]^b}$$

The **equilibrium constant, *K***, is the ratio of the concentration of the products to the concentration of the reactants, raised to powers that are their coefficients in the equation. The ratio of products to reactants can be calculated at any time during the reaction and is called the **reaction quotient, *Q***, or concentration fraction.

- If $Q > K$, the reaction shifts left to establish equilibrium.
- If $Q < K$, the reaction shifts right to establish equilibrium.
- If $Q = K$, the reaction is at equilibrium.

At equilibrium, for the reaction:

$$N_2(g) + 3H_2(g) \rightleftharpoons 2NH_3(g)$$

$$Q = K = \frac{[NH_3]^2}{[N_2][H_2]^3}$$

### The extent of reaction

The value of the equilibrium constant, *K*, indicates how far the reaction proceeds in the direction of the products. The value of *K* gives an indication of the extent of the chemical reaction, the position of equilibrium. It is specific for an equilibrium system at a specified temperature.

- $K > 10^4$ indicates an extensive forward reaction; the equilibrium mixture consists mostly of products. The position of equilibrium is 'to the right'.
- $K < 10^{-4}$ indicates a negligible forward reaction; the equilibrium mixture consists mostly of reactants. The position of equilibrium is 'to the left'.
- $10^{-4} < K < 10^4$ equilibrium mixture consists of significant amounts of both reactants and products.

Note that the term **extent of reaction** is used to indicate how far the reaction has proceeded when equilibrium is achieved, whereas the **rate of reaction** indicates how fast the reaction occurred. A reaction can have a large equilibrium constant, *K*, but a very low rate, and vice versa.

The units for the equilibrium constant, *K*, can be determined by substituting M (or mol $L^{-1}$) into the expression for the equilibrium law. Units are dependent on the equation for the equilibrium system.

### Factors affecting the value of *K*

The value of the equilibrium constant for a reaction depends on the:

- coefficients in the equation: doubling the coefficients will square the original value of *K*; halving the coefficients will cause *K* to be the square root of the original
- direction of the equation: reversing the equation causes *K* to have the inverse value
- temperature: the effect of change of temperature on the value of *K* depends on whether the reaction is exothermic or endothermic (see Effect of temperature change on equilibria on page 68).

### Calculations involving stoichiometry and equilibrium

Stoichiometry enables you to calculate the equilibrium concentrations in an equilibrium system, and therefore calculate the equilibrium constant, *K*. The use of an **ICE** (Initial, Change, Equilibrium) **table** is recommended, as the following worked example shows. Work in moles in the ICE table and then convert to concentration when required later in the question.

**Example:** In the equilibrium system:

$$2A + B \rightleftharpoons 2C$$

0.64 mol of A and 0.55 mol of B react to produce C. The volume of the container is 4 L and the equilibrium concentration of C is measured to be 0.12 M. Calculate the equilibrium concentrations of A and B and then calculate the equilibrium constant for this system at 25°C.

| | 2A + | B ⇌ | 2C |
|---|---|---|---|
| I | 0.64 mol | 0.55 mol | 0.00 mol |
| C | −2*x* | −*x* | +2*x* |
| E | 0.64 − 2*x* | 0.55 − *x* | 0.00 + 2*x*<br>= 0.12 × 4 = 0.48 mol |

ISBN 978 0 6557 0027 2 

KEY KNOWLEDGE

| Thinking | Working |
|---|---|
| 1 Use the ICE table to calculate the number moles of each chemical at equilibrium. | The equilibrium concentration of C is given as 0.12 M. Hence $n(C) = 0.12 \times 4 = 0.48$ mol. Since $2x = 0.48$ mol, therefore, $x = 0.24$ mol. For A: $0.64 - 2x = 0.64 - 0.48 = 0.16$ mol; therefore, equilibrium concentration of $A = \frac{0.16}{4} = 0.04$ M. For B: $0.55 - x = 0.55 - 0.24 = 0.31$ mol; therefore, equilibrium concentration of $B = \frac{0.31}{4} = 0.08$ M |
| 2 Calculate the value of $K$. | $K = \frac{[C]^2}{[A]^2[B]}$ $= \frac{0.12^2}{0.04^2 \times 0.08}$ $= 112.5\ M^{-1}$ |

- **You will now be able to complete Worksheet 14.**

## LE CHÂTELIER'S PRINCIPLE

**Le Châtelier's principle** states that if a change is imposed on a system at equilibrium, the system will adjust itself to partially oppose the effect of the change. The equilibrium will not completely return to its original state, but it will tend to oppose the effect.

Adding a catalyst speeds up the forward and backward reactions equally, so no change in the equilibrium concentrations of the reactants or products occurs. Hence, addition of a catalyst causes no change to the position of the equilibrium and no change in $K$.

The addition of an inert gas to an equilibrium system does not have any effect on the position of equilibrium or the value of $K$ because the inert gas is not part of the equilibrium reaction and does not appear in the expression for the equilibrium law.

For changes where Le Châtelier's principle predicts more product will be formed, the yield of the reaction will increase.

Table 3.2.2 summarises the effect of different changes on an equilibrium system when the temperature is constant. (The value of $K$ will remain unchanged.)

**Table 3.2.2** Effect of changes on an equilibrium system at constant temperature; hence no change in $K$

| Change | Effect of change on equilibrium position |
|---|---|
| adding extra reactant | shifts to the right |
| removing product | shifts to the right |
| adding a catalyst | no change |
| adding an inert gas | no change |
| decreasing pressure by increasing volume (for gases) | shifts in the direction of most gaseous particles |
| adding water (dilution of solutions) | shifts in the direction of most solute particles |

### Effect of temperature change on equilibria

Only a change in temperature changes the value of $K$. A rise in temperature decreases the amount of product at equilibrium in an exothermic reaction and increases the amount of product in an endothermic reaction (Table 3.2.3).

**Table 3.2.3** Relationship between $K$ and $\Delta H$ as temperature increases

| $\Delta H$ | Effect on $K$ as temperature increases |
|---|---|
| – (exothermic) | decreases |
| + (endothermic) | increases |

- **You will now be able to complete Worksheets 15 and 16 and conduct Practical activities 9–11.**

### Optimising industrial yield

When developing a process for synthesising a chemical for use in society, the aim is for a high yield of a pure product that is produced quickly and efficiently at a marketable price. Reactants are selected that produce a product with specific properties suitable for its intended purpose.

Factors considered when designing a chemical include availability of raw materials, reaction conditions (rate and equilibrium), yield and energy use as well as environmental, social and economic issues.

Inevitably, during a synthesis, there will be some loss of reactants and products during transfers between reaction vessels and in the separation and purification stages. This results in fewer products being obtained than expected. The formation of an equilibrium can also limit the yield of product from a chemical synthesis as the factors that shift an equilibrium in the forward direction to maximise the amount of product formed can, at times, be in conflict with the factors affecting rate of reaction. Temperature, pressure and concentration need to be selected carefully and it may be necessary to make compromises to ensure optimum yield.

The Haber process, which produces ammonia, is an example of such an industrial process (Table 3.2.4 on the following page). The reaction in this process involves the equilibrium:

$$N_2(g) + 3H_2(g) \rightleftharpoons 2NH_3(g) \qquad \Delta H = -91\text{ kJ}$$

Reagents and reaction conditions are chosen to optimise yield and rate, and green chemistry principles are applied to reduce energy costs and optimise energy efficiency. In particular, there is a conflict between the conditions needed for a high rate of reaction and the conditions needed for a high equilibrium yield, and in the selection of a suitable catalyst where cost can be in conflict with reusability.


ISBN 978 0 6557 0027 2

## KEY KNOWLEDGE

| **Table 3.2.4** The Haber process: an example of the need to compromise to maximise yield and energy efficiency | |
|---|---|
| To increase the equilibrium yield:<br>• decrease *T*, because $\Delta H$ is negative, so *K* increases when *T* decreases<br>• increase *p*, because the equilibrium position moves forward in the direction of the least number of molecules. | To increase the rate of reaction:<br>• increase *T*<br>• increase *p*<br>• use a catalyst: usually porous $Fe/Fe_3O_4$, which is a compromise between cost and reusability. |
| *Conflict:* A higher rate occurs at high temperatures, but the higher equilibrium yield occurs at low temperatures. | |
| *Compromise:* Use a moderate temperature of 350–500°C and a pressure of 25 000 kPa. Use a porous $Fe/Fe_3O_4$ catalyst. The gas is passed quickly over the catalyst beds several times. It is cooled between passes to keep the temperature below 500°C. | |

- **You will now be able to complete Worksheet 18.**

# Production of chemicals using electrolysis

In the process of **electrolysis**, electrical energy is converted to chemical energy. This is the reverse of the energy transformation that occurs in galvanic cells. Table 3.2.5 compares electrolytic and galvanic electrochemical cells. The reactions that occur in electrolytic cells are **non-spontaneous** and require an external power supply to provide energy.

## FEATURES AND PRINCIPLES OF ELECTROLYTIC CELLS

### Electrolysis of molten liquids

Figure 3.2.4 shows an electrolytic cell containing molten salt, NaCl. Inert electrodes, such as graphite or platinum, are used. No water is present. Electrons are 'pumped' by a power supply onto the negative electrode and withdrawn from the positive electrode. The reactions that occur in the cell are described in Table 3.2.6 on the following page. If operating as a commercial cell (Downs cell), the products $Cl_2(g)$ and Na(l) would be collected separately. In the Downs cell, chlorine gas is collected near the top of the cell using a screen to prevent mixing in the cell; whereas molten sodium, being less dense than molten sodium chloride, is collected separately from the side of the cell.

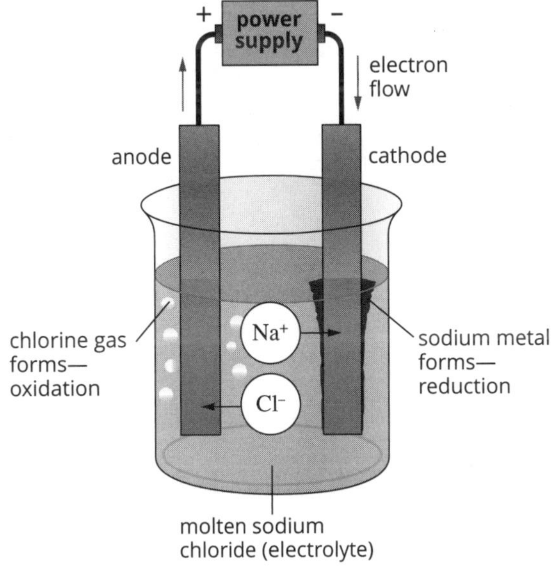


$2Cl^-(l) \rightarrow Cl_2(g) + 2e^-$ $\quad Na^+(l) + e^- \rightarrow Na(l)$

**Figure 3.2.4** Electrolysis of molten NaCl

**Table 3.2.5** A comparison of galvanic cells and electrolytic cells

| Galvanic cell | Electrolytic cell |
|---|---|
| produces energy | consumes energy |
| chemical energy → electrical energy | electrical energy → chemical energy |
| spontaneous reaction occurs | non-spontaneous reaction occurs |
| Oxidation occurs at the anode; reduction occurs at the cathode. | Oxidation occurs at the anode; reduction occurs at the cathode. |
| Anode is negative; cathode is positive (polarity of the electrode is the result of the half-reaction occurring at the electrode). | Anode is positive; cathode is negative (polarity of the electrode is imposed by external power supply). |
| Reactants are not allowed to be in contact, preventing a spontaneous reaction. Laboratory cells can be made up of separate half-cells connected by a salt bridge. | Products are not allowed to be in contact, preventing a spontaneous reaction. Different methods are used to keep products separate, e.g. semipermeable plastic membranes, iron mesh screens. |

# KEY KNOWLEDGE

| Table 3.2.6 The electrode half-equations and overall equation for the electrolysis of molten NaCl | | |
|---|---|---|
| | **Description** | **Equation** |
| at cathode (negative) | Electrons are 'pumped' by the power supply onto the negative electrode, where they are consumed and cause a reduction reaction to occur. | $Na^+(l) + e^- \rightarrow Na(l)$ |
| at anode (positive) | Electrons are withdrawn from the positive electrode, causing an oxidation reaction to occur. | $2Cl^-(l) \rightarrow Cl_2(g) + 2e^-$ |
| overall | The overall equation is obtained by combining the half-equations in the usual way, ensuring that electrons balance. | $2Na^+(l) + 2Cl^-(l) \rightarrow 2Na(l) + Cl_2(g)$ |

## Electrolysis of aqueous solutions

Figure 3.2.5 shows an electrolytic cell containing an aqueous solution of $KNO_3$ with inert electrodes. In this cell, the water in the electrolyte, $KNO_3(aq)$, reacts rather than the $K^+$ or $NO_3^-$ ions. The reactions that occur in the cell are described in Table 3.2.7.

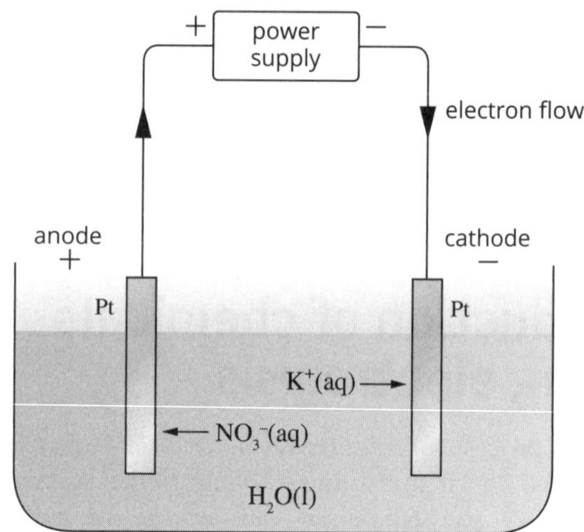


Oxidation reaction (anode):

$$2H_2O(l) \rightarrow O_2(g) + 4H^+(aq) + 4e^-$$

Reduction reaction (cathode):

$$2H_2O(l) + 2e^- \rightarrow H_2(g) + 2OH^-(aq)$$

**Figure 3.2.5** The electrolysis of 1 M $KNO_3(aq)$

| Table 3.2.7 The electrode half-equations and overall equation in electrolysis of $KNO_3(aq)$ | | |
|---|---|---|
| | **Description** | **Equation** |
| at cathode (negative) | Reduction occurs at the cathode. Water is a stronger oxidising agent than potassium ions, so water is reduced. | $2H_2O(l) + 2e^- \rightarrow H_2(g) + 2OH^-(aq)$ |
| at anode (positive) | Oxidation occurs at the anode. Water is the strongest reducing agent in the cell and is oxidised. | $2H_2O(l) \rightarrow O_2(g) + 4H^+(aq) + 4e^-$ |
| overall | The overall equation for the cell is determined by multiplying the cathode half-equation by 2 and then adding the two half-equations. | $2H_2O(l) \rightarrow 2H_2(g) + O_2(g)$ |


ISBN 978 0 6557 0027 2

# KEY KNOWLEDGE

## Using the electrochemical series

At the electrodes of some electrolytic cells, there is more than one chemical that might react. The electrochemical series (see Appendix 2) can be used to predict the electrode reactions in electrolytic cells at standard conditions. The following worked example shows the steps involved in determining the electrode reactions.

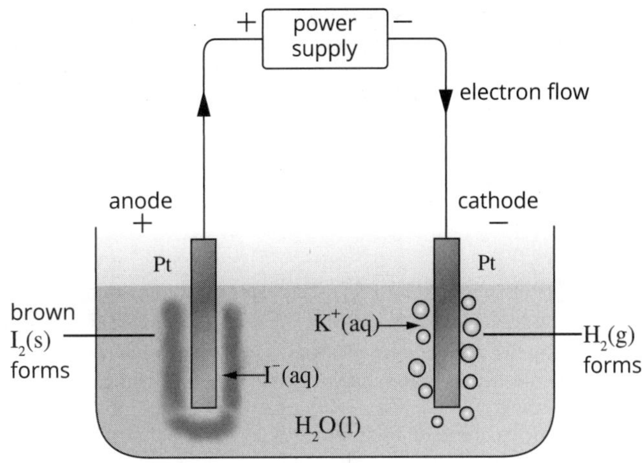


Oxidation reaction: $2I^-(aq) \rightarrow I_2(aq) + 2e^-$

Reduction reaction: $2H_2O(l) + 2e^- \rightarrow H_2(g) + 2OH^-(aq)$

**Figure 3.2.6** The electrolysis of 1 M KI solution using inert electrodes

**Example:** Predict the electrode reactions that occur during the electrolysis of 1 M potassium iodide solution using unreactive platinum electrodes (Figure 3.2.6). The negative $I^-$ ions are attracted to the positive electrode and the positive $K^+$ ions are attracted to the negative electrode.

| Thinking | Working |
|---|---|
| 1 Circle or bold all species in the electrochemical series that are present in the cell when the reaction begins. Remember to include water if it is present. | $O_2(g) + 4H^+(aq) + 4e^- \rightarrow$ **$2H_2O(l)$**<br>$I_2(s) + 2e^- \rightarrow$ **$2I^-(aq)$**<br>**$2H_2O(l)$** $+ 2e^- \rightarrow H_2(g) + 2OH^-(aq)$<br>**$K^+(aq)$** $+ e^- \rightarrow K(s)$<br>Identify the strongest oxidising agent (highest bolded species on the left of the series, $H_2O(l)$) and the strongest reducing agent (lowest bolded species on the right of the series, $I^-(aq)$). |
| 2 Copy the half-equations from the series, writing the circled (bolded) species as reactants (i.e. the half-equation with the strongest reducing agent is reversed). | Positive electrode (anode): $2I^-(aq) \rightarrow I_2(s) + 2e^-$<br>Negative electrode (cathode): $2H_2O(l) + 2e^- \rightarrow H_2(g) + 2OH^-(aq)$<br>The equation involving oxidation will occur at the positive electrode in the cell. The cathode is the site of reduction and is the negative electrode. Although the $K^+$ ions are attracted to the negative electrode, $H_2O$ is a stronger oxidising agent than $K^+$ and so $H_2O$ is reduced in preference to $K^+$. |
| 3 Write the overall equation. | $2I^-(aq) + 2H_2O(l) \rightarrow I_2(s) + H_2(g) + 2OH^-(aq)$ |

## Limitations of predictions using the electrochemical series

Note that the electrochemical series applies to reactions under standard conditions (1 M and 100 kPa and at 25°C). For non-standard conditions, the series cannot be used reliably to predict electrode reactions.

For example, consider the electrolysis of sodium chloride solution using inert electrodes. At low concentrations, the predicted products of $O_2$ gas and $H_2$ gas are formed, but when 4 M NaCl solution (brine) is electrolysed, the major product at the anode becomes $Cl_2$ gas instead of $O_2$ gas.

## Choosing electrodes and electrolyte materials

The production of chemicals, electroplating, recharging secondary batteries and the refining of copper are some examples of the use of electrolysis. The types of electrodes and electrolytes chosen for a commercial cell depend on several factors, including the relative strength of the oxidising and reducing agents compared to water and the costs of materials and energy.

Highly reactive metals, such as Al, Mg, Na, Ca, K and Li, are produced by electrolysis of their molten salts. These metals are so reactive that, if aqueous electrolytes were used, the water would be more readily reduced than their metal ions (which are weaker oxidising agents). Other chemicals are produced commercially by electrolysis using aqueous electrolytes, including $Cl_2$, NaOH and $H_2$ (Table 3.2.8 on the following page).

# KEY KNOWLEDGE

| Table 3.2.8 Choosing appropriate electrodes and electrolytes for commercial electrolytic cells | |
|---|---|
| **Cells with an aqueous electrolyte** | **Cells with molten electrolyte** |
| • Water is a possible reactant at both the anode and the cathode.<br>• The use of aqueous solutions reduces energy costs associated with molten electrolytes. | • Used for the production of metals whose ions are weaker oxidising agents than water in the electrochemical series.<br>• A non-reacting substance mixed with the molten electrolyte can be used to lower the melting point (e.g. $CaCl_2$ is added to molten NaCl in the commercial version of the cell in Figure 3.2.4). This reduces energy costs. |
| **Cells with inert electrodes** | **Cells with non-inert electrodes** |
| • Cathodes are always inert. Carbon (graphite) or iron are often used. Iron is less expensive than graphite and more easily shaped.<br>• Inert anodes are used in some cells. | • The material used to make the anode may be a reactant. In such cases, the anode will be consumed and need replacing. |

**Electroplating** involves the deposition of a layer of metal on the surface of another metal by electrolysis. When an object is plated, it is connected to the negative terminal of a power supply, making it the cathode. The positive electrode is usually made of the metal to be plated and the electrolyte contains ions of the metal (Figure 3.2.7). The metal ions are reduced to form a layer of metal on the object.

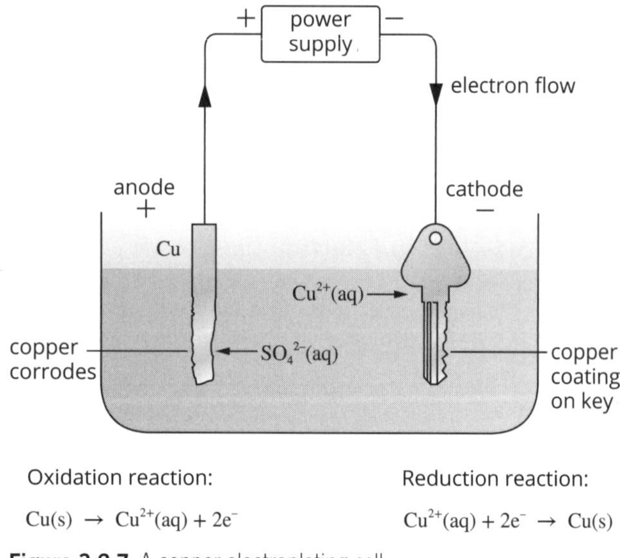


Oxidation reaction:

$Cu(s) \rightarrow Cu^{2+}(aq) + 2e^-$

Reduction reaction:

$Cu^{2+}(aq) + 2e^- \rightarrow Cu(s)$

**Figure 3.2.7** A copper electroplating cell

● **You will now be able to complete Worksheet 17 and conduct Practical activity 12.**

## RECHARGEABLE BATTERIES (SECONDARY CELLS)

Rechargeable batteries, such as car batteries and computer batteries, are examples of secondary cells. Secondary cells, like primary cells, can act as galvanic cells and discharge electrical energy. However, in secondary cells, the products of the cell reaction remain in contact with the electrodes, so that recharging is possible. When a cell is recharged, electrical energy from a battery charger is converted to chemical energy by reversing the normal discharge reaction of the cell. The products in the recharging cell become reactants again.

The polarity of the electrodes does not change for the discharge or recharge reactions, but the process at each electrode is reversed. The behaviour of a secondary cell is summarised in Table 3.2.9.

● **You will now be able to complete Worksheet 19.**

**Table 3.2.9** The operation of secondary cells while discharging and recharging

| | During discharge | During recharge |
|---|---|---|
| type of cell | acts as a galvanic cell | acts as an electrolytic cell |
| nature of reaction | spontaneous | non-spontaneous |
| energy transformation | chemical to electrical | electrical to chemical |
| at negative electrode | • electrons spontaneously flow from this electrode<br>• anode—oxidation occurs | • electrons are pumped onto this electrode by the charger<br>• cathode—reduction occurs |
| at positive electrode | • electrons spontaneously flow to this electrode<br>• cathode—reduction occurs | • electrons are withdrawn from this electrode by the charger<br>• anode—oxidation occurs |

 ISBN 978 0 6557 0027 2

# KEY KNOWLEDGE

## FARADAY'S LAWS

Faraday's laws were described in Area of Study 1. They can also be used to determine the mass of product formed at an electrode during electrolysis, as shown in the following worked example. The formulas used in calculations are:

- $Q = I \times t$
- $Q = n(e^-) \times F = n(e^-) \times 96\,500$

**Example:** Copper is plated onto a nickel medal using a copper-plating cell. A current of 10.5 A runs through the cell for 25.0 minutes. What mass of copper will be deposited onto the medal?

| Thinking | Working |
|---|---|
| 1 Write the half-equation for the reaction at the cathode. | $Cu^{2+}(aq) + 2e^- \rightarrow Cu(s)$ |
| 2 Use the formula $Q = I \times t$ to calculate the charge, $Q$, which flows through the cell. | $Q = 10.5 \times 25.0 \times 60$<br>$= 15\,750$ C |
| 3 Use the formula $Q = n(e^-) \times F$ to calculate the number of moles of electrons flowing through the cell. | $Q = n(e^-) \times 96\,500$<br>$n(e^-) = \frac{Q}{96500}$<br>$= \frac{15750}{96500} = 0.163$ mol |
| 4 Use the half-equation and the mole ratio to determine $n$(product) at the cathode. | $\frac{n(Cu)}{n(e^-)} = \frac{1}{2}$<br>$n(Cu) = \frac{1}{2} \times 0.163$<br>$= 0.0815$ mol |
| 5 Calculate the mass of product at the cathode. | $m(Cu) = n \times M$<br>$= 0.0815 \times 63.5 = 5.18$ g |

## CHALLENGING AND INNOVATIVE CELLS PRODUCING 'GREEN' HYDROGEN

As previously discussed, the use of hydrogen as a fuel in fuel cells promises to be an environmentally friendly source of energy because the only products are water and energy. However, conventional methods of producing hydrogen release greenhouse gases, so the use of new methods for producing hydrogen is critical in the development of a 'green' hydrogen economy. **Green hydrogen** is produced when water is split to form hydrogen and oxygen gases by using renewable energy supplies.

New technologies for hydrogen production use a device called an **electrolyser**. An electrolyser is an electrolytic cell with an anode and a cathode separated by an electrolyte. There are many types of electrolysers and they operate differently due to the different electrolyte materials used.

### Polymer electrolyte membrane electrolyser

An example of an electrolyser that uses a polymer electrolyte is shown in Figure 3.2.8.

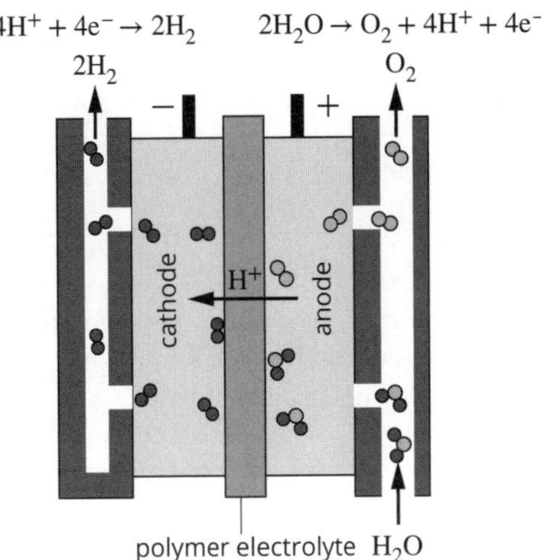


**Figure 3.2.8** A polymer electrolyte membrane electrolyser

Polymer electrolyte membrane (PEM) electrolysers are powered by either solar energy or wind energy. The reactions at the electrodes are:

Anode: $2H_2O \rightarrow 4H^+ + O_2 + 4e^-$

Cathode: $4H^+ + 4e^- \rightarrow 2H_2$

Water reacts at the anode, producing $H^+$ ions and $O_2$ gas. The $H^+$ ions migrate through an electrolyte made of a specially-designed polymer membrane, to the cathode where $H_2$ is formed.

Although PEM electrolysers have high production rates, the process requires expensive catalysts in the electrodes, such as gold or iridium.

### Artificial photosynthesis

Other new technologies include artificial photosynthesis, which converts carbon dioxide and water into oxygen and hydrocarbons. The hydrocarbons can be stored and used as fuels. A cleaner fuel, $H_2$, can be produced when the same process oxidises water at the anode, forming oxygen. The oxidation reaction produces protons and electrons for the reduction reaction, which occurs at the cathode, making hydrogen.

Artificial photosynthesis has been developed to copy the natural photosynthesis process, but uses solar cells instead of chlorophyll and is significantly more efficient. The process captures light for its energy source and, besides producing hydrocarbons, it can split water into $H_2$ and $O_2$, but requires a catalyst. Significant advances have been made in this area in the past decade. Three of the more successful catalysts are manganese, dye-sensitised titanium oxide and cobalt oxide.

- **You will now be able to complete Worksheets 20 and 21 and conduct Practical activity 13.**

# WORKSHEET 12



## Knowledge review—ionisation reactions and Faraday's laws

**1** Define the following terms to check your knowledge and understanding of the key ideas involved in acid–base theory and the dilution of solutions.

**a** weak acid: ____________________

**b** strong base: ____________________

**c** dilute solution: ____________________

**d** concentrated solution: ____________________

**2** Show the balanced equations for the reactions that occur when the following substances dissolve in water. Include states of matter in the equations. Use different arrow types to show which species fully ionise and which partially ionise.

**a** $HCl(g)$: ____________________

**b** $NaOH(s)$: ____________________

**c** $CH_3COOH(l)$: ____________________

**d** $NH_3(g)$: ____________________

**3** Calculate the concentration of a solution prepared by dissolving 3.65 g of sodium carbonate, $Na_2CO_3$, in water and making the volume up to 250.0 mL in a volumetric flask.

**4** Determine the mass of sodium carbonate, $Na_2CO_3$, you would need to dissolve in a 500 mL volumetric flask to make up a 1.00 mol $L^{-1}$ solution.

**5** To check your knowledge and understanding of redox reactions, galvanic cells and Faraday's laws, complete the following statements or formulas with the appropriate terms from the box.

| strongest | weakest | cathode | anode | time in seconds |
|---|---|---|---|---|
| electrical | 96 500 | chemical | current in amps | |

**a** Oxidation always occurs at the ____________________.

**b** Reduction always occurs at the ____________________.

**c** The strongest oxidising agent reacts with the ____________________ reducing agent.

**d** Galvanic cells convert ____________________ energy to ____________________ energy.

**e** $Q = n(e^-) \times$ ____________________

**f** $Q = I$ (where $I$ is ____________________) $\times\ t$ (where $t$ is ____________________)


ISBN 978 0 6557 0027 2

# WORKSHEET 13



## Reaction routes—rate of reaction

**1** Use the terms in the box to help you complete this question. Some terms may be used more than once and others not at all.

| collision | frequency | activation | kinetic energy |
|---|---|---|---|
| catalyst | orientation | energy | |

Factors that affect the rate of reaction can be explained in terms of the ____________________ theory. Reactions occur when particles of the reactants collide with sufficient energy to produce a reaction. This energy needs to be greater than the ____________________ energy. Two factors determine whether a reaction occurs. They are the:

- ____________________ of the colliding particles
- ____________________ of the reacting particles when they collide.

The overall rate of reaction also depends on the ____________________ of collisions.

**2** Use the terms in the box below to help you complete the flow chart that summarises the factors that affect the rate of reaction. Some terms may be used more than once and others not at all.

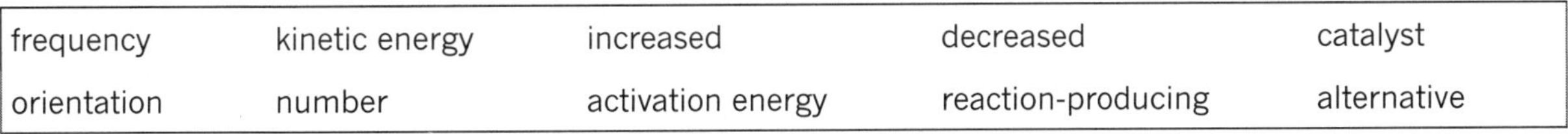

| frequency | kinetic energy | increased | decreased | catalyst |
|---|---|---|---|---|
| orientation | number | activation energy | reaction-producing | alternative |

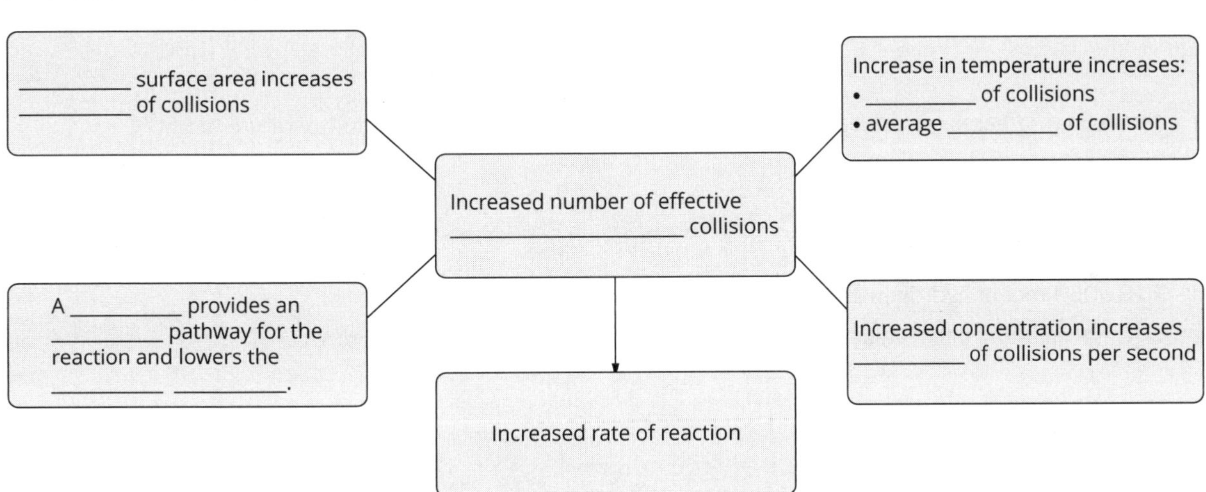


**3** The effect of a catalyst on energy during a chemical reaction can be seen in the energy profile diagram below. Use all the terms from the box to label the diagram.

| reactants | products | activation energy without catalyst |
|---|---|---|
| activation energy with catalyst | progress of reaction | energy |
| The energy needed to break bonds (activation energy) is less if a catalyst is used. | | |

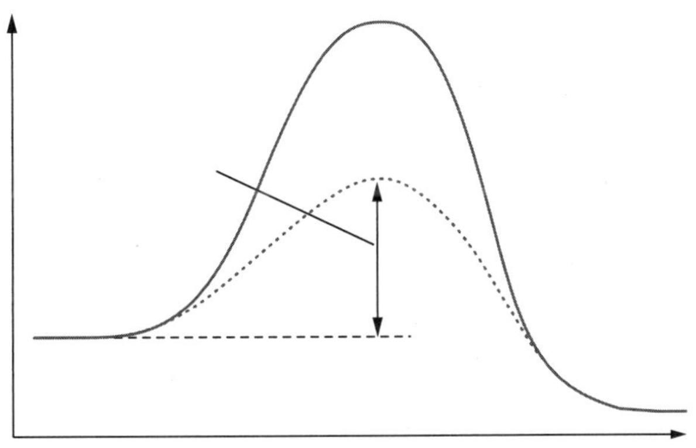

ISBN 978 0 6557 0027 2

# WORKSHEET 14

Simulation

## Calculations—equilibrium constants and concentrations

This worksheet allows you to practise doing calculations involving chemical equilibria.

Steam reforming is used for the large-scale industrial production of hydrogen gas. In this process, methane gas is converted to synthesis gas, which is a mixture of carbon monoxide gas and hydrogen gas. The thermochemical equation is:

$CH_4(g) + H_2O(g) \rightleftharpoons 3H_2(g) + CO(g) \quad \Delta H = +206$ kJ

The equilibrium constant at 650°C is $2.4 \times 10^{-4}$ $M^2$.

1 Write the expression for the equilibrium constant for the equation above.

2 Calculate the equilibrium constant at 650°C and the $\Delta H$ value for the equation:

$6H_2(g) + 2CO(g) \rightleftharpoons 2CH_4(g) + 2H_2O(g)$

3 The following gases are added to a sealed 2.00 L container at a constant temperature of 650°C.

- 0.012 mol of methane gas
- 0.0080 mol of water vapour
- 0.016 mol of carbon monoxide gas
- 0.0060 mol of hydrogen gas

**a** Determine the initial value of the reaction quotient (concentration fraction), $Q$.

**b** In which direction would the reaction move to establish equilibrium? Explain.

**c** The temperature of the equilibrium is increased to 850°C.

i Will the equilibrium constant increase, decrease or remain unchanged? Explain your answer.

 ISBN 978 0 6557 0027 2

ii How will the increased temperature affect the rate and equilibrium yield of the reaction? Explain your answer.

______________________________________________

______________________________________________

______________________________________________

______________________________________________

4 At a different temperature, an equilibrium mixture of $CH_4$, $H_2O$, $H_2$ and CO was prepared in a 2.00 L flask by initially adding only 0.600 mol $CH_4$ and 0.400 mol of $H_2O$ to the reaction vessel. At equilibrium, 0.110 mol of CO was present.

a Complete the following table to determine the amounts of $CH_4$, $H_2O$ and $H_2$ at equilibrium. Let *x* be the amount of $CH_4$ used.

| | $CH_4$ + | $H_2O$ ⇌ | $3H_2$ + | CO |
|---|---|---|---|---|
| **Initial mol** | 0.600 | 0.400 | 0.000 | 0.000 |
| **Change in mol** | | | | |
| **Equilibrium mol** | | | | 0.110 |

b Calculate the equilibrium concentrations of $CH_4$, $H_2O$, $H_2$ and CO.

$[CH_4]$ = ______________ M

$[H_2O]$ = ______________ M

$[H_2]$ = ______________ M

[CO] = ______________ M

c Calculate the equilibrium constant, *K*, for the reaction at this temperature.

# WORKSHEET 15

Case study

## Equilibrium reactions

1 Chemical equilibria can be investigated by mixing different amounts of reactants and products together and measuring the concentrations of all species present at equilibrium. The table gives the results of four of these experiments for the following equilibrium:

$H_2(g) + I_2(g) \rightleftharpoons 2HI(g)$

| Equilibrium concentrations (M) for $H_2(g) + I_2(g) \rightleftharpoons 2HI(g)$ at 458°C | | | | | | |
|---|---|---|---|---|---|---|
| Mixture | $[H_2]$ | $[I_2]$ | $[HI]$ | $[H_2] + [I_2] + [HI]$ | $\frac{[HI]}{[H_2][I_2]}$ | $\frac{[HI]^2}{[H_2][I_2]}$ |
| 1 | 0.0056170 | 0.0005940 | 0.0126990 | | | |
| 2 | 0.0045800 | 0.0009930 | 0.0148580 | | | |
| 3 | 0.0038420 | 0.0015240 | 0.0168710 | | | |
| 4 | 0.0046670 | 0.0010580 | 0.0154450 | | | |

a Calculate the missing entries in the blank columns in the spreadsheet. Which expression was almost constant?

b The ratio that was almost constant is the equilibrium constant, *K*, for this reaction at the specified temperature. Write the expression for *K* for this reaction.

c Write a general expression for the equilibrium law for the following reaction: $aA(g) + bB(g) \rightleftharpoons cC(g) + dD(g)$.

2 a Write the equation $H_2(g) + I_2(g) \rightleftharpoons 2HI(g)$ in reverse.

b Write the expression for the equilibrium constant for this reverse reaction.

c Use the data for one of the mixtures in Question 1 to calculate the value of *K* for this reverse reaction.

d What is the relationship between the two constants you have calculated?

e Calculate the value of *K* for the reaction:

$\frac{1}{2}H_2(g) + \frac{1}{2}I_2(g) \rightleftharpoons HI(g)$

f What is the relationship between this constant and the first constant you calculated?


ISBN 978 0 6557 0027 2

3 **a** Consider the following solution equilibrium:

$Fe^{3+}(aq) + SCN^{-}(aq) \rightleftharpoons FeSCN^{2+}(aq)$

Complete the concentration–time graphs on the grids provided if each of the following changes were made to this system at constant temperature.

i A small amount of $Fe^{3+}(aq)$ is added. (Assume the volume change is negligible.)

ii A volume of water is added to the system.

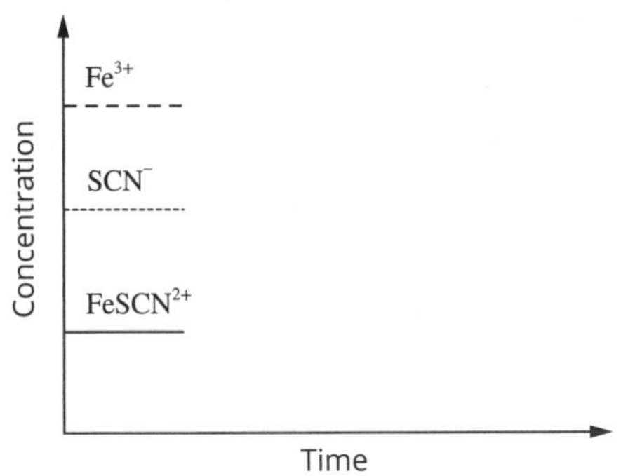


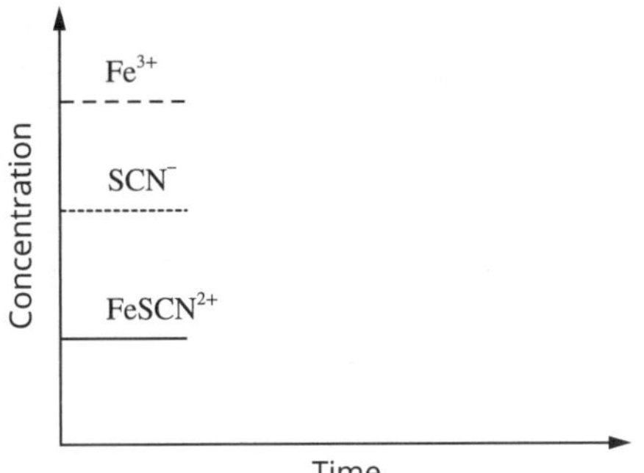


**b** Consider the following gaseous equilibrium:

$N_2O_4(g) \rightleftharpoons 2NO_2(g)$

Complete the graph if the volume of the container is increased at constant temperature.

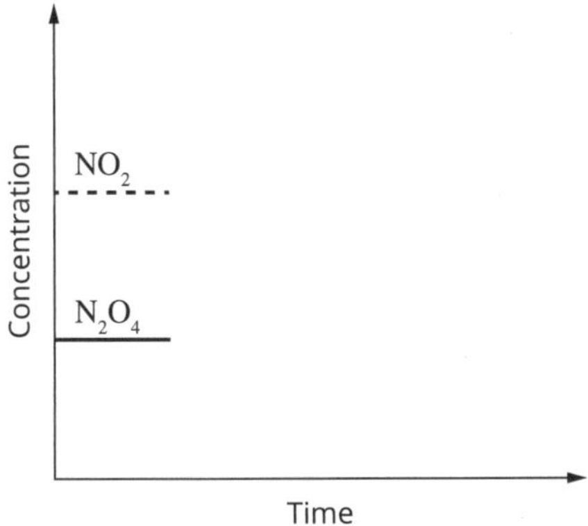


4 The following table shows the results of a simple experiment involving temperature change on two equilibria at constant volume.

| **Effects of temperature on two equilibria** | | | |
|---|---|---|---|
| **Equilibrium system** | **Type of reaction** | **Colour in ice bath** | **Colour in hot water bath** |
| $Fe^{3+}(aq) + SCN^{-}(aq) \rightleftharpoons FeSCN^{2+}(aq)$<br>pale yellow red | exothermic | red | pale yellow |
| $N_2O_4(g) \rightleftharpoons 2NO_2(g)$<br>colourless brown | endothermic | light brown | dark brown |

Complete the table below.

| **Effect of temperature on equilibrium system** | | | | |
|---|---|---|---|---|
| **Equilibrium system** | **Effect of heating on concentration of products** | **After heating, position of equilibrium has moved ...** | **Effect of heating on value of *K*** | **Sign of $\Delta H$** |
| $Fe^{3+}(aq) + SCN^{-}(aq) \rightleftharpoons FeSCN^{2+}(aq)$<br>pale yellow red | | | | |
| $N_2O_4(g) \rightleftharpoons 2NO_2(g)$<br>colourless brown | | | | |

5 Examine the concentration–time graph below and answer the questions for the system:

$N_2(g) + 3H_2(g) \rightleftharpoons 2NH_3(g)$

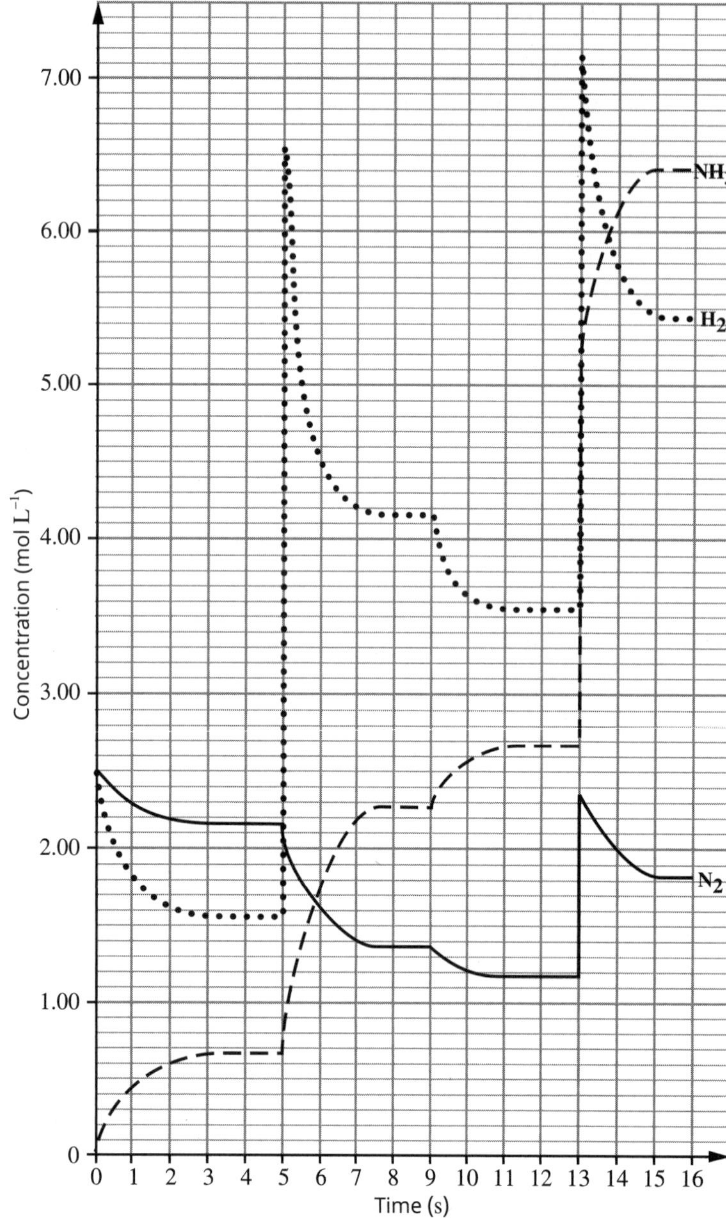


a During which time intervals was the system at equilibrium? ______

b Calculate the equilibrium constant at:

i 4 seconds ______

ii 8 seconds ______

iii 12 seconds ______

iv 16 seconds ______

c Using the answers in part **b** and your knowledge of equilibrium, explain what changes were made to the system at:

i 5 seconds ______

ii 9 seconds ______

iii 13 seconds ______

 ISBN 978 0 6557 0027 2

# WORKSHEET 16

Case study

## Equilibrium—Le Châtelier's principle and the equilibrium law

**1** State Le Châtelier's principle.

**2** Consider the following equilibrium system:

$4HCl(g) + O_2(g) \rightleftharpoons 2H_2O(g) + 2Cl_2(g) \quad \Delta H = -116\ kJ$

Complete the following table by predicting the effect (increase, decrease, no change) of the following changes to the equilibrium system on each quantity. Briefly explain the reason for your answer by using Le Châtelier's principle.

| Predicted effects on equilibrium system | | | | |
|---|---|---|---|---|
| | Quantity | Change | Effect (compared to initial equilibrium) | Reason |
| **a** | $[Cl_2]$ | adding $O_2$ at constant volume and temperature | | |
| **b** | [HCl] | adding $O_2$ at constant volume and temperature | | |
| **c** | amount of $H_2O$ (in mol) | decreasing the volume of the container at constant temperature | | |
| **d** | [HCl] | decreasing the pressure by increasing the volume of the container at constant temperature | | |
| **e** | *K* | increasing the volume of the container at constant temperature | | |
| **f** | *K* | increasing the temperature at constant volume | | |
| **g** | $[O_2]$ | increasing the temperature at constant volume | | |
| **h** | [HCl] | adding argon gas at constant volume and temperature | | |
| **i** | *K* | adding a catalyst at constant volume and temperature | | |

3 a Consider the following reaction:

$2H_2(g) + S_2(g) \rightleftharpoons 2H_2S(g)$ $K = 9.4 \times 10^5\ M^{-1}$ at 750°C

A mixture of $H_2$, $S_2$ and $H_2S$ was allowed to come to equilibrium in a closed 2.0 L container at 750°C. The equilibrium concentrations of $H_2$ and $H_2S$ gases were analysed and found to be 0.234 M and 0.442 M respectively.

i What does the value of the equilibrium constant for this reaction tell you about the extent of the reaction?

ii Write an expression for the equilibrium constant for this reaction.

iii Calculate the equilibrium concentration of $S_2(g)$ in the mixture.

b 1.364 mol of $H_2$, 0.682 mol of $S_2$ and 0.680 mol of $H_2S$ were mixed in another 2.00 L container at 550°C. At equilibrium, the concentration of $H_2S$ was measured as 1.02 M.

i What is the value of the equilibrium constant at this temperature?

ii Compare your answer to the value of *K* for this reaction in part **a**. Is this reaction exothermic or endothermic? Give a reason for your answer.

4 Consider the following reaction at equilibrium:

$N_2O_4(g) \rightleftharpoons 2NO_2(g)$ $\Delta H = +57$ kJ $K = 3.62 \times 10^2$ M at 327°C

a Calculate the equilibrium constant at 327°C for the following reaction:

$NO_2(g) \rightleftharpoons \frac{1}{2} N_2O_4(g)$

b In terms of Le Châtelier's principle, explain the effect on the position of equilibrium when the:

i temperature is decreased at constant volume

ii volume of the container is increased at constant temperature.


ISBN 978 0 6557 0027 2

5 The value of $\Delta H$ for a chemical reaction depends on the direction of the equation for the reaction and the coefficients and states of species in the equation. These factors also affect the value of an equilibrium constant. However, when an equation is written in a different way, the effects on $\Delta H$ and $K$ are different.

Complete the following table.

| **Effects on $\Delta H$ and $K$** | | |
|---|---|---|
| **Equation** | **$\Delta H$** | **$K$ (at 527°C)** |
| $N_2(g) + 3H_2(g) \rightleftharpoons 2NH_3(g)$ | –92 kJ | 0.051 $M^{-2}$ |
| $2NH_3(g) \rightleftharpoons N_2(g) + 3H_2(g)$ | | |
| $2N_2(g) + 6H_2(g) \rightleftharpoons 4NH_3(g)$ | | |
| $NH_3(g) \rightleftharpoons \frac{1}{2} N_2(g) + \frac{3}{2} H_2(g)$ | | |
| **Equation** | **$\Delta H$** | **$K$ (at 600°C)** |
| $2SO_2(g) + O_2(g) \rightleftharpoons 2SO_3(g)$ | –198 kJ | 1.56 $M^{-1}$ |
| | –99 kJ | |
| | | 2.43 $M^{-2}$ |
| | +198 kJ | |

6 Use your results from Question 5 to complete the summary statements.

For $\Delta H$: When equations are reversed, $\Delta H$ has the ___________ sign and ___________ magnitude. When coefficients are doubled, $\Delta H$ has the ___________ sign and the magnitude is ___________. When coefficients are halved, $\Delta H$ has the ___________ sign and the magnitude is ___________.

For $K$: When equations are reversed, the value of $K$ is ___________. When coefficients are doubled, the value of $K$ is ___________. When the coefficients are halved, the value of $K$ is ___________.

# WORKSHEET 17

Simulation

# Electrolytic cells—predicting reactions and comparing electrolytes

**1** Use the electrochemical series to predict the products of the following electrolysis reactions.

| Predicting electrode reactions | | | |
|---|---|---|---|
| **Cell no.** | **Solution and electrodes** | **Anode half-equation** | **Cathode half-equation** |
| 1 | 1 M $CuSO_4(aq)$, inert electrodes | | |
| | **Overall equation:** | | |
| 2 | 0.5 M $Pb(NO_3)_2(aq)$, inert electrodes | | |
| | **Overall equation:** | | |
| 3 | 1 M NaCl(aq), inert electrodes | | |
| | **Overall equation:** | | |
| 4 | 1 M $CuSO_4(aq)$, Cu electrodes | | |
| | **Overall equation:** | | |
| 5 | Molten KI, inert electrodes | | |
| | **Overall equation:** | | |

**2** In some of the cells in Question 1, the reactions observed are not the same as predicted by the electrochemical series. Explain the limitations associated with the predictions you have made.

______________________________________________

______________________________________________

**3** Two solutions are electrolysed in series using the arrangement shown. The following experiment was carried out.

At 0 minutes, the power was switched on; at 5 minutes, electrode 3 was replaced with a copper electrode; and at 10 minutes, the power stopped.

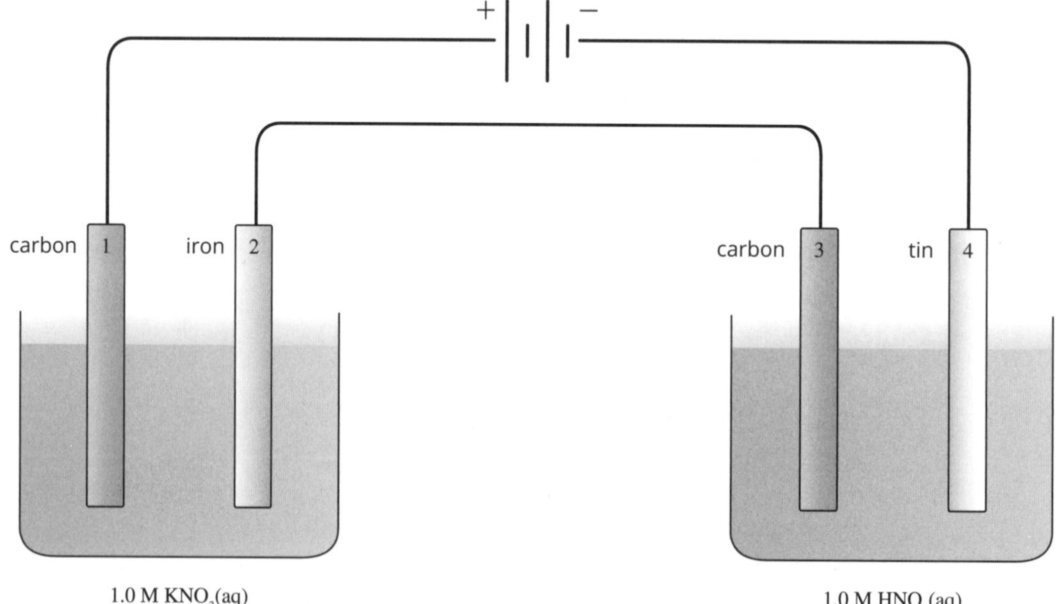


 ISBN 978 0 6557 0027 2

**a** Complete the half-equations for the reactions as indicated below.

| Electrode | Half-equation for reaction at 2 min | Half-equation for reaction at 8 min |
|---|---|---|
| 1 | | |
| 2 | | |
| 3 | | |
| 4 | | |

**b** Explain the changes that occur to the half-reactions over time in part a.

**4** In industrial electrolytic cells, the cathode is often made of iron. Explain why this is so and why the products of the cells are not contaminated with iron ions.

**5** Select one of the two electrolytic cells that use molten electrolytes from the left column of Table 1. Compare it with the cell using aqueous electrolyte (right column) by completing Table 2.

**Table 1**

| Electrolytic cells with molten electrolyte | Electrolytic cell with aqueous electrolyte |
|---|---|
| Downs cell (for production of sodium and chlorine)<br>Features:<br>• $NaCl(l)/CaCl_2(l)$ electrolyte<br>• carbon anode<br>• iron cathode<br>• iron screen around cathode | Membrane cell (for production of sodium hydroxide, chlorine and hydrogen)<br>Features:<br>• concentrated NaCl(aq) electrolyte (called brine)<br>• carbon anode<br>• iron cathode<br>• semipermeable membrane separates electrode regions |
| Hall-cell (production of aluminium)<br>Features:<br>• electrolyte is $Al_2O_3$ dissolved in molten cryolite<br>• carbon anode<br>• carbon cathode | |

**Table 2** Comparing cells with molten and aqueous electrolytes

| | Cell using molten electrolyte | Cell using aqueous electrolyte |
|---|---|---|
| Anode equation | | |
| Cathode equation | | |
| Overall cell equation | | |
| How are the products from each electrode kept separate and why? | | |
| Explain the choice of electrodes. | | |
| Explain why each electrolyte is used.<br>As appropriate, discuss:<br>• why compounds are mixed to form the electrolyte<br>and/or<br>• the choice of concentration.<br>In each case explain the reason for choosing a molten or aqueous electrolyte. | | |

# WORKSHEET 18

Case study

# Industrial applications—finding a compromise

In industrial processes, the conditions required for a high rate of reaction and a high equilibrium yield are often in conflict with each other. This worksheet examines the conditions chosen for a particular industrial process.

Dimethyl ether (DME) is used as a propellant in spray cans of paints. It has similar properties to LPG and has been proposed as an alternative fuel for power generation, domestic uses and transportation. It does not contain any sulfur or nitrogen, is harmless to humans, biodegradable and non-corrosive to metals. The following table contains some information about DME.

| Dimethyl ether (DME) data | |
|---|---|
| Molecular formula | $CH_3OCH_3$ |
| Recommended purity of DME as vehicle fuel | <0.01% methanol and <0.05% water |
| Energy value compared to LPG | Similar |
| Cetane number (A fuel with a high cetane number starts to burn quickly after it is injected.) | 10 times LPG |
| Boiling point of DME compared to LPG | –25°C compared with –42°C |
| Liquid density at 20°C compared to LPG | 0.67 g $mL^{-1}$ compared with 0.49 g $mL^{-1}$ |

There are three ways of producing DME, but the most recent and environmentally friendly process is a three-step synthesis involving a specially developed liquid slurry reactor containing a powdered catalyst. Like many industrial processes, the reaction conditions must be carefully adjusted to ensure a relatively high yield of DME in a reasonable time.

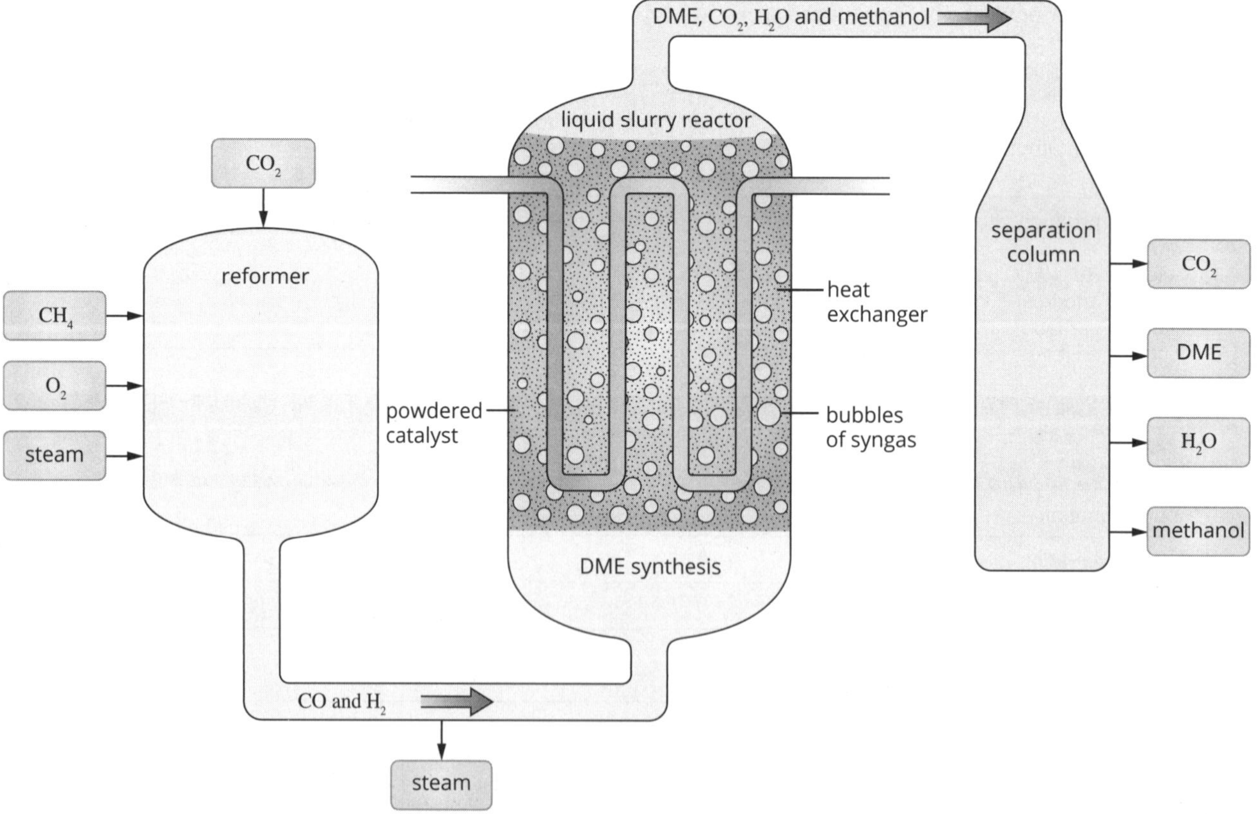


**Figure 3.2.9** The three-step synthesis involved in producing DME (dimethyl ether)


ISBN 978 0 6557 0027 2

The steps in this process are:

- a highly efficient conversion of methane, oxygen and carbon dioxide into syn-gas (CO and $H_2$) and steam in a reformer
- DME synthesis in the liquid slurry reactor where bubbles of syn-gas react as they rise through the catalytic slurry in an exothermic, temperature-sensitive process that produces DME and carbon dioxide (some methanol is also formed by a side reaction)
- purification to produce DME, $CO_2$, $H_2O$ and methanol.

**1** Use Figure 3.2.9 and the information provided to write an equation for the main reaction that occurs in the liquid slurry reactor.

**2** Suggest why the heat exchanger is present in the reactor.

**3** Suggest how DME might be separated from other compounds in the reaction mixture in the separation column.

In a pilot plant, the exothermic reaction producing DME in the liquid slurry reactor takes place at 260°C and 5 MPa. At this temperature, the conversion of syn-gas to DME is about 90%. The DME produced is 99.9% pure and contains less than 0.01% water and methanol. The catalyst is water sensitive and gradually deactivates at temperatures higher than 300°C.

**4** With reference to reaction rates and equilibrium, discuss why the particular temperatures and pressure in the reactor would have been chosen.

**5** In the reactor, what would be the effect on the yield of DME if you were to:

**a** decrease the pressure at constant temperature?

**b** increase the temperature at constant pressure?

**6** Identify the wastes likely to be produced from this process and suggest a management process for each one.

**7** What health and safety issues might be involved in this method of DME production? You can use the internet to investigate this further.

**8** Given the information in this worksheet, state an advantage or disadvantage you can foresee for the use of DME as an alternative fuel to petrol. You can use the internet to investigate this further.

**9** In what way(s) does this process meet the aims of green chemistry and in what way(s) doesn't it?

# WORKSHEET 19

Case study

## Secondary cells—storing and producing energy

Li-ion cells are an advanced type of battery technology that last longer, charge faster and have a higher power density than conventional batteries. One electrode is composed of a lithium–cobalt oxide compound ($LiCoO_2$) and the other electrode is a compound made from lithium and graphite ($LiC_6$). While the processes that occur within a Li-ion cell are complex, the overall cell reaction during use can be written as:

$LiC_6 + CoO_2 \rightarrow LiCoO_2 + C_6$

The half-reactions when the cell is generating electricity are:

$LiC_6 \rightarrow Li^+ + C_6 + e^-$

and:

$CoO_2 + Li^+ + e^- \rightarrow LiCoO_2$

1 Write the half-equation for the reaction that occurs at the cathode when the cell is in use.

______

2 Write the half-equation for the reaction that occurs at the negative electrode when it is being recharged.

______

3 Write the equation for the overall reaction that occurs in the cell when it is being recharged.

______

4 Describe the energy transformation that occurs when the cell is being recharged.

______

5 A battery charger can be used to recharge the cell.

**a** Is the cathode of the cell connected to the positive or negative terminal of the charger? ______

**b** Is the cathode of the cell positive or negative when it is being recharged? ______

6 Explain why a secondary cell, such as this, can be recharged whereas primary cells cannot.

______

______

7 During the operation of the cell, $Li^+$ ions flow through the electrolyte. In which direction do they flow when the cell is in use?

______

8 A lithium-ion cell in a particular mobile phone operates with a peak current of 0.850 A. Calculate the maximum mass of lithium ions that could be transferred between the electrodes during a 5-minute phone call.


ISBN 978 0 6557 0027 2

# WORKSHEET 20

Case study

# Commercial cells and Faraday's laws

1 Three important chemicals, hydrogen gas, chlorine gas and sodium hydroxide solution, are manufactured by the electrolysis of brine (concentrated salt solution) in a membrane cell. The cell usually operates at 4.5 V and 200 A. A diagram of the cell is shown in Figure 3.2.10.

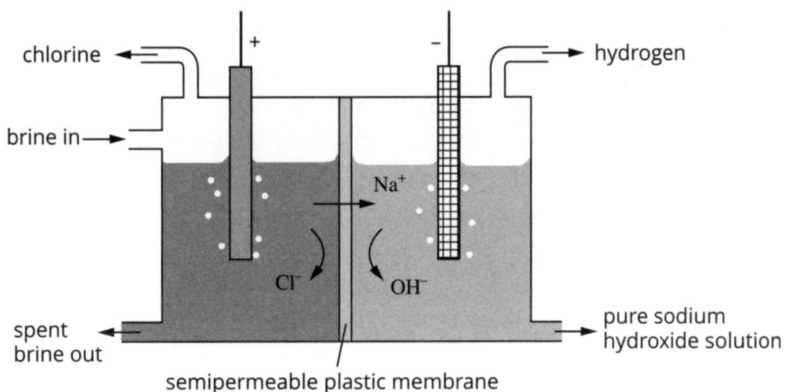


**Figure 3.2.10** A diagram of the electrolysis of brine

**a** Use your knowledge of the construction of electrolytic cells to explain the function of the following components of this cell.

**i** semipermeable membrane

**ii** brine

**b** Write equations for the reactions occurring at each electrode and the overall equation.

anode:

cathode:

overall:

**c** Calculate the maximum volume of hydrogen gas, in litres, that would be produced at SLC when 950.0 g of sodium hydroxide is produced in the cell.

**d** The energy released as electricity passes through a cell is given by the formula:

$E$ (energy in joules) = $V$ (voltage in V) × $I$ (current in A) × $t$ (time in seconds) = $V$ × $Q$ (charge in C).

Calculate the electrical energy, in joules, required to produce this volume of hydrogen if the cell operates at 100% efficiency.

**2** The 'blister' copper produced from a copper smelter is about 99.4% pure. For copper to conduct electricity well, this level of purity must be increased to 99.98%. Purification is achieved by electrolysis. Gold and silver are often found in blister copper as impurities.

Consider the simplified diagram of a copper electrorefining cell. In such a cell, the anode is made of blister copper and the cathode is made of pure copper.

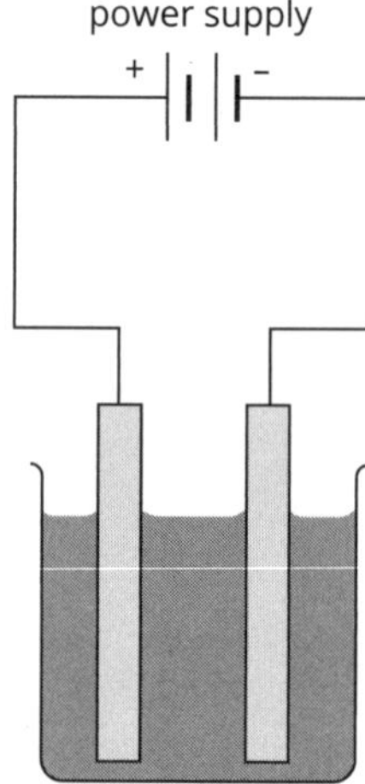


**a** Label the anode, cathode, electrolyte, polarity of the electrodes and direction of flow of the electrons in the wires.

**b** Write half-equations for the electrode reactions.

anode: ______

cathode: ______

**c** Would the gold and silver impurities in the blister copper ultimately become ions in the electrolyte or be found as pieces of metal at the bottom of the cell? Explain your answer by referring to the electrochemical series.

______

______

**3** Two students were given the chemicals and equipment needed to zinc plate a metal toy car. They constructed an electroplating cell using a large beaker containing 60 mL of 1 M $ZnSO_4$ solution and a zinc strip. Their results and other data are given in the table.

| | |
|---|---|
| **Voltage used** | 6.0 V |
| **Time cell operated** | 100 min |
| **Mass of toy car before plating** | 105.0 g |
| **Mass of toy car after plating** | 115.0 g |
| **Molar mass of Zn** | 65.4 g mol$^{-1}$ |
| **1 faraday** | 96 500 C mol$^{-1}$ |

**a** Write the equation for the reaction occurring at the cathode in the zinc-plating cell.

______

 ISBN 978 0 6557 0027 2

**b** Using the mass of zinc that was deposited on the toy car, calculate the amount of zinc deposited, in mol.

**c** Calculate the number of moles of electrons needed to deposit this mass of zinc.

**d** Calculate the current, in A, that was needed to deposit this mass of zinc.

**e** The energy released as electricity passes through a cell is given by the formula:
$E$ (energy in J) = $V$ (voltage in V) × $I$ (current in A) × $t$ (time in seconds) = $V \times Q$ (charge in C).
Calculate the energy, in J, consumed by the cell to deposit this mass of zinc.

**4** **a** Briefly explain how an electrolyser with a polymer electrolyte (a PEM electrolyser) works and give the electrode half-equations.

**b** Explain the term 'green hydrogen'.

# WORKSHEET 21

Classification and identification

## Literacy review—equilibrium and electrolysis

1 Complete the following table to review your understanding of some of the terms in equilibrium.

| Term | Meaning |
|---|---|
| equilibrium law | |
| $K$ | |
| $Q$ | |
| open system | |
| dynamic equilibrium | |
| Le Châtelier's principle | |

2 The following table contains statements about galvanic and electrolytic cells. Complete the table by indicating whether each statement is true for galvanic cells, electrolytic cells or both types of cells.

| Statement | Galvanic, electrolytic or both | Statement | Galvanic, electrolytic or both |
|---|---|---|---|
| **a** Oxidation occurs at the anode. | | **b** Chemical energy is converted to electrical energy. | |
| **c** Electrical energy is converted to chemical energy. | | **d** Reaction is spontaneous. | |
| **e** Reduction occurs at the cathode. | | **f** Reaction is non-spontaneous. | |
| **g** Cell produces electrical energy. | | **h** The strongest oxidising agent reacts with the strongest reducing agent. | |
| **i** Cathode is positive. | | **j** Cathode is negative. | |
| **k** Cell consumes electrical energy. | | **l** Reactants must be kept separate. | |
| **m** Anode is negative. | | **n** Anode is positive. | |
| **o** Electrons flow in the wires in the external circuit from the negative to the positive electrodes. | | **p** Cell can be used to run an appliance. | |

3 The following summary of chemical equilibrium contains five errors. Highlight the errors and correct them below.

Dynamic equilibrium occurs when there is a reversible reaction in an open system. In an equilibrium system, the rates of the forward and reverse reactions are equal and there is no change in the concentrations of the reactant and product molecules. However, it can be shown that there is continual conversion of reactants to products and vice versa. Only when a system is open will it achieve equilibrium because only energy is lost to the surroundings.

The equilibrium law is a mathematical expression that shows the ratio of the concentrations of products and reactants at equilibrium. At a specified temperature, this ratio is constant and equal to $K$, the equilibrium constant. If this concentration ratio, $Q$, is less than $K$ the system will favour the reverse reaction to restore equilibrium.

When the temperature is decreased for an exothermic reaction, the value of $K$ decreases. When the temperature is increased for an endothermic reaction, $K$ increases.

Le Châtelier's principle predicts that when the pressure of a system decreases by increasing the volume, the system will move in the direction of the least particles to re-establish equilibrium.

Corrections:


ISBN 978 0 6557 0027 2

# WORKSHEET 22

## Reflection—How can the rate and yield of chemical reactions be optimised?

The following table lists the key knowledge covered in this area of study.

1 Reflect on how well you understand the concepts listed. Rate your learning by shading the circle that corresponds to your current level of understanding for each one.

| Key knowledge | Not confident ◄ | | | ► Very confident |
|---|---|---|---|---|
| Factors affecting the rate of a chemical reaction, including the role of catalysts | ○ | ○ | ○ | ○ |
| Distinction between reversible and irreversible reactions, and between rate and extent of a reaction | ○ | ○ | ○ | ○ |
| The dynamic nature of equilibria | ○ | ○ | ○ | ○ |
| Changes in position of equilibrium that can occur when changes occur; representation of changes using concentration–time graphs | ○ | ○ | ○ | ○ |
| Applications of Le Châtelier's principle | ○ | ○ | ○ | ○ |
| Calculations involving equilibrium expressions; dependence of *K* value on the temperature and equation | ○ | ○ | ○ | ○ |
| The reaction quotient (*Q*) as a measure of the extent of a chemical reaction | ○ | ○ | ○ | ○ |
| Responses to the conflict between optimal rate and temperature; green chemistry principles of catalysis and designing for energy efficiency | ○ | ○ | ○ | ○ |
| Use and limitations of the electrochemical series to explain or predict the products of electrolysis | ○ | ○ | ○ | ○ |
| Common design features and operating principles of commercial electrolytic cells | ○ | ○ | ○ | ○ |
| Common design features and operating principles of secondary cells | ○ | ○ | ○ | ○ |
| The role of innovation in meeting society's energy needs for producing 'green' hydrogen | ○ | ○ | ○ | ○ |
| Applications of Faraday's laws and stoichiometry | ○ | ○ | ○ | ○ |

2 Consider the points you have shaded from 'Not confident' to 'Very confident'. List specific ideas you can identify that were challenging.

______________________________

______________________________

3 Write down two different strategies that you will apply to help further your understanding of these ideas.

______________________________

______________________________

# PRACTICAL ACTIVITY 8

Controlled experiment

# Factors affecting the rate of reaction

## SUGGESTED DURATION

- 50 minutes

## MATERIALS

- 100 mL of 0.25 M sodium thiosulfate solution, $Na_2S_2O_3(aq)$
- 40 mL of 2 M hydrochloric acid, HCl(aq)
- 10 mL measuring cylinder
- 100 mL measuring cylinder
- 6 × 100 mL beakers
- thermometer, –10 to 110°C
- Bunsen burner, tripod stand and gauze mat
- bench mat
- stopwatch
- black marking pen
- cling film
- safety glasses

## INTRODUCTION

Hydrochloric acid reacts with sodium thiosulfate to produce sulfur, sulfur dioxide gas and water:

$S_2O_3^{2-}(aq) + 2H^+(aq) \rightarrow S(s) + SO_2(g) + H_2O(l)$

The sulfur precipitate causes the reaction mixture to become cloudy. The rate at which the solution becomes cloudy indicates the rate of reaction: the shorter the period before the cloudiness obscures a cross on a sheet of paper placed under the beaker, the more rapid the reaction. In this practical activity the effects of temperature and concentration of one of the reactants on the rate of reaction are studied.

## AIM

To explore the ways in which changes in temperature and concentration of reactants affect the rate of a reaction

As directed by your teacher, complete the pre-lab safety information by referring to the safety data sheets (SDSs) or your teacher's risk assessment for the activity.

| PRE-LAB SAFETY INFORMATION | | |
|---|---|---|
| **Material used** | **Hazard** | **Control** |
| sulfur dioxide gas, $SO_2(g)$* | • toxic by inhalation; causes irritation to the eyes, nose and throat | Perform the reactions in a fume cupboard if possible. Avoid breathing the gas: work in a well-ventilated area and dispose of reaction mixtures in a sink in a fume cupboard. Wear safety glasses. |
| 2 M hydrochloric acid, HCl(aq) | | |
| Complete and indicate that you have understood the information in the safety table. | | |
| Name (print): | | |
| I understand the safety information (signature): | | |

*Although concentrations of $SO_2$ gas should be low in this activity, the gas can exacerbate breathing difficulties in people with conditions such as asthma. The laboratory should be well ventilated and the neck of the conical flask covered with cling film.

## METHOD

### PART A • EFFECT OF TEMPERATURE

1. Mark a cross on a sheet of paper with a black marking pen.
2. Place a 100 mL beaker on top of the cross. Pour 10 mL of 0.25 M sodium thiosulfate solution and 35 mL of water into the beaker.
3. Record the temperature of the solution in Table 1.
4. Add 5 mL of 2 M hydrochloric acid and commence timing. Immediately cover the top of the beaker with cling film. Measure and record in the table the time taken for the cross to disappear when viewed from above the beaker.
5. Pour 10 mL of 0.25 M sodium thiosulfate solution and 35 mL of water into another beaker. Heat the solution to 40°C. Remove the beaker from the source of heat, place it on top of the cross, and immediately add 5 mL of hydrochloric acid. Cover the top of the beaker with cling film. Again, measure and record the time taken for the cross to disappear.
6. Repeat step 5 with a third beaker, this time heating the solution to 60°C.


ISBN 978 0 6557 0027 2

## PART B • EFFECT OF CONCENTRATION OF REACTANTS

1 ▪ Place a 100 mL beaker on top of a cross marked on a sheet of paper. Pour 10 mL of 0.25 M sodium thiosulfate solution and 35 mL of water into the beaker. Add 5 mL of 2 M hydrochloric acid and commence timing. Cover the top of the beaker with cling film. Measure and record in Table 2 the time taken for the cross to disappear when viewed from above the beaker.

2 ▪ Place another 100 mL beaker on top of the cross. Pour 25 mL of the sodium thiosulfate solution and 20 mL of water into the beaker. Add 5 mL of hydrochloric acid and commence timing. Cover the top of the beaker with cling film. Measure and record the time taken for the cross to disappear.

3 ▪ Place another 100 mL beaker on top of the cross. Pour 40 mL of sodium thiosulfate solution and 5 mL of water into the beaker. Add 5 mL of hydrochloric acid and commence timing. Cover the top of the beaker with cling film. Measure and record the time taken for the cross to disappear.

## RESULTS

**Table 1** Results for effect of temperature

| Temperature (°C) | Time for the cross to disappear (s) |
|---|---|
| | |
| | |
| | |
| | |

**Table 2** Results for effect of concentration of reactants

| Solution | Concentration of sodium thiosulfate in solution (M) | Time for the cross to disappear (s) |
|---|---|---|
| 10 mL $Na_2S_2O_3$ solution, 35 mL water and 5 mL HCl | | |
| 25 mL $Na_2S_2O_3$ solution, 20 mL water and 5 mL HCl | | |
| 40 mL $Na_2S_2O_3$ solution, 5 mL water and 5 mL HCl | | |

## DISCUSSION

1 What effect does an increase in temperature have on the rate of this reaction?

2 Explain why an increase in temperature leads to a change in the rate of reaction.

3 Calculate the concentrations of the sodium thiosulfate in each of the three reaction mixtures made up in Part B and enter these in Table 2.

4 What effect does increasing the concentration of sodium thiosulfate have on the rate of the reaction?

## PRACTICAL ACTIVITY 8

**5** Explain why the rate of a reaction depends upon the concentration of the reactants.

**6** Discuss the validity of your experiment. Include a definition of validity in your answer.

### CONCLUSION

Summarise the effects of a change in temperature and in the concentration of the reactants on the rate of reaction.

 ISBN 978 0 6557 0027 2

# PRACTICAL ACTIVITY 9

Controlled experiment

# Effect of concentration changes on equilibrium yields

## SUGGESTED DURATION

- 50 minutes

## INTRODUCTION

You will be given a solution of $Fe(SCN)^{2+}$ that contains the ions $Fe^{3+}$, $SCN^-$ and $Fe(SCN)^{2+}$ at equilibrium:

$Fe^{3+}(aq) + SCN^-(aq) \rightleftharpoons Fe(SCN)^{2+}(aq)$

pale yellow colourless red

The intense, blood-red colour of the solution is due to the presence of the $Fe(SCN)^{2+}$ ion. The colour of the solution in each test tube, when viewed down the tube, is a measure of the amount of $Fe(SCN)^{2+}$ ions, in mol, present in the tube. By noting how the intensity of this colour changes, it is possible to deduce the effect of each of the tests performed in this experiment on the equilibrium. For instance, if the colour deepens, the number of $Fe(SCN)^{2+}$ ions has increased and the number of $Fe^{3+}$ and $SCN^-$ ions must have simultaneously decreased because they are used up to form more $Fe(SCN)^{2+}$. The equilibrium would be described as having a net forward reaction (its position would have 'shifted to the right').

In Tests A to E, the amount (number of moles) of $Fe^{3+}$ or of $SCN^-$ ions present in the solution is initially changed as follows.

- In Test A, addition of $Fe(NO_3)_3$ increases the amount of $Fe^{3+}$.
- In Test B, addition of KSCN increases the amount of $SCN^-$.
- In Test C, addition of NaF decreases the amount of $Fe^{3+}$ because $F^-$ ions react with $Fe^{3+}$ ions to form $FeF_6^{3-}(aq)$.
- In Test D, addition of $AgNO_3$ decreases the amount of $SCN^-$ because $Ag^+$ ions react with $SCN^-$ ions to form a white precipitate of AgSCN.
- In Test E, addition of water has no effect.

## MATERIALS

- 25 mL of $5 \times 10^{-4}$ M iron(III) thiocyanate solution, $Fe(SCN)^{2+}(aq)$*
- 0.1 M iron(III) nitrate solution, $Fe(NO_3)_3(aq)$
- 0.1 M potassium thiocyanate solution, KSCN(aq)
- 0.1 M sodium fluoride solution, NaF(aq)
- 0.1 M silver nitrate solution, $AgNO_3(aq)$
- 6 semi-micro test tubes
- semi-micro test-tube rack
- bench mat
- marking pen
- dropping pipette
- white tile or white sheet of paper
- safety glasses

*20 mL of 0.1 M $Fe(NO_3)_3$ and 20 mL of 0.1 M KSCN per litre

## AIM

To explore the effect of concentration changes on an aqueous equilibrium

As directed by your teacher, complete the pre-lab safety information by referring to the safety data sheets (SDSs) or your teacher's risk assessment for the activity.

| PRE-LAB SAFETY INFORMATION | | |
|---|---|---|
| **Material used** | **Hazard** | **Control** |
| 0.1 M iron(III) nitrate solution, $Fe(NO_3)_3(aq)$ | | |
| 0.1 M sodium fluoride solution, NaF(aq) | | |
| 0.1 M silver nitrate solution, $AgNO_3(aq)$ | • can stain skin, clothing and bench surfaces | |
| Complete and indicate that you have understood the information in the safety table.<br>Name (print):<br>I understand the safety information (signature): | | |

## PRACTICAL ACTIVITY 9

### METHOD

1 ▪ Fill each of six semi-micro test tubes to one-third of its volume with $Fe(SCN)^{2+}$ solution. Check that the liquid in each tube has the same intensity of colour when you look down the tube, using a white tile or sheet of paper as a background. If necessary, add more solution so that the liquid in each tube is the same colour. Label the tubes A–F.

2 ▪ Using test tube F for the purposes of comparison, perform each of the tests described in Table 1 and record the change that occurs in the colour of the solution when viewed down the test tube.

### RESULTS

**Table 1** Colour results

| Test tube | Test | Colour change |
|---|---|---|
| A | 1 drop of $Fe(NO_3)_3(aq)$ added | |
| B | 1 drop of KSCN(aq) added | |
| C | 1 drop of NaF(aq) added | |
| D | 1 drop of $AgNO_3(aq)$ added | |
| E | equal volume of water added | |
| F | none | no change |

### DISCUSSION

**1** Write an expression for the equilibrium constant of the reaction that is the subject of this practical activity.

**2** Complete Table 2 for each test by stating the:

**a** initial effect on the concentration of $Fe^{3+}$ or $SCN^-$ of each test

**b** concentration change of $Fe(SCN)^{2+}$ after the test

**c** direction in which the position of equilibrium has shifted.

The first part has been completed for you.

**Table 2** Equilibrium results

| Test-tube | Initial effect | $[Fe(SCN)^{2+}]$ change | Direction of equilibrium shift |
|---|---|---|---|
| A | $[Fe^{3+}]$: increases | increases | forward |
| B | $[SCN^-]$: | | |
| C | $[Fe^{3+}]$: | | |
| D | $[SCN^-]$: | | |
| E | $[Fe^{3+}]$: | | |

**3** Sketch graphs to show how the concentration of each ion changed during each test. (The graph for Test A has been completed for you.)

 ISBN 978 0 6557 0027 2

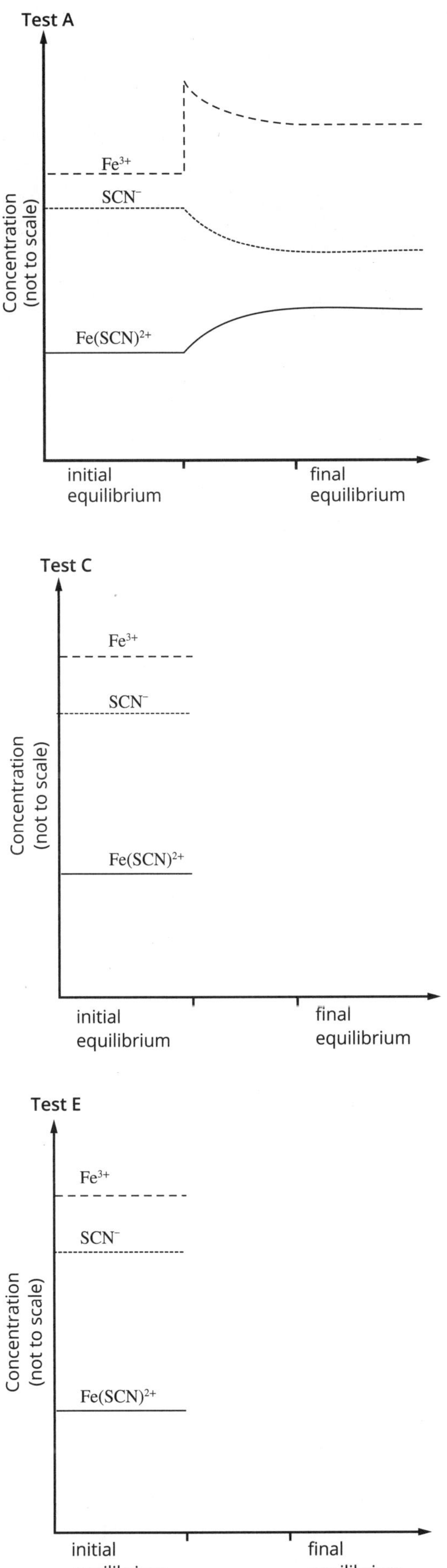
Test A
Fe3+
SCN−
Fe(SCN)2+
Concentration (not to scale)
initial equilibrium
final equilibrium
Test C
Fe3+
SCN−
Fe(SCN)2+
Concentration (not to scale)
initial equilibrium
final equilibrium

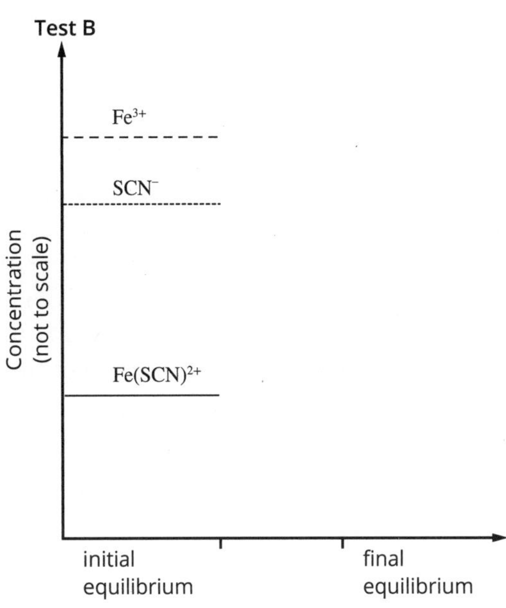
Test B
Fe3+
SCN−
Fe(SCN)2+
Concentration (not to scale)
initial equilibrium
final equilibrium

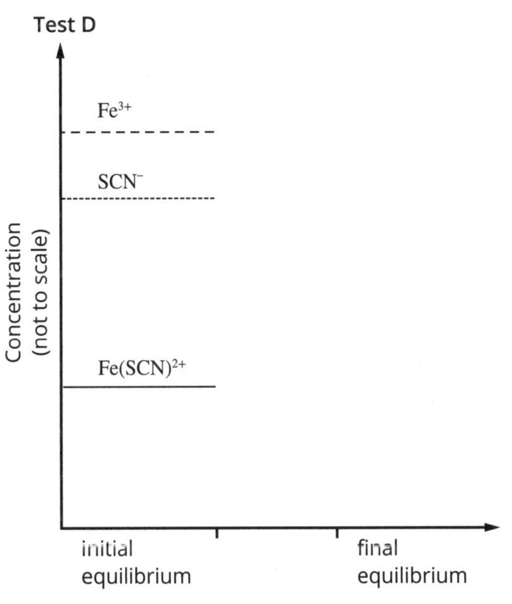
Test D
Fe3+
SCN−
Fe(SCN)2+
Concentration (not to scale)
initial equilibrium
final equilibrium

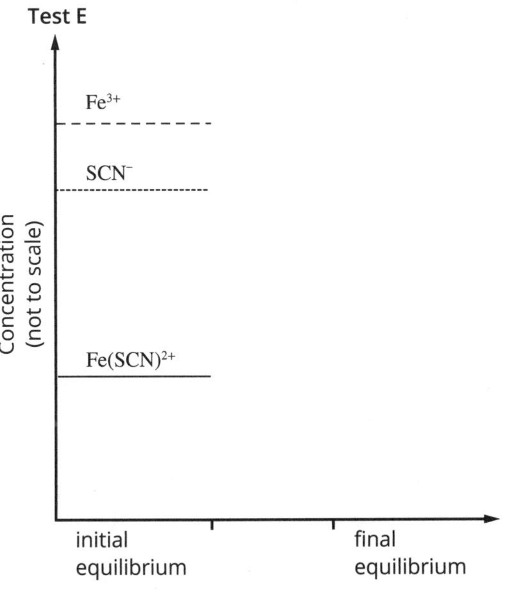
Test E
Fe3+
SCN−
Fe(SCN)2+
Concentration (not to scale)
initial equilibrium
final equilibrium

4 Use Le Châtelier's principle to account for the way in which the position of equilibrium shifts in Test A.

5 Account for the way in which the position of equilibrium shifts in Test E.

6 Does this method enable these results to be reproducible? Explain briefly how Le Châtelier's principle allows you to be confident your results are accurate.

## CONCLUSION

Summarise the effects of changes in concentration on an equilibrium system.

 ISBN 978 0 6557 0027 2

# PRACTICAL ACTIVITY 10

Controlled experiment

## Effect of volume changes on equilibrium yields (teacher demonstration only)

### SUGGESTED DURATION

- 15 minutes

### INTRODUCTION

The syringe used in this experiment contains the gases nitrogen dioxide ($NO_2$) and dinitrogen tetroxide ($N_2O_4$) in equilibrium:

$N_2O_4(g) \rightleftharpoons 2NO_2(g)$

colourless    brown

Changes in the concentration of $NO_2$ can be monitored by observing changes in the intensity of the colour of the gas mixture.

### AIM

To investigate the effect of changes in volume on a gaseous equilibrium mixture

**PRE-LAB SAFETY INFORMATION**

| Material used | Hazard | Control |
|---|---|---|
| nitrogen dioxide, $NO_2(g)$ | • very toxic if inhaled; causes burns; risk of serious damage to the eyes | Use a fume hood; wear gloves, glasses and a laboratory coat. |

Please indicate that you have understood the information in the safety table.

Name (print): ______________________

I understand the safety information (signature): ______________________

### MATERIALS

- sealed gas syringe containing $NO_2/N_2O_4$ in equilibrium (The gas mixture is obtained by pouring a small volume of concentrated $HNO_3$ onto copper turnings in a conical flask in a fume cupboard. A one-hole stopper and rubber delivery tube fitted to the neck of the flask should be used to fill the syringes.)
- safety glasses
- disposable gloves

### METHOD

Both Test 1 and Test 2 must be performed in a fume hood as a demonstration by the teacher.

**Test 1**

Hold a syringe containing a mixture of $NO_2$ gas and $N_2O_4$ gas in equilibrium and rapidly withdraw the plunger, holding it in position once you have done so*. Note and record in Table 1 the change that occurs in the intensity of the brown colour immediately after the volume is increased, and the change in colour that occurs a moment later. (You may need to do the test a few times to identify both of these changes.)

**Test 2**

Hold the syringe securely, this time placing one finger firmly* over the sealed end of the syringe. Rapidly push in the plunger to decrease the volume of gas. Hold the plunger in the new position. Again, record the change that occurs in the intensity of the brown colour immediately after the volume is decreased and the change that occurs a moment later.

* It may be preferable to have a rubber stopper on the end of the syringe (make a hole in the stopper that does not go more than halfway) and use your finger to hold the stopper in place.

## PRACTICAL ACTIVITY 10

### RESULTS

**Table 1** Colour results

| Test | Immediate colour change | Later colour change |
|---|---|---|
| 1 Volume increase | | |
| 2 Volume decrease | | |

### DISCUSSION

**1** Write an expression for the equilibrium constant, *K*, for the reaction being investigated.

______________________________________________

______________________________________________

**2** For each test complete Table 2.

**Table 2** Test results

| Reason for initial colour change | Change in concentration of $NO_2$ as system returns to equilibrium after change in volume | Direction of change in position of equilibrium |
|---|---|---|
| Test 1 | | |
| Test 2 | | |

**3** Circle the correct word in each pair of terms.

The increase in volume initially causes the concentration of $NO_2$ to *increase/decrease* and the concentration of $N_2O_4$ to *increase/decrease* so that the reaction quotient is *greater/less* than the equilibrium constant, *K*.

To regain equilibrium, the reaction quotient must *increase/decrease* and so the position of equilibrium shifts *left/right*, causing the amount of $NO_2/N_2O_4$ to increase and the amount of $NO_2/N_2O_4$ to decrease.

**4** Use Le Châtelier's principle to explain what you observed when the volume of the gas mixture was increased.

______________________________________________

______________________________________________

______________________________________________

______________________________________________

______________________________________________

______________________________________________


ISBN 978 0 6557 0027 2

5 Complete the following graphs to show how the concentrations of $NO_2$ and of $N_2O_4$ change when the volume of the equilibrium mixture is:

a increased

b decreased.

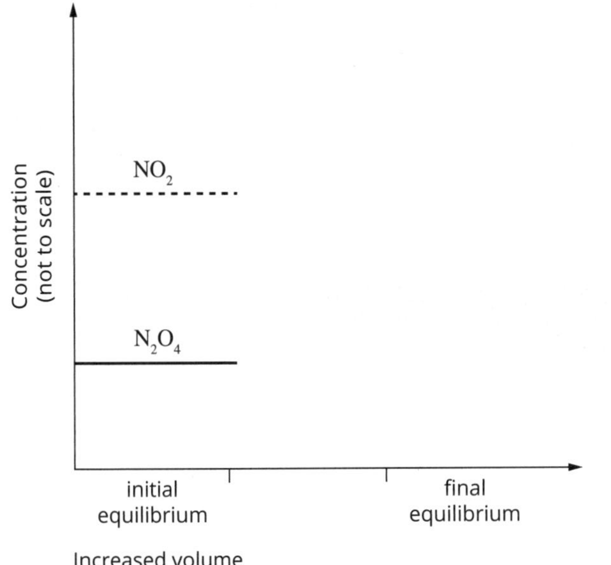


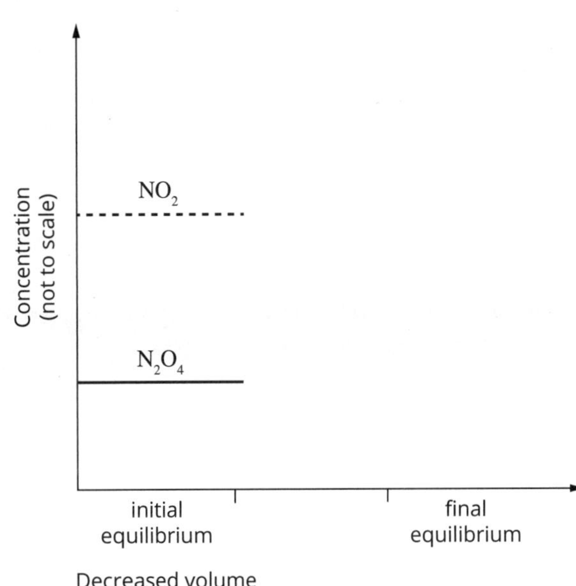


6 From your understanding of equilibrium, were the results of this experiment as you would have expected? Discuss the reproducibility of these results.

______________________________

______________________________

______________________________

## CONCLUSION

Summarise the effects of changes in volume, and therefore of concentration, on a gaseous equilibrium system.

______________________________

______________________________

______________________________

______________________________

______________________________

______________________________

# PRACTICAL ACTIVITY 11

Controlled experiment

# Effect of temperature on equilibrium yields

## SUGGESTED DURATION

- 20 minutes

## AIM

To explore the way in which the temperature of an equilibrium mixture influences the position of the equilibrium

As directed by your teacher, complete the pre-lab safety information by referring to the safety data sheets (SDSs) or your teacher's risk assessment for the activity.

| PRE-LAB SAFETY INFORMATION | | |
|---|---|---|
| **Material used** | **Hazard** | **Control** |
| 1 M phosphoric acid, $H_3PO_4(aq)$ | • irritating to the eyes and skin | |
| methyl violet indicator | • may irritate the eyes and skin | |
| Complete and indicate that you have understood the information in the safety table. Name (print): ______ I understand the safety information (signature): ______ | | |

## MATERIALS

- 10 mL of $5 \times 10^{-4}$ M iron(III) thiocyanate solution, $Fe(SCN)^{2+}(aq)$*
- 10 mL of 1 M phosphoric acid, $H_3PO_4(aq)$
- methyl violet indicator
- 250 mL beaker of ice–water
- 4 semi-micro test tubes
- semi-micro test-tube rack
- semi-micro test-tube holder
- 250 mL beaker
- Bunsen burner, gauze mat and tripod stand
- bench mat
- safety glasses

*20 mL of 0.1 M $Fe(NO_3)_3$ and 20 mL of 0.1 M KSCN per litre

## METHOD

### PART A

$Fe^{3+}(aq) + SCN^-(aq) \rightleftharpoons Fe(SCN)^{2+}(aq)$ exothermic

1 ▪ Half-fill two semi-micro test tubes with $Fe(SCN)^{2+}$ solution.

2 ▪ Place one test tube in ice–water. Carefully heat the other test tube until the solution almost boils.

3 ▪ Compare the colour of the mixtures in the two test tubes. Record your observations in Table 1.

### PART B

$H_3PO_4(aq) \rightleftharpoons H_2PO_4^-(aq) + H^+(aq)$ exothermic

1 ▪ Pour 1 M phosphoric acid into each of two semi-micro test tubes, to a depth of about 3 cm. Add two drops of methyl violet indicator to each test tube.

2 ▪ Place one test tube in ice–water. Carefully heat the other test tube until the solution almost boils. Record in Table 1 the colour of the indicator in the heated and cooled test tubes.

3 ▪ Note the acidity of each solution. Methyl violet indicator is yellow in solutions with a high concentration of $H^+(aq)$. Its colour changes through green, blue to violet as the $H^+(aq)$ concentration decreases.

## RESULTS

**Table 1** Colour results

| Test | Colour when cold | Colour when hot |
|---|---|---|
| Part A. $Fe^{3+}(aq) + SCN^-(aq) \rightleftharpoons Fe(SCN)^{2+}(aq)$ | | |
| Part B. $H_3PO_4(aq) \rightleftharpoons H_2PO_4^-(aq) + H^+(aq)$ | | |



ISBN 978 0 6557 0027 2

## PRACTICAL ACTIVITY 11

### VARIABLES

Independent variable: ____________________

Dependent variable: ____________________

Controlled variables: ____________________

### DISCUSSION

1 In Part A of this experiment, the red colour of the equilibrium mixture is caused by the ion $Fe(SCN)^{2+}$.

a How does the concentration of $Fe(SCN)^{2+}$ ions in the equilibrium mixture change as the temperature is increased?

b What does this indicate about the value of the equilibrium constant, *K*, for this reaction as the temperature increases? Explain.

c Describe how the position of equilibrium has shifted.

2 Account for your observations in Part B of this experiment.

3 Comment on the validity of this experiment and the accuracy of the results in terms of Le Châtelier's principle.

### CONCLUSION

Summarise the effects of changes in temperature on an equilibrium system.

# PRACTICAL ACTIVITY 12

Controlled experiment

## Electrolysis of aqueous solutions

### SUGGESTED DURATION

- 50 minutes

### INTRODUCTION

During the operation of an electrolytic cell, the electrode polarity is determined by the external power supply. The half-cell connected to the negative electrode of the power supply accepts electrons and hence a reduction reaction occurs in the half-cell. This electrode is the cathode and the negative electrode. The half-cell connected to the positive electrode of the power supply donates electrons and hence an oxidation reaction occurs. This electrode is the anode and the positive electrode.

### AIM

To investigate the products formed at electrodes during the electrolysis of various aqueous solutions

As directed by your teacher, complete the pre-lab safety information by referring to the safety data sheets (SDSs) or your teacher's risk assessment for the activity.

| PRE-LAB SAFETY INFORMATION | | |
|---|---|---|
| **Material used** | **Hazard** | **Control** |
| 0.5 M copper(II) chloride solution, $CuCl_2(aq)$ | | Avoid contact. Wear safety glasses and a laboratory coat. Wash hands at the conclusion of the experiment. |
| 0.5 M potassium iodide solution, KI(aq) | | Avoid contact. Wear safety glasses and a laboratory coat. Wash hands at the conclusion of the experiment. |
| 0.5 M sulfuric acid, $H_2SO_4(aq)$ | | Avoid contact. Wear safety glasses and a laboratory coat. Wash hands at the conclusion of the experiment. |
| 0.5 M zinc nitrate solution, $Zn(NO_3)_2(aq)$ | | Avoid contact. Wear safety glasses and a laboratory coat. Wash hands at the conclusion of the experiment. |
| electrical equipment | • electric shock or damage to instruments from incorrect connection of circuit | Use electrical equipment carefully. Check all connections. |
| Complete and indicate that you have understood the information in the safety table.<br>Name (print):<br>I understand the safety information (signature): | | |

### MATERIALS

- 60 mL of 0.5 M each of:
  - copper(II) chloride solution, $CuCl_2(aq)$
  - potassium iodide solution, KI(aq)
  - potassium chloride solution, KCl(aq)
  - sulfuric acid, $H_2SO_4(aq)$
  - zinc nitrate solution, $Zn(NO_3)_2$
- 2 carbon electrodes
- U-tube
- sheet of emery paper
- cotton wool
- litmus paper or pH paper
- starch paper or starch indicator solution
- DC power supply
- 2 wire leads with alligator clips
- retort stand and clamp
- safety glasses

 ISBN 978 0 6557 0027 2

## PRACTICAL ACTIVITY 12

### METHOD

1 ▪ For each of the solutions to be electrolysed, predict the products at the anode and cathode by referring to the electrochemical series in Appendix 2. Complete Table 1.

2 ▪ Insert a plug of cotton wool in the base of a U-tube. The plug will reduce the extent to which the two solutions in the sides of the U-tube mix during electrolysis. Do not make the plug extremely tight, as this can prevent any reaction occurring. Hold the U-tube in position using a retort stand and clamp.

3 ▪ Fill the U-tube with 0.5 M copper(II) chloride solution until the solution is within 3 cm of the top. Insert the carbon electrodes, one into each side of the U-tube, securing them by attaching the alligator clips on the wire leads at right angles to the electrodes.

4 ▪ Connect a power supply to the electrodes and observe and record in Table 2 the change at each electrode using a voltage of 4–8 V.

5 ▪ Depending on the electrolyte in use and what you have predicted will be the products of the electrolysis, test for the presence of $H^+$ or $OH^-$ ions using litmus paper, test for iodine using starch paper (which turns blue in the presence of iodine), or test for chlorine using litmus paper, which is bleached if chlorine is present.

6 ▪ Empty and wash the U-tube, discarding the cotton wool plug. If a deposit has formed on an electrode, remove it using emery paper.

7 ▪ Repeat steps 2–6 for each of the other solutions.

### RESULTS

**Table 1** Electrode predictions

| Solutions | Predicted reaction at anode | Predicted reaction at cathode |
|---|---|---|
| $CuCl_2(aq)$ | | |
| $KI(aq)$ | | |
| $KCl(aq)$ | | |
| $H_2SO_4(aq)$ | | |
| $Zn(NO_3)_2(aq)$ | | |

**Table 2** Electrode observations and equations

| Solutions | Observations and equations at anode | Observations and equations at cathode |
|---|---|---|
| $CuCl_2(aq)$ | | |
| $KI(aq)$ | | |
| $KCl(aq)$ | | |
| $H_2SO_4(aq)$ | | |
| $Zn(NO_3)_2(aq)$ | | |

**Variables**

Independent variables: ____________________

Dependent variable: ____________________

Controlled variables: ____________________

### DISCUSSION

1 Explain why electrolysis sometimes yields different products from those predicted using the electrochemical series.

____________________

____________________

____________________

2 Discuss any modifications or improvements you could possibly make to this experiment to improve the reproducibility and therefore the accuracy of the results.

## CONCLUSION

Summarise how to determine the electrode products during electrolysis of aqueous solutions.

 ISBN 978 0 6557 0027 2

# PRACTICAL ACTIVITY 13

Controlled experiment

# Determination of Faraday's constant and Avogadro's constant

## SUGGESTED DURATION

- 50 minutes

## INTRODUCTION

In 1909, Robert Millikan, an American scientist, determined the size of the charge on an electron. Millikan's method involved measuring the size of the electric field required to suspend charged oil droplets between two parallel plates. The value he obtained for the electron charge was $1.60 \times 10^{-19}$ C.

In this practical activity, this value for the charge on an electron is used, in conjunction with data obtained from a nickel electroplating cell, to determine values for two important constants: Faraday's constant and Avogadro's constant. A copper cathode is used, rather than a nickel one, so that the nickel plating that forms on the cathode is clearly visible.

## AIM

To determine a value for Faraday's constant ($F$) and for Avogadro's constant ($N_A$), using data obtained from an electroplating cell

As directed by your teacher, complete the pre-lab safety information by referring to the safety data sheets (SDSs) or your teacher's risk assessment for the activity.

## MATERIALS

- copper strip or mesh, approximately 3 cm × 7 cm
- nickel strip, approximately 3 cm × 7 cm
- 60 mL of 0.1 M nickel(II) sulfate solution, $NiSO_4(aq)$
- 50 mL acetone (propanone), $CH_3COCH_3(l)$
- 3 × 100 mL beakers
- paper towelling
- 4 wire leads with alligator clips
- scourer or sheet of emery paper
- ammeter
- variable resistor
- DC power supply
- stopwatch
- marking pen
- safety glasses
- gloves

**PRE-LAB SAFETY INFORMATION**

| Material used | Hazard | Control |
|---|---|---|
| 0.1 M nickel(II) sulfate solution, $NiSO_4(aq)$ | • harmful; possible sensitisation by contact with skin and by inhalation | Wear gloves, safety glasses and a laboratory coat. |
| acetone (propanone), $CH_3COCH_3(l)$ | • highly flammable; irritating to the eyes | |
| electrical equipment | • electric shock and damage to equipment from incorrect connection of circuit | Use electrical equipment carefully. Check all connections carefully. |

Complete and indicate that you have understood the information in the safety table.

Name (print): ____________________

I understand the safety information (signature): ____________________

## METHOD

1. Clean a copper strip using a scourer and wipe it with paper towelling. Hold the strip by the edges after it has been cleaned. The strip will act as the cathode of an electroplating cell. Use a marking pen to label it 'C'.
2. Weigh the copper strip and record its mass in Table 1.
3. Place this copper electrode and a nickel electrode opposite each other in a 100 mL beaker. Secure them by bending the top of each metal strip over the glass rim.
4. Wearing gloves, pour 60 mL of 0.1 M nickel(II) sulfate solution into the beaker.
5. Construct the electric circuit shown in Figure 3.2.11. Turn on the power supply and quickly adjust the voltage and the variable resistor to obtain a current of 1.0 A. At the same time, using a stopwatch, start timing. Maintain a constant current throughout the experiment by adjusting the variable resistor or the voltage control on the power supply.

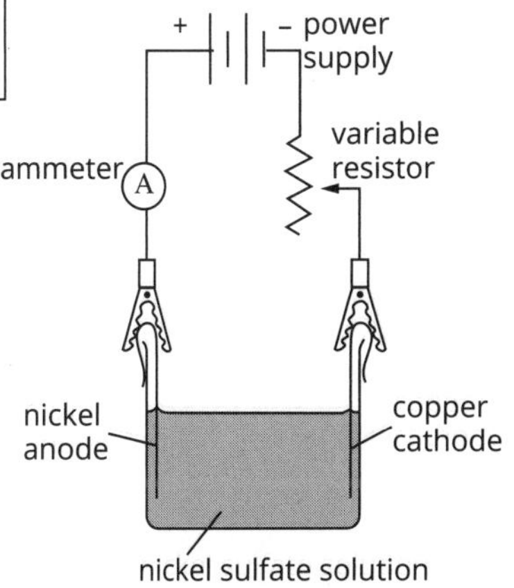


**Figure 3.2.11** The experimental circuit set-up

6 ▪ When exactly 10 minutes has elapsed, turn off the power supply. Record the current that passed through the cell.

7 ▪ Carefully remove the cathode from the cell and wash it by dipping it first into a beaker half-filled with water and then into a beaker half-filled with acetone (propanone). Allow it to dry in air and reweigh it, recording its mass in Table 1.

## RESULTS

| Table 1 Mass and current results | |
|---|---|
| Mass of copper strip (g) | |
| Mass of copper strip and nickel plating (g) | |
| Current through cell (A) | |

## DISCUSSION

1 Write a half-equation for the reaction that occurs at the cathode.

2 Use your measurement results of current and time to calculate the charge, $Q$, that flowed through the cell during the electroplating.

3 From the measurements of mass, calculate the amount of nickel, in mol, deposited at the cathode.

4 What amount of electrons ($n(e^-)$), in mol, must have combined with $Ni^{2+}$ ions at the cathode?

5 Use the values of $Q$ and $n(e^-)$ to calculate the charge on 1 mole of electrons (the Faraday constant, $F$).

6 Given that the charge on an electron is $1.60 \times 10^{-19}$ C, calculate the number of electrons in 1 mole of electrons. This is Avogadro's constant.

7 Compare your calculated value from the measured results with the published true value for these constants. Discuss their accuracy and precision and therefore their repeatability and reproducibility.

## CONCLUSION

Summarise your results for the values of Faraday's constant and Avogadro's constant.

 ISBN 978 0 6557 0027 2

# EXAM QUESTIONS

The following exam questions reflect the wording, presentation and symbols used in past VCAA exam questions. Occasionally there will be discrepancies with the VCE Chemistry 2024–2027 Study Design; in particular, the use of $K_c$ to express the equilibrium constant instead of $K$, and the use of units for $\Delta H$ in thermochemical equations.

## Multiple-choice questions

**Question 1** VCE Chemistry 2017 (A) 1

A catalyst

**A.** slows the rate of reaction.

**B.** ensures that a reaction is exothermic.

**C.** moves the chemical equilibrium of a reaction in the forward direction.

**D.** provides an alternative pathway for the reaction with a lower activation energy.

*Use the following information to answer Questions 2 and 3.*

The magnitude of the equilibrium constant, $K_c$, at 25°C for the following reaction is 640.

$N_2(g) + 3H_2(g) \rightleftharpoons 2NH_3(g)$ $\qquad \Delta H = -92.3 \text{ kJ mol}^{-1}$

**Question 2** VCE Chemistry 2020 (A) 14

For the reaction

$\frac{1}{3} N_2(g) + H_2(g) \rightleftharpoons \frac{2}{3} NH_3(g)$, the magnitude of $K_c$ at 25°C is

**A.** 9 and $\Delta H = -30.8 \text{ kJ mol}^{-1}$

**B.** 213 and $\Delta H = -30.8 \text{ kJ mol}^{-1}$

**C.** 640 and $\Delta H = -30.8 \text{ kJ mol}^{-1}$

**D.** 640 and $\Delta H = -92.3 \text{ kJ mol}^{-1}$

**Question 3** VCE Chemistry 2020 (A) 15

For the reaction $N_2(g) + 3H_2(g) \rightleftharpoons 2NH_3(g)$

**A.** a catalyst increases the number of collisions between the reactants.

**B.** the rate of the forward reaction increases when the temperature increases.

**C.** a catalyst reduces the activation energy of the forward and backward reactions by the same proportion.

**D.** the activation energy of the forward reaction is greater than the activation energy of the reverse reaction.

**Question 4** VCE Chemistry 2020 (A) 17

The following equation represents the reaction between chlorine gas, $Cl_2$, and carbon monoxide gas, CO.

$Cl_2(g) + CO(g) \rightleftharpoons COCl_2(g)$ $\qquad \Delta H = -108 \text{ kJ mol}^{-1}$

The concentration–time graph at right represents changes to the system.

Which of the following identifies the changes to the system that took place at 1 minute and at 7 minutes?

| | 1 minute | 7 minutes |
|---|---|---|
| **A.** | increase in temperature | increase in volume |
| **B.** | decrease in temperature | decrease in volume |
| **C.** | decrease in temperature | increase in volume |
| **D.** | increase in temperature | decrease in volume |

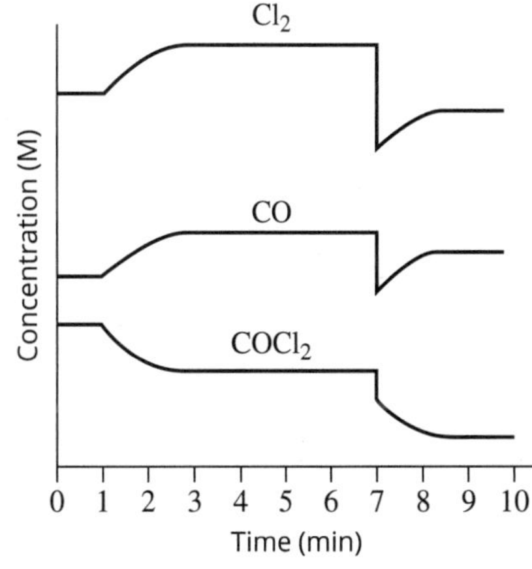

# EXAM QUESTIONS

**Question 5** VCE Chemistry 2021 (A) 28

Hydrogen, $H_2$, and iodine, $I_2$, react to form hydrogen iodide, HI.

$\frac{1}{2}H_2(g) + \frac{1}{2}I_2(g) \rightleftharpoons HI(g)$ $\Delta H = +25.9$ kJ mol$^{-1}$

The graph below shows the concentrations of $H_2$, $I_2$ and HI in a sealed container.

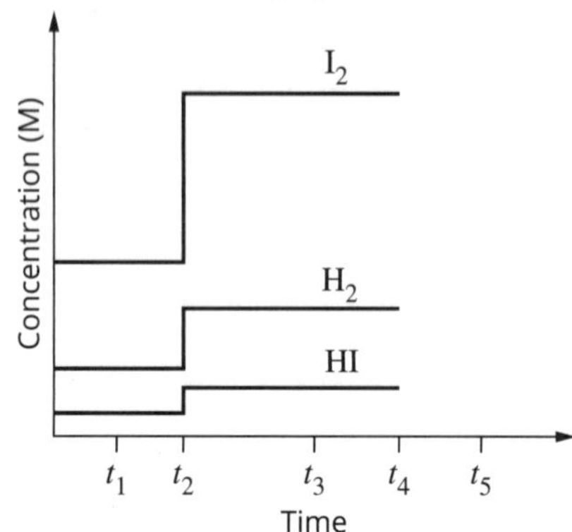


One change was made to the equilibrium system at time $t_4$, which altered the equilibrium constant. Equilibrium was re-established at time $t_5$. The rate of the reverse reaction at time $t_5$ was higher than at time $t_3$.

Which of the following options correctly shows the change in the equilibrium system from time $t_3$ to time $t_5$?

Change from time $t_3$ to time $t_5$

| | Equilibrium constant | Total chemical energy |
|---|---|---|
| **A.** | increase | increase |
| **B.** | increase | decrease |
| **C.** | decrease | increase |
| **D.** | decrease | decrease |

**Question 6** VCE Chemistry 2019 (A) 20

The oxidation of sulfur dioxide, $SO_2$, to sulfur trioxide, $SO_3$, can be represented by the following equation.

$2SO_2(g) + O_2(g) \rightleftharpoons 2SO_3(g)$ $K_c = 1.75$ M$^{-1}$ at 1000°C

An equilibrium mixture has a concentration of 0.12 M $SO_2$ and 0.16 M oxygen gas, $O_2$. The temperature of the container is 1000°C.

The equilibrium concentration of $SO_3$ at 1000°C is

**A.** $1.5 \times 10^{-4}$ M

**B.** $4.0 \times 10^{-3}$ M

**C.** $1.2 \times 10^{-2}$ M

**D.** $6.3 \times 10^{-2}$ M

**Question 7** VCE Chemistry 2018 (A) 24

The four equations below represent different equilibrium systems.

| | | |
|---|---|---|
| Equation 1 | $2SO_2(g) + O_2(g) \rightleftharpoons 2SO_3(g)$ | $\Delta H = -180$ kJ mol$^{-1}$ |
| Equation 2 | $CO(g) + H_2O(g) \rightleftharpoons CO_2(g) + H_2(g)$ | $\Delta H = -46$ kJ mol$^{-1}$ |
| Equation 3 | $PCl_5(g) \rightleftharpoons PCl_3(g) + Cl_2(g)$ | $\Delta H = 93$ kJ mol$^{-1}$ |
| Equation 4 | $CH_4(g) + H_2O(g) \rightleftharpoons CO(g) + 3H_2(g)$ | $\Delta H = 205$ kJ mol$^{-1}$ |


ISBN 978 0 6557 0027 2

# EXAM QUESTIONS

After equilibrium was established in each system, the temperature was decreased and the pressure was increased.

In which equilibrium system would both changes result in an increase in yield?

**A.** Equation 1

**B.** Equation 2

**C.** Equation 3

**D.** Equation 4

**Question 8** VCE Chemistry 2021 (A) 9

An electrolysis cell consumed a charge of 4.00 C in 5.00 minutes.

This represents a consumption of

**A.** $4.15 \times 10^{-5}$ mol of electrons.

**B.** $2.07 \times 10^{-4}$ mol of electrons.

**C.** $1.93 \times 10^{4}$ mol of electrons.

**D.** $2.41 \times 10^{4}$ mol of electrons.

**Question 9** VCE Chemistry 2019 (A) 7

A molten mixture of equal parts aluminium fluoride, $AlF_3$, and sodium chloride, NaCl, undergoes electrolysis.

Which one of the following statements about this reaction is correct?

**A.** Sodium metal will be produced at the cathode and fluorine gas will be produced at the anode.

**B.** Sodium metal will be produced at the anode and chlorine gas will be produced at the cathode.

**C.** Aluminium metal will be produced at the cathode and chlorine gas will be produced at the anode.

**D.** Aluminium metal will be produced at the anode and fluorine gas will be produced at the cathode.

**Question 10** VCE Chemistry 2018 (A) 16

The silver oxide–zinc battery is rechargeable and utilises sodium hydroxide, NaOH, solution as the electrolyte. The battery is used as a back-up in spacecraft, if the primary energy supply fails.

The overall reaction during discharge is

$Zn + Ag_2O \rightarrow ZnO + 2Ag$

When the silver oxide–zinc battery is being recharged, the reaction at the anode is

**A.** $2Ag + 2OH^- \rightarrow Ag_2O + H_2O + 2e^-$

**B.** $Ag_2O + H_2O + 2e^- \rightarrow 2Ag + 2OH^-$

**C.** $ZnO + H_2O + 2e^- \rightarrow Zn + 2OH^-$

**D.** $Zn + 2OH^- \rightarrow ZnO + H_2O + 2e^-$

# EXAM QUESTIONS

## Short-answer questions

**Question 1** (9 marks) VCE Chemistry 2020 (B) 1

Methanol is a very useful fuel. It can be manufactured from biogas.

The main reaction in methanol production from biogas is represented by the following equation.

$CO(g) + 2H_2(g) \rightleftharpoons CH_3OH(g) \qquad \Delta H < 0$

This reaction requires the use of a catalyst to maximise the yield of methanol produced in optimum conditions. The energy profile diagram below represents the uncatalysed reaction.

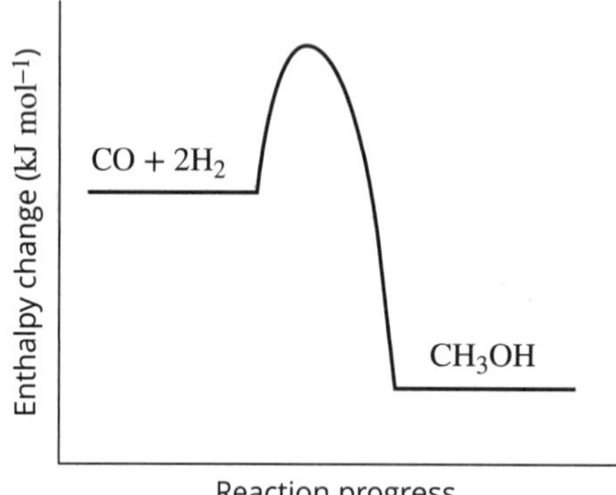


**a.** On the energy profile diagram above, sketch how the catalyst would alter the reaction pathway. 1 mark

**b.** **i.** How does the reaction temperature affect the yield of methanol from biogas? In your answer, refer to Le Châtelier's principle. 2 marks

**ii.** How does the reaction pressure affect the yield of methanol from biogas? In your answer, refer to Le Châtelier's principle. 2 marks

**c.** Write the expression for the equilibrium constant, $K_c$, for this reaction. 1 mark

 ISBN 978 0 6557 0027 2

**d.** 0.760 mol of carbon monoxide, CO, and 0.525 mol of hydrogen, $H_2$, were allowed to reach equilibrium in a 500 mL container. At equilibrium the mixture contained 0.122 mol of methanol.

Calculate the equilibrium constant, $K_c$. 3 marks

**Question 2** (8 marks) VCE Chemistry 2017 (B) 4

Sulfur trioxide, $SO_3$, is made by the reaction of sulfur dioxide, $SO_2$, and oxygen, $O_2$, in the presence of a catalyst, according to the equation below.

$2SO_2(g) + O_2(g) \rightleftharpoons 2SO_3(g) \qquad \Delta H < 0$

In a closed system in the presence of the catalyst, the reaction quickly achieves equilibrium at 1000 K.

**a.** A mixture of 2.00 mol of $SO_2$(g) and 2.00 mol of $O_2$(g) was placed in a 4.00 L evacuated, sealed vessel and kept at 1000 K until equilibrium was reached. At equilibrium, the vessel was found to contain 1.66 mol of $SO_3$(g).

Calculate the equilibrium constant, $K_c$, at 1000 K. 4 marks

**b.** A manufacturer of $SO_3$ investigates changes to the reaction conditions used in part a in order to increase the percentage yield of the product in a closed system, where the volume may be changed, if required.

What changes would the manufacturer make to the temperature and volume of the system in order to increase the percentage yield of $SO_3$? Justify your answer. 4 marks

# EXAM QUESTIONS

**Question 3** (8 marks) VCE Chemistry 2020 (B) 2

The electrolysis of carbon dioxide gas, $CO_2$, in water is one way of making ethanol, $C_2H_5OH$.

The diagram below shows a $CO_2$–$H_2O$ electrolysis cell. The electrolyte used in the electrolysis cell is sodium bicarbonate solution, $NaHCO_3(aq)$.

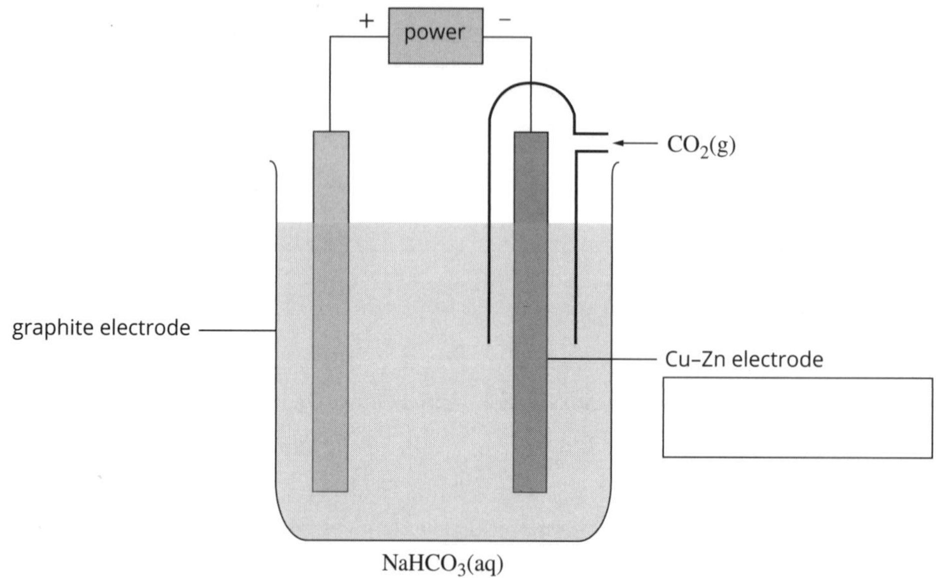


The following half-cell reactions occur in the $CO_2$–$H_2O$ electrolysis cell.

$O_2(g) + 2H_2O(l) + 4e^- \rightleftharpoons 4OH^-(aq)$ $\quad E° = +0.40$ V

$2CO_2(g) + 9H_2O(l) + 12e^- \rightleftharpoons C_2H_5OH(l) + 12OH^-(aq)$ $\quad E° = -0.33$ V

**a.** Identify the Cu–Zn electrode as either the anode or the cathode in the box provided in the diagram above. 1 mark

**b.** Determine the applied voltage required for the electrolysis cell to operate. 1 mark

**c.** Write the balanced equation for the overall electrolysis reaction. 1 mark

**d.** Identify the oxidising agent in the electrolysis reaction. Give your reasoning using oxidation numbers. 2 marks

**e.** A current of 2.70 A is passed through the $CO_2$–$H_2O$ electrolysis cell. The cell has an efficiency of 58%. Calculate the time taken, in minutes, for this cell to consume $6.05 \times 10^{-3}$ mol of $CO_2(g)$. 3 marks


ISBN 978 0 6557 0027 2

# EXAM QUESTIONS

**Question 4** (9 marks) VCE Chemistry 2017 (B) 8

Fluorine, $F_2$, gas is the most reactive of all non-metals. Anhydrous liquid hydrogen fluoride, HF, can be electrolysed to produce $F_2$ and hydrogen, $H_2$, gases. Potassium fluoride, KF, is added to the liquid HF to increase electrical conductivity. The equation for the reaction is

$2HF(l) \rightarrow F_2(g) + H_2(g)$

$F_2$ is used to make a range of chemicals, including sulfur hexafluoride, $SF_6$, an excellent electrical insulator, and xenon difluoride, $XeF_2$, a strong fluorinating agent.

The diagram below shows an electrolytic cell used to prepare $F_2$ gas.

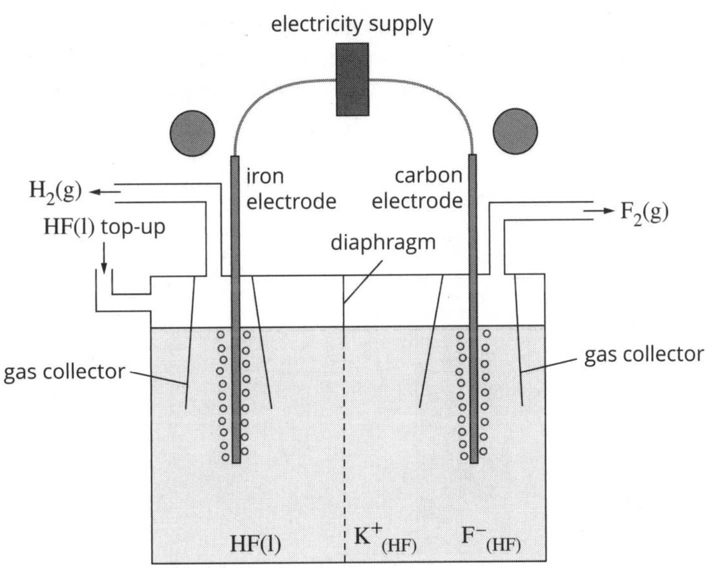


Liquid HF, like water, is an excellent solvent for ionic compounds. In the same way that water molecules in an aqueous solution form the ions $K^+(aq)$ and $F^-(aq)$, when KF is dissolved in HF, the $K^+$ and $F^-$ ions form ions that are written as $K^+_{(HF)}$ and $F^-_{(HF)}$.

**a.** Label the polarities of each electrode in the circles provided on the diagram above. 1 mark

**b.** Write the equation for the half-reaction occurring at the anode. 1 mark

______________________________________________

**c.** Suggest why the diaphragm, shown in the diagram above, is important for the safe operation of the electrolytic cell. 1 mark

______________________________________________

______________________________________________

**d.** Explain why the carbon electrode cannot be replaced with an iron electrode. 3 marks

______________________________________________

______________________________________________

______________________________________________

______________________________________________

______________________________________________

**e.** Calculate the volume of $F_2$ gas, measured at standard laboratory conditions (SLC), that would be produced when a current of 1.50 A is passed through the cell for 2.00 hours. 3 marks

**Question 5** (8 marks) VCE Chemistry 2017 (B) 6

Submarines operate both on the surface and under water. When operating under water, the submarine acts as a closed system, where there is no interaction with the atmosphere. Most types of submarines use both batteries and diesel engines to provide their energy requirements. A new type of submarine uses proton exchange membrane (PEM) fuel cells and diesel engines.

Below is a diagram of a PEM fuel cell.

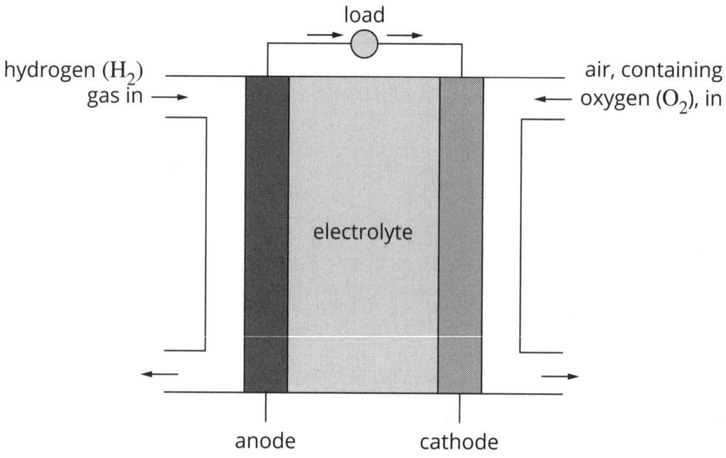

**a. i.** State the function of the electrolyte in a fuel cell. 1 mark

**ii.** Write the balanced overall redox reaction that occurs in this PEM fuel cell. 1 mark

**iii.** Give two safety considerations for the safe storage of hydrogen, $H_2$, gas on a submarine. 2 marks

**b. i.** State two advantages of using a PEM fuel cell compared to a diesel engine when a submarine is under water. 2 marks

**ii.** Most submarines generate more $H_2$ gas for their fuel cells when travelling on the surface. Explain how the $H_2$ gas could be generated. 2 marks

 ISBN 978 0 6557 0027 2

# UNIT 4 How are carbon-based compounds designed for purpose?

## AREA OF STUDY 1

## How are organic compounds categorised and synthesised?

### Outcome 1

On completion of this unit the student should be able to analyse the general structures and reactions of the major organic families of compounds, design reaction pathways for organic synthesis, and evaluate the sustainability of the manufacture of organic compounds used in society.

### Key knowledge

**Structure, nomenclature and properties of organic compounds**

- characteristics of the carbon atom that contribute to the diversity of organic compounds formed, with reference to valence electron number, relative bond strength, relative stability of carbon bonds with other elements, degree of unsaturation, and the formation of structural isomers
- molecular, structural and semi-structural (condensed) formulas and skeletal structures of alkanes (including cyclohexane), alkenes, benzene, haloalkanes, primary amines, primary amides, alcohols (primary, secondary and tertiary), aldehydes, ketones, carboxylic acids and non-branched esters
- the International Union of Pure and Applied Chemistry (IUPAC) systematic naming of organic compounds up to C8, with no more than two functional groups for a molecule, limited to non-cyclic hydrocarbons, haloalkanes, primary amines, alcohols (primary, secondary and tertiary), aldehydes, ketones, carboxylic acids and non-branched esters
- trends in physical properties within homologous series (boiling point and melting point, viscosity), with reference to structure and bonding

**Reactions of organic compounds**

- organic reactions and pathways, including equations, reactants, products, reaction conditions and catalysts (specific enzymes not required):
  - synthesis of primary haloalkanes and primary alcohols by substitution
  - addition reactions of alkenes
  - the esterification between an alcohol and a carboxylic acid
  - hydrolysis of esters
  - pathways for the synthesis of primary amines and carboxylic acids
  - transesterification of plant triglycerides using alcohols to produce biodiesel
  - hydrolytic reactions of proteins, carbohydrates and fats and oils to break down large biomolecules in food to produce smaller molecules
  - condensation reactions to synthesise large biologically important molecules for storage as proteins, starch, glycogen and lipids (fats and oils)
- calculations of percentage yield and atom economy of single-step or overall reaction pathways, and the advantages for society and for industry of developing chemical processes with a high atom economy
- the sustainability of the production of chemicals, with reference to the green chemistry principles of use of renewable feedstocks, catalysis and designing safer chemicals

VCE Chemistry Study Design extracts © VCAA (2022); reproduced by permission.

KEY KNOWLEDGE

# Structure, nomenclature and properties of organic compounds

In this topic, you will explore the structure, naming, properties and reactions of organic compounds. These concepts will be used as the basis for examining the chemical reactions associated with the metabolism of food. You will also discuss how synthetic organic compounds can be produced more sustainably for use in society.

- **As revision, you will now be able to complete Worksheet 23.**

## CHARACTERISTICS OF CARBON COMPOUNDS

Carbon is a unique element. It has the unusual property of bonding strongly to itself, by covalent bonds, to form long chains or rings of carbon atoms. It also bonds covalently to other non-metals such as hydrogen, nitrogen, oxygen, sulfur and the halogens. As a consequence, millions of carbon-containing compounds are known.

**Hydrocarbons** are commonly found in petroleum deposits in the form of crude oil and natural gas. They were formed millions of years ago from the decomposition of marine plants and animals. Crude oil is composed mainly of hydrocarbons, in particular **alkanes**, which can be separated by fractional distillation before being used. Since the boiling points of the alkanes depend on the strength of the dispersion forces between molecules, fractional distillation of crude oil results in the separation of molecules of different molecular mass. Although some of the hydrocarbon fractions obtained from the fractional distillation of crude oil are used as raw materials for manufacturing other chemicals, their main use is as fuels to provide energy for electricity production and transport.

A carbon atom can form four single covalent bonds to complete its outer shell and the atom is said to have a valency of 4. As a result of electron pair repulsion, four single electron pairs are arranged in a tetrahedral shape around a carbon atom. Alternatively, carbon atoms can form double or triple bonds with other carbon atoms or atoms of other elements.

Different covalent bonds have different bond strengths (Table 4.1.1). **Bond energy** is a measure of the strength of a chemical bond. Carbon–carbon triple bonds (C≡C) are stronger and shorter than carbon–carbon double bonds (C=C), which are stronger and shorter than carbon–carbon single bonds (C–C) (Figure 4.1.1). The single covalent bond between two carbon atoms (C–C) is much stronger then bonds between other elements of the same type. This is one of the reasons carbon tends to form the stable chain-like structures that are the basis for life.

**Table 4.1.1** Bond energies of different covalent bonds

| Covalent bond | Bond energy (kJ $mol^{-1}$) |
|---|---|
| C≡C | 839 |
| C=C | 614 |
| C–C | 348 |
| F–F | 155 |
| N–N | 163 |
| O–O | 146 |
| S–S | 226 |

### Saturated hydrocarbons

**Saturated hydrocarbons** have only C–C single bonds and belong to a **homologous series** called alkanes, with a general chemical formula of $C_nH_{2n+2}$ (Table 4.1.2). A homologous series is a series of organic compounds whose successive members differ by one $-CH_2-$ unit. The simplest saturated hydrocarbon is methane, $CH_4$.

Branches on a chain formed from alkanes have the formula $C_nH_{2n+1}$ and are called **alkyl groups** (Table 4.1.2). Alkyl groups (alkanes with one hydrogen atom removed) are often found in organic molecules. They are named according to the parent alkane, but with a -yl suffix.

**Table 4.1.2** The alkane homologous series and alkyl groups

| Alkane | Formula | Alkyl groups | Formula |
|---|---|---|---|
| methane | $CH_4$ | methyl | $-CH_3$ |
| ethane | $CH_3CH_3$ | ethyl | $-CH_2CH_3$ |
| propane | $CH_3CH_2CH_3$ | propyl | $-(CH_2)_2CH_3$ |
| butane | $CH_3(CH_2)_2CH_3$ | butyl | $-(CH_2)_3CH_3$ |
| pentane | $CH_3(CH_2)_3CH_3$ | pentyl | $-(CH_2)_4CH_3$ |
| hexane | $CH_3(CH_2)_4CH_3$ | hexyl | $-(CH_2)_5CH_3$ |
| heptane | $CH_3(CH_2)_5CH_3$ | heptyl | $-(CH_2)_6CH_3$ |
| octane | $CH_3(CH_2)_6CH_3$ | octyl | $-(CH_2)_7CH_3$ |

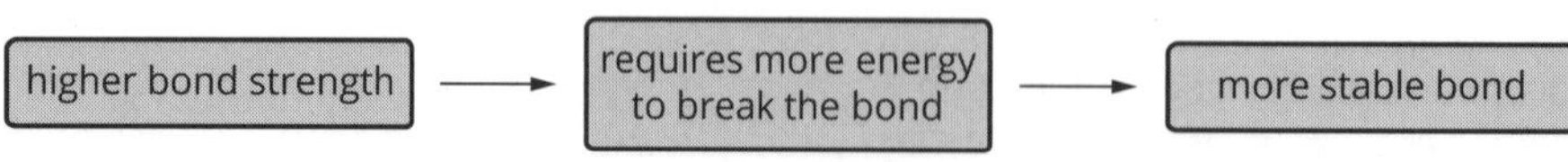


**Figure 4.1.1** The relationship between bond strength and bond stability

 ISBN 978 0 6557 0027 2

# KEY KNOWLEDGE

## Unsaturated hydrocarbons

**Unsaturated hydrocarbons** have at least one carbon–carbon double bond or a carbon–carbon triple bond. Hydrocarbons with one double bond belong to the homologous series of **alkenes** (formula $C_nH_{2n}$) (Table 4.1.3). Some of the properties of hydrocarbons are compared in Table 4.1.4.

**Table 4.1.3** The homologous series of alkenes

| Name of alkene | Formula |
|---|---|
| ethene | $C_2H_4$ |
| propene | $C_3H_6$ |
| butene | $C_4H_8$ |
| pentene | $C_5H_{10}$ |
| hexene | $C_6H_{12}$ |
| heptene | $C_7H_{14}$ |
| octene | $C_8H_{16}$ |

**Table 4.1.4** A comparison of alkanes and alkenes

| Property | Alkanes | Alkenes |
|---|---|---|
| type of bond between molecules | dispersion forces | dispersion forces |
| polarity of molecules | non-polar | non-polar |
| solubility in water | insoluble | insoluble |
| general formula | $C_nH_{2n+2}$ | $C_nH_{2n}$ |
| saturated or unsaturated | saturated | unsaturated |
| main reaction type | substitution | addition |

## Degree of unsaturation

When analysing organic compounds, the **degree of unsaturation** (also called index of hydrogen deficiency or double bond equivalent) is a calculation that determines the number of rings or double bonds present in the molecule. For hydrocarbons, this is equal to the number of pairs of hydrogen atoms that are needed to make the compound saturated.

There are several different formulas, but all will give the same answer.

One formula is:

For example, consider the molecule with formula $C_5H_8$. An alkane with 5 carbon atoms has the formula $C_5H_{(2\times5+2)} = C_5H_{12}$. The degree of unsaturation and the number of double bonds in the $C_5H_8$ molecule $= \frac{(12-8)}{2} = 2$, i.e. there are two double bonds or rings present.

## Cyclic hydrocarbons

Carbon forms rings of all sizes and with different combinations of single and double bonds between the carbon atoms. Cyclohexane and benzene are two examples of cyclic hydrocarbons.

- Cyclohexane is saturated and has the molecular formula $C_6H_{12}$. Its structure is shown in Figure 4.1.2a. Although the figure shows the molecule in two dimensions, in reality there is a tetrahedral arrangement of bonds around each carbon atom (bond angles are about 109°), and therefore the ring is puckered. The carbon atoms in the cyclohexane molecule usually adopt a 'chair' shape (Figure 4.1.2b).

(a)

(b)

**Figure 4.1.2** Structural formula and shape of cyclohexane

- Benzene is unsaturated and has the molecular formula $C_6H_6$. There are six carbon atoms in a planar ring in a benzene molecule. As Figure 4.1.3 on the following page shows, benzene can be represented in different ways. One electron from each carbon atom (six electrons in total) is delocalised over the ring, giving significant stability to the molecule.

$$\text{degree of unsaturation} = \text{number of double bond or ring equivalents}$$
$$= \frac{\text{maximum number of H possible per C} - \text{actual number of H per C}}{2}$$

(a)

The shaded ring represents the six delocalised electrons.

(b)

The delocalised electrons can be represented by a circle inside the hexagon.

(c)

This representation of benzene indicates that each carbon–carbon bond is equal to one and a half bonds on average.

**Figure 4.1.3** Representations of a molecule of benzene

## ISOMERS

**Isomers** are compounds with the same molecular formula (same number of each type of atom) that differ in the way the atoms are arranged. There are two main types of isomers: **structural isomers** and **stereoisomers**.

### Structural isomers

Structural isomers differ in the way the atoms are bonded to each other. They may have similar, but not identical, properties (Figure 4.1.4). For example, butane, $CH_3CH_2CH_2CH_3$, and 2-methylpropane, $(CH_3)_3CH$, have the same number of atoms of each element but the connectivity is different.

### Stereoisomers

In stereoisomers, all atoms are connected to the same atoms but the orientation in space of some of the atoms is different. There are two types of stereoisomers:

- **optical isomers**—called chiral molecules, which you will study in Unit 4, Area of Study 2
- geometric isomers—you will not study these in this course.

### Structures and formulas

A **structural formula** shows the arrangement of atoms in space and all the bonds between atoms. The differences between **molecular**, **semi-structural** (or condensed) and structural formulas as well as **skeletal formulas** (or structures), are shown in Table 4.1.5 on the following page.

(a)

butane

2-methylpropane

(b)

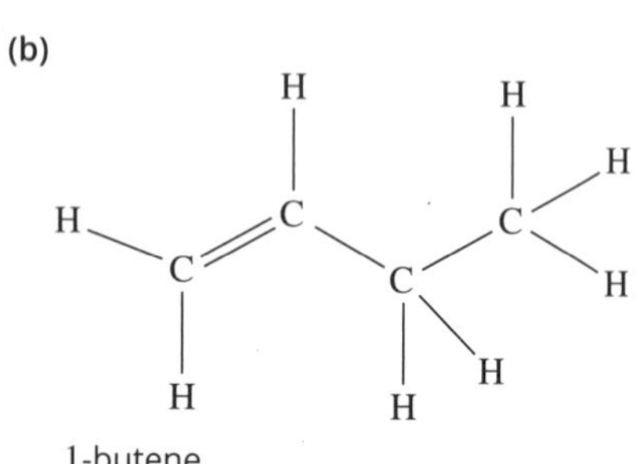


2-butene

2-methylpropene

**Figure 4.1.4** Structural isomers of (a) butane and (b) butene


ISBN 978 0 6557 0027 2

# KEY KNOWLEDGE

**Table 4.1.5** Formulas of some organic compounds

| Name | Molecular formula | Semi-structural formula | Structural formula | Skeletal structure |
|---|---|---|---|---|
| propan-1-ol | $C_3H_8O$ | $CH_3CH_2CH_2OH$ | 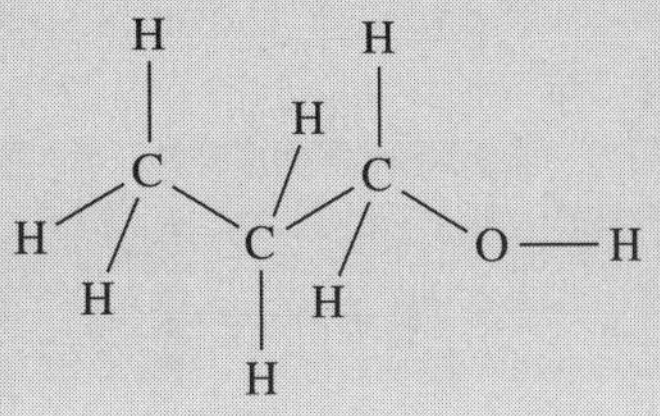 | 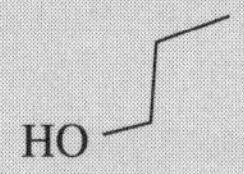 |
| propan-2-ol | $C_3H_8O$ | $CH_3CH(OH)CH_3$ | 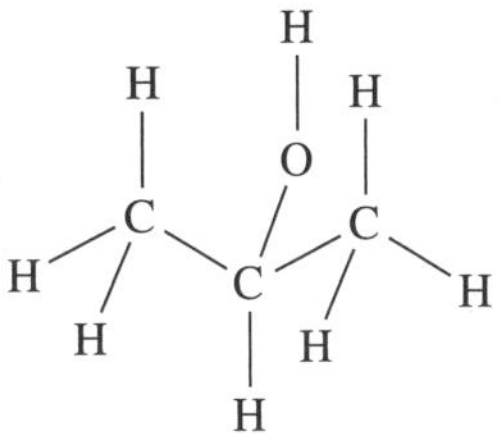 | 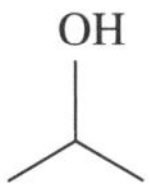 |
| but-2-ene | $C_4H_8$ | $CH_3CHCHCH_3$ | 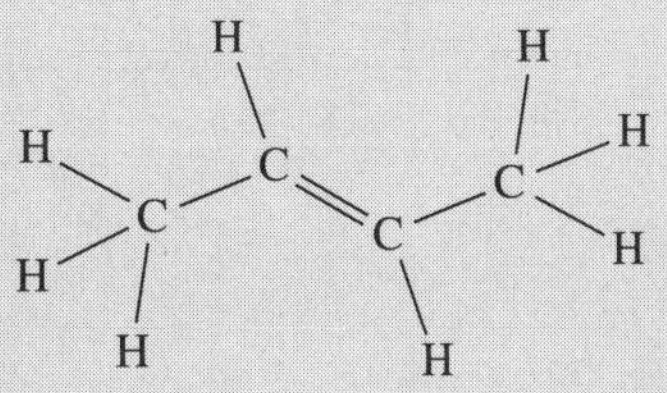 |  |

## SYSTEMATIC RULES FOR NAMING ORGANIC COMPOUNDS

To name organic compounds unambiguously, the International Union of Pure and Applied Chemistry (IUPAC) has developed a set of rules. Some of these rules for naming hydrocarbons are listed in Table 4.1.6. For some homologous series, particular prefixes or suffixes are used (Table 4.1.7 on the following page).

**Table 4.1.6** IUPAC rules for naming hydrocarbons

| Rule | Example | Name |
|---|---|---|
| 1 Choose the longest carbon chain and name according to the alkane with the same number of carbon atoms. If the hydrocarbon contains a C=C or C≡C bond, this should be included in the longest chain. | $CH_3CH_2CH_2CH_2CH_2CH_3$ | hexane |
| 2 If there is a C=C, replace -ane with -ene and number from the end that gives the smallest number for the first carbon atom involved in the bond. | $CH_3CH{=}CHCH_2CH_2CH_3$ | hex-2-ene (not hex-4-ene) |
| 3 When there is a branch and numbers have not already been established (such as by a double bond), number from the end that gives the smallest number to the branch. | $CH_3CH_2CH(CH_3)_2$ | 2-methylbutane (not 3-methylbutane) |
| 4 Number all branches. Use prefixes such as di-, tri- and tetra- if branches are identical. | $CH_3CH(CH_3)CH(CH_3)CH_3$ | 2,3-dimethylbutane |
| 5 If different alkyl branches are present, write them in alphabetical order. | $CH_3CH(CH_3)CH(C_2H_5)CH_2CH_3$ | 3-ethyl-2-methylpentane |

# KEY KNOWLEDGE

## FUNCTIONAL GROUPS

**Functional groups** are atoms or groups of atoms attached to a molecule whose presence determines the properties of the molecule (Table 4.1.7). The systematic naming of compounds with functional groups follows the same general rules for hydrocarbons and uses specific prefixes or suffixes for different functional groups (Table 4.1.8 on the following page).

**Table 4.1.7** Common functional groups

| Name of functional group | Name of homologous series to which it belongs | Structure | Prefix or suffix used in the name |
|---|---|---|---|
| double bond | alkenes | 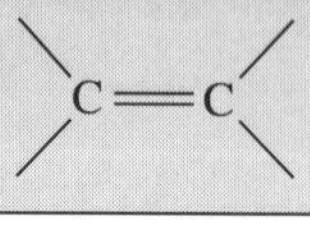 >C=C< | -ene |
| hydroxyl | alcohols | –OH | -ol (the number of the carbon atom to which the –OH group is bonded is given before the -ol suffix) |
| carboxyl | carboxylic acids | –COOH 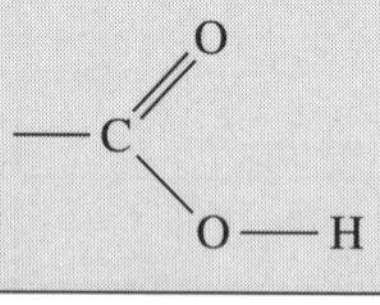 | -oic acid (the C=O carbon atom is always numbered as carbon 1) |
| halo | halohydrocarbons | –F, –Cl, –Br, –I | fluoro-, chloro-, bromo-, iodo- |
| amino | amines | $-NH_2$ 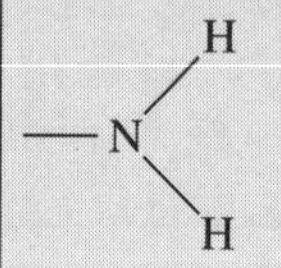 | -amine (the number of the carbon atom to which the $-NH_2$ group is bonded is given before the amine suffix) |
| ester | esters | –COOC– 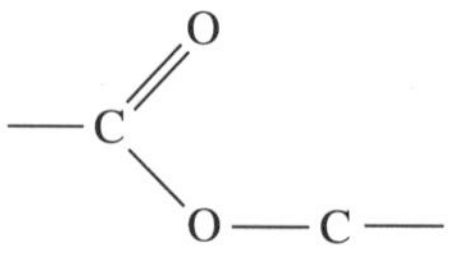 | The name has two parts; the part singly bonded to oxygen is written in front of the parent name and ends in -yl and the parent name ends in -oate. |
| ether | ethers | –COC– 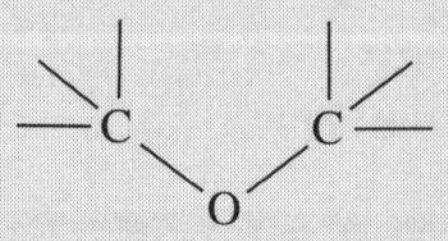 | (Naming of ethers is not required for VCE Chemistry.) |
| amide | amides (also present in proteins where it is called peptide) | –CONH– 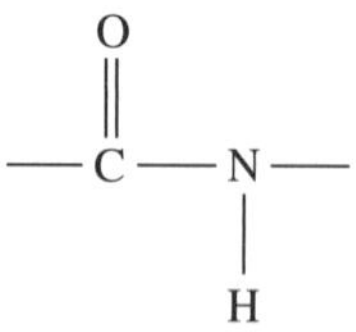 | (Naming of amides is not required for VCE Chemistry.) |
| carbonyl | aldehydes (the carbonyl group, C=O, is bonded to a hydrogen atom, which is at the end of the chain) | –CO 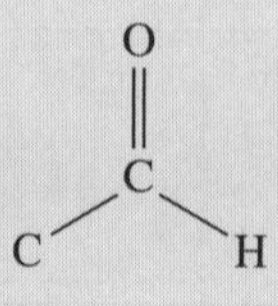 | -al (the C bonded to O is always numbered as carbon 1) |
| carbonyl | ketones (the carbonyl group is in the middle of the chain) | –CO 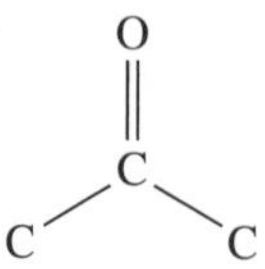 | -one (the number of the carbon in the carbonyl group is given before the -one suffix) |


ISBN 978 0 6557 0027 2

## KEY KNOWLEDGE

**Table 4.1.8** Rules for naming organic compounds containing functional groups

| Rule | Example | Name |
|---|---|---|
| 1 For alcohols, replace the -e of the hydrocarbon with -ol at the end of the hydrocarbon name. | $CH_3CH_2CH_2OH$ | propan-1-ol |
| 2 For alcohols, when the –OH is positioned along the chain, assign the smallest number possible to the –OH group. Place the number before the -ol ending. | $CH_3CH_2(OH)CH_3$ | propan-2-ol |
| 3 For amines, replace the -e of the hydrocarbon with -amine at the end of the hydrocarbon name. | $CH_3CH_2CH_2NH_2$ | propan-1-amine |
| 4 For amines, indicate the position of the amine group by the numbering system. | $CH_3CH(NH_2)CH_3$ | propan-2-amine |
| 5 For carboxylic acids, replace -e with -oic acid. Remember, the carbon atom doubly bonded to oxygen is always counted as carbon 1. | $CH_3CH_2COOH$ | propanoic acid |
| 6 Esters can be regarded as being made from an alcohol and a carboxylic acid. To name an ester, the first part of the name comes from the alcohol, so replace -anol with -yl, and the second part of the name comes from the carboxylic acid, so replace -oic acid with -oate. | $\underbrace{CH_3CH_2COO}_{\text{carboxylic acid part}}\underbrace{CH_2CH_2CH_2CH_3}_{\text{alcohol part}}$ | butyl propanoate |

When there are two functional groups in a molecule, IUPAC nomenclature indicates an order of priority (Table 4.1.9).

- The functional group with the higher priority determines the parent name of the molecule.
- The presence of a functional group with the lower priority is shown by a prefix in front of the parent name; a number indicates its position in a molecule.

**Table 4.1.9** IUPAC priority for functional groups (from highest to lowest)

| Functional group priority order | Suffix | Alternative name (when needed) |
|---|---|---|
| carboxyl | -oic acid | – |
| carbonyl on end carbon (aldehyde) | -al | oxo- |
| carbonyl not on end carbon (ketone) | -one | oxo- |
| hydroxyl | -ol | hydroxy- |
| amine | -amine | amino- |
| alkene | -ene | -an becomes -en |
| halo | halo- (used as a prefix) | halo- |

For example, the name of $CH_3CH(OH)CH_2COOH$ is 3-hydroxybutanoic acid and $CH_3CHClCH(OH)CH_3$ is 3-chlorobutan-2-ol.

### Types of alcohols

The term 'alcohol' is used for any hydrocarbon containing an –OH group. There are three types of alcohols, depending on the position of the –OH group (Figure 4.1.5).

- A **primary alcohol** is one in which the carbon atom bonded to the –OH group is bonded to no more than one other carbon; for example, butan-1-ol.
- A **secondary alcohol** is one in which the carbon atom bonded to the –OH group is bonded to two other carbons; for example, butan-2-ol.
- A **tertiary alcohol** is one in which the carbon atom bonded to the –OH group is bonded to three other carbons; for example, 2-methylpropan-2-ol.

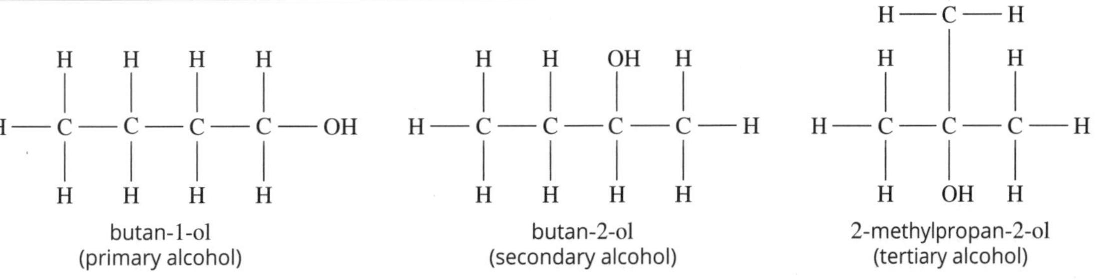


**Figure 4.1.5** Three types of alcohols. They have the same molecular formula and are structural isomers.

- **You will now be able to complete Worksheet 24.**

# KEY KNOWLEDGE

## TRENDS IN PHYSICAL PROPERTIES OF HOMOLOGOUS SERIES

The physical properties of organic compounds are directly related to their structure and bonding.

### Boiling point and melting point

As the number of carbon atoms in hydrocarbon molecules increases, the strength of the **dispersion forces** between molecules increases and their boiling points increase as a consequence. Alkenes show the same trend as alkanes, although they have slightly lower boiling points.

Organic compounds containing functional groups have higher boiling points than alkanes of similar molar mass. This is due to the stronger intermolecular forces that exist. In the examples in Table 4.1.10, the higher boiling points of carboxylic acids and amides are due to the presence of hydrogen bonds between their molecules. Carboxylic acids form dimers (Figure 4.1.6) and their presence further increases the strength of the intermolecular forces in carboxylic acids.

**Table 4.1.10** A comparison of boiling points and melting points of similar molar mass alkane, alcohol, carboxylic acid and amine compounds

| Compound | Formula | Molar mass (g mol$^{-1}$) | Boiling point (°C) | Melting point (°C) |
|---|---|---|---|---|
| butane | $C_4H_{10}$ | 58 | −1 | −138 |
| propan-1-ol | $C_3H_7OH$ | 60 | 97 | −126 |
| ethanoic acid | $CH_3COOH$ | 60 | 118 | 16.6 |
| propan-1-amine | $C_3H_7NH_2$ | 59 | 50 | −83 |

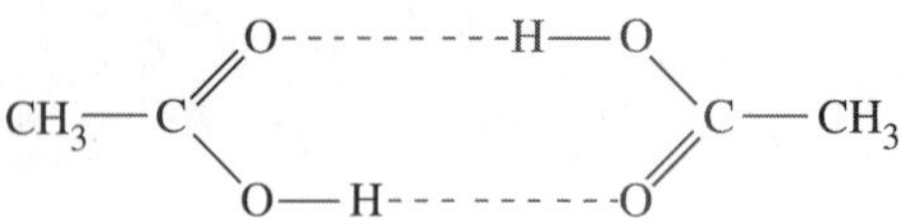


**Figure 4.1.6** When two ethanoic acid molecules form hydrogen bonds with each other, a dimer results. This effectively doubles the size of the molecules, increasing the strength of the dispersion forces between particles.

Melting points follow the same trends as boiling points, for the same reasons. However, more energy is needed to break the stronger forces between the particles in a solid. Other variations in melting point trends are due to how well the molecules pack together, which is a consequence of the number of carbon atoms in the chain and the degree of branching.

### Viscosity

The resistance of a liquid to flowing is called **viscosity**. Honey has a higher viscosity than water. As for boiling points, the stronger the intermolecular forces between molecules in a liquid, the higher the viscosity. So, as the length of the hydrocarbon chain in organic molecules increases, the viscosity of the substance also increases.

- **You will now be able to complete Worksheet 25.**

## Reactions of organic compounds

Some of the reactions of organic compounds are shown in Table 4.1.11.

**Table 4.1.11** Reactions of organic compounds

| Type of reaction | Description/examples |
|---|---|
| Combustion | **Description:** A type of oxidation reaction involving oxygen<br>**Examples:** Complete combustion: $CH_4(g) + 2O_2(g) \rightarrow CO_2(g) + 2H_2O(l)$<br>Incomplete combustion: $2CH_4(g) + 3O_2(g) \rightarrow 2CO(g) + 4H_2O(l)$ |
| **Substitution reactions** to form haloalkanes, amines and alcohols | **Description:** Replacement of an atom or group of atoms by another atom or group of atoms. At least two new products are formed.<br>**Examples:** $CH_3CH_3(g) + Cl_2(g) \xrightarrow{\text{light or heat}} CH_3CH_2Cl(g) + HCl(g)$<br>A H atom is substituted by a Cl atom to form a chloroalkane.<br>$CH_3Cl(aq) + OH^-(aq) \rightarrow CH_3OH(aq) + Cl^-(aq)$<br>The Cl atom is substituted by the OH group to form an alcohol.<br>$CH_3OH(aq) + NH_3(aq) \rightarrow CH_3NH_2(aq) + H_2O(aq)$<br>The OH group is substituted by the $NH_2$ group to form an amine. |

*(Continued)*

 ISBN 978 0 6557 0027 2

# KEY KNOWLEDGE

**Table 4.1.11** Reactions of organic compounds *(Continued)*

| Type of reaction | Description/examples |
|---|---|
| **Addition reactions** of alkenes to form haloalkanes, alcohols and alkanes | **Description:** Addition to both ends of a C=C double bond. One product is formed (i.e. there are no by-products). Alkenes undergo addition reactions.<br>**Examples:**<br>$H_2C{=}CH_2 + Br{-}Br \longrightarrow H_2C(Br){-}CH_2(Br)$<br>$CH_2{=}CH_2 + Br_2 \longrightarrow CH_2Br{-}CH_2CH_2Br$<br>1, 2 -dibromoethane<br>$H_2C{=}CH_2 + H_2O \xrightarrow[300°C]{H_3PO_4 \text{ catalyst}} H_3C{-}CH_2(OH)$<br>$CH_2{=}CH_2 + H_2O \xrightarrow[300°C]{H_3PO_4 \text{ catalyst}} CH_3CH_2OH$<br>ethanol<br>$CH_2{=}CH_2 + H_2 \xrightarrow{\text{Ni catalyst}} CH_3CH_3$ |
| Oxidation of primary alcohols to carboxylic acids | **Description:** Involves a change in oxidation number (in organic chemistry, oxidation is defined as an increase in the number of C–O bonds and a simultaneous decrease in the number of C–H bonds at the same C atom)<br>**Example:** Primary (1°) alcohols are oxidised to a carboxylic acid.<br>$CH_3CH_2OH(aq) \xrightarrow[H^+(aq)]{Cr_2O_7^{2-} \text{ or } MnO_4^-} CH_3COOH(aq)$<br>(During this reaction an aldehyde is formed initially, which is then oxidised to the carboxylic acid; the VCE Chemistry course does not require knowledge of the aldehyde formation.) |
| Esterification | **Description:** A type of condensation reaction involving an alcohol and a carboxylic acid in the presence of a catalyst, concentrated $H_2SO_4$<br>**Examples:**<br>$CH_3C(=O)OH + HO{-}CH_2{-}CH_3 \xrightarrow{H_2SO_4} CH_3C(=O){-}O{-}CH_2{-}CH_3 + H_2O$<br>ethanoic acid + ethanol → ethyl ethanoate + water<br>$CH_3COOH(l) + CH_3CH_2OH(l) \xrightarrow{\text{conc. } H_2SO_4} CH_3COOCH_2CH_3(l) + H_2O(l)$ |
| Hydrolysis of esters | **Description:** Water splits an ester link, producing an alcohol and a carboxylic acid. (Acid or alkali catalysts are often used.)<br>**Example:**<br>$CH_3{-}C(=O){-}O{-}CH_2{-}CH_3 \xrightarrow[H^+ \text{ catalyst}]{H_2O} CH_3{-}C(=O){-}OH + HO{-}CH_2{-}CH_3$<br>ethyl ethanoate → ethanoic acid + ethanol<br>$CH_3COOCH_2CH_3 + H_2O \longrightarrow CH_3COOH + CH_3CH_2OH$ |

*(Continued)*

# KEY KNOWLEDGE

**Table 4.1.11** Reactions of organic compounds *(Continued)*

| Type of reaction | Description/examples |
| --- | --- |
| **Trans-esterification** of plant triglycerides | **Description:** A reaction between vegetable oils (composed of triglycerides, which contain three ester groups) and an alcohol, commonly methanol, to form glycerol and biodiesel (KOH solution is often used as a catalyst.)<br>**Example:** 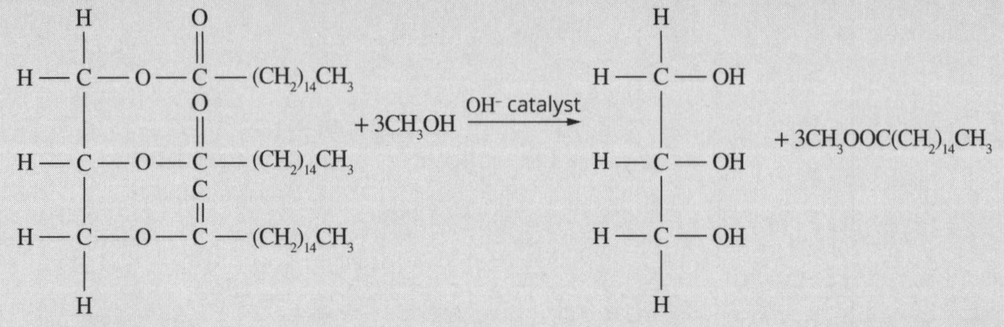<br>$CH_2(OOC(CH_2)_{14}CH_3)CH(OOC(CH_2)_{14}CH_3)CH_2(OOC(CH_2)_{14}CH_3) + 3CH_3OH \xrightarrow{OH^- \text{ catalyst}} CH_2(OH)CH(OH)CH_2(OH) + 3CH_3OOC(CH_2)_{14}CH_3$<br>triglyceride (palm oil) — methanol — glycerol — fatty acid methyl esters (biodiesel) |
| **Condensation polymerisation** reactions to form proteins and carbohydrates | The reactions to synthesise large biologically important molecules such as proteins and carbohydrates are called condensation polymerisation reactions. Such reactions are enzyme catalysed.<br>**Description: Protein**: formed by reactions between amino acids, whose general structure is: 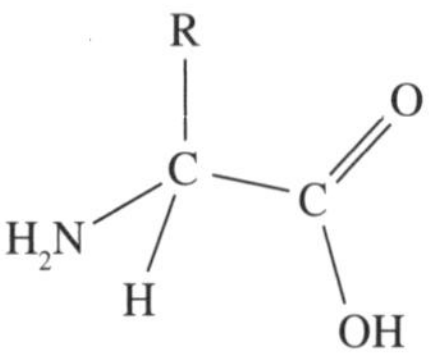<br>$H_2N{-}CH(R){-}COOH$<br>It is the COOH and $NH_2$ groups that are bonded to the same C atom that react. When the COOH group of one amino acid reacts with the $NH_2$ group of another amino acid, a CONH amide (peptide) bond is formed.<br>**Example: Formation of proteins** 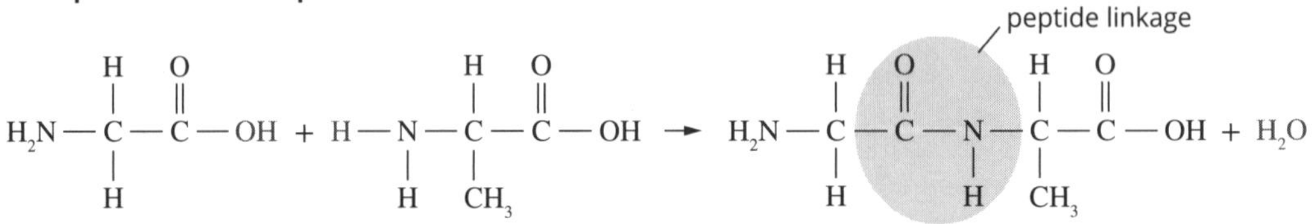<br>$H_2N{-}CH_2{-}COOH + H{-}NH{-}CH(CH_3){-}COOH \rightarrow H_2N{-}CH_2{-}CO{-}NH{-}CH(CH_3){-}COOH + H_2O$<br>glycine — alanine — a dipeptide (peptide linkage) — water<br>Glycine and alanine are two amino acids that can undergo a condensation reaction. Multiple amino acids (monomers) react to form a protein polymer. Water is released each time an amide (peptide) bond is formed.<br>**Description: Carbohydrate**, e.g. **starch** or **glycogen**: formed by reactions between –OH groups on glucose molecules. Ether (glycosidic) functional groups are formed, resulting in the polymers starch or glycogen.<br>**Example: Formation of carbohydrates** 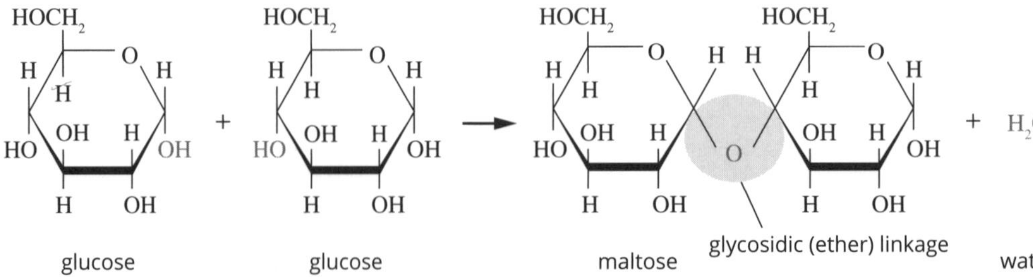<br>glucose + glucose → maltose + $H_2O$<br>glucose — glucose — maltose (glycosidic (ether) linkage) — water<br>Multiple glucose molecules (monomers) undergo condensation reactions repeatedly to produce a carbohydrate polymer. Starch and glycogen are similar polymers that are formed by these condensation polymerisation reactions. They have different degrees of branching. As shown above, water is produced each time an ether bond is formed. |

*(Continued)*


ISBN 978 0 6557 0027 2

# KEY KNOWLEDGE

| **Table 4.1.11** Reactions of organic compounds *(Continued)* | |
|---|---|
| **Type of reaction** | **Description/examples** |
| Condensation reactions to form lipids (fats and oils) | **Description:** Large biological molecules such as fats and oils are formed by enzyme-catalysed condensation reactions. These reactions occur between each of the three –OH groups in the tri-alcohol glycerol and three carboxylic acid groups on fatty acids (long hydrocarbon chains with a COOH group at one end of the chain). Ester groups are formed, resulting in a large lipid molecule, a triglyceride.<br>**Example:**<br>$CH_2(OH)-CH(OH)-CH_2(OH) + 3HO-CO-C_{17}H_{35} \longrightarrow CH_2(O-CO-C_{17}H_{35})-CH(O-CO-C_{17}H_{35})-CH_2(O-CO-C_{17}H_{35}) + 3H_2O$<br>three ester linkages<br>glycerol; stearic acid (a fatty acid); tristearin (a fat); water |
| Hydrolytic (hydrolysis) reactions of biomolecules | **Description:** Hydrolytic reactions of proteins, carbohydrates and fats and oils break down large biomolecules in food to produce smaller molecules. These are enzyme-catalysed reactions in which water splits the amide (peptide), glycosidic (ether) or ester link to produce smaller molecules. The hydrolytic reactions can be regarded as the reverse of the condensation (polymerisation) reactions that formed these biomolecules.<br>**Example: Proteins**<br>$H_2N-CH_2-CO-NH-CH(CH_3)-COOH \xrightarrow{H_2O} H_2N-CH_2-COOH + H-NH-CH(CH_3)-COOH$<br>peptide linkage<br>a dipeptide; glycine; alanine<br>Multiple amino acids are formed by hydrolysis of proteins, when the covalent bonds (peptide links) between amino acids are broken. |

*(Continued)*

# KEY KNOWLEDGE

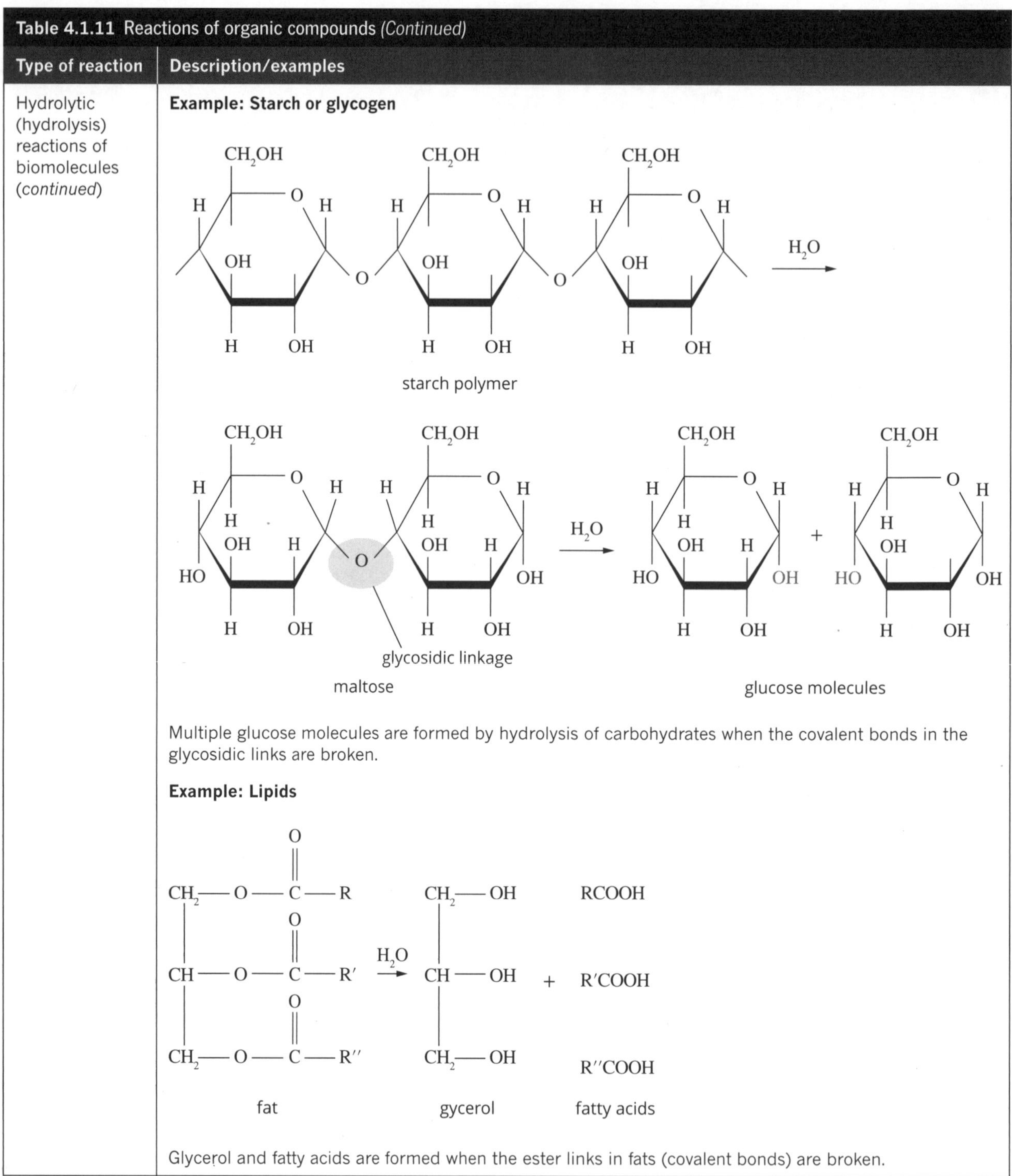


**Table 4.1.11** Reactions of organic compounds *(Continued)*

| Type of reaction | Description/examples |
|---|---|
| Hydrolytic (hydrolysis) reactions of biomolecules (*continued*) | **Example: Starch or glycogen**<br>starch polymer<br>maltose → glucose molecules<br>Multiple glucose molecules are formed by hydrolysis of carbohydrates when the covalent bonds in the glycosidic links are broken.<br>**Example: Lipids**<br>fat → gycerol + fatty acids<br>Glycerol and fatty acids are formed when the ester links in fats (covalent bonds) are broken. |

- **You will now be able to conduct Practical activities 17 and 18.**

 ISBN 978 0 6557 0027 2

# KEY KNOWLEDGE

## ORGANIC REACTION SEQUENCES

Organic chemicals are often made in a sequence of reactions (a pathway), rather than in a single step from reactants to products. For example, an amine or a carboxylic acid can be produced from an alkane or an alkene by the reaction pathways shown in Figure 4.1.7.

## CALCULATIONS OF ATOM ECONOMY AND PERCENTAGE YIELD

The efficiency of single-step or overall reaction pathways for the synthesis of a chemical can be measured by calculating atom economy and percentage yield.

### Atom economy

**Atom economy** measures the mass of reactant atoms that are used to make a final product. It is calculated using the formula:

$$\text{atom economy} = \frac{\text{molar mass of desired product}}{\text{molar mass of all reactants}} \times 100\%$$

Ideally, for environmental and financial reasons, all or almost all of the atoms from the reactants should be incorporated in the final product and there will be as few 'wasted' atoms as possible. If all the atoms in the reactants are converted to the desired product, the atom economy is 100%.

### Theoretical yield

The **theoretical yield** is the maximum mass of product that could be formed if the reactants reacted completely. It can be calculated from the stoichiometry of the reaction after determining the limiting reactant.

### Percentage yield

Inevitably, during a synthesis, there will be some loss of reactants and products during transfers between reaction vessels and in separation and purification stages, resulting in less product being obtained than expected. The formation of equilibria can further reduce the amount of product obtained.

**Percentage yield** compares the **actual yield** to the theoretical yield and indicates the **efficiency** of the reaction:

$$\text{percentage yield} = \frac{\text{actual yield}}{\text{theoretical yield}} \times 100\%$$

The yield for each step in a multi-step reaction has an effect on the overall yield. The overall percentage yield helps scientists decide on the best pathway to use to produce a desired product with minimal waste. The overall yield is determined by multiplying the percentage yield for each step and then expressing the answer as a percentage.

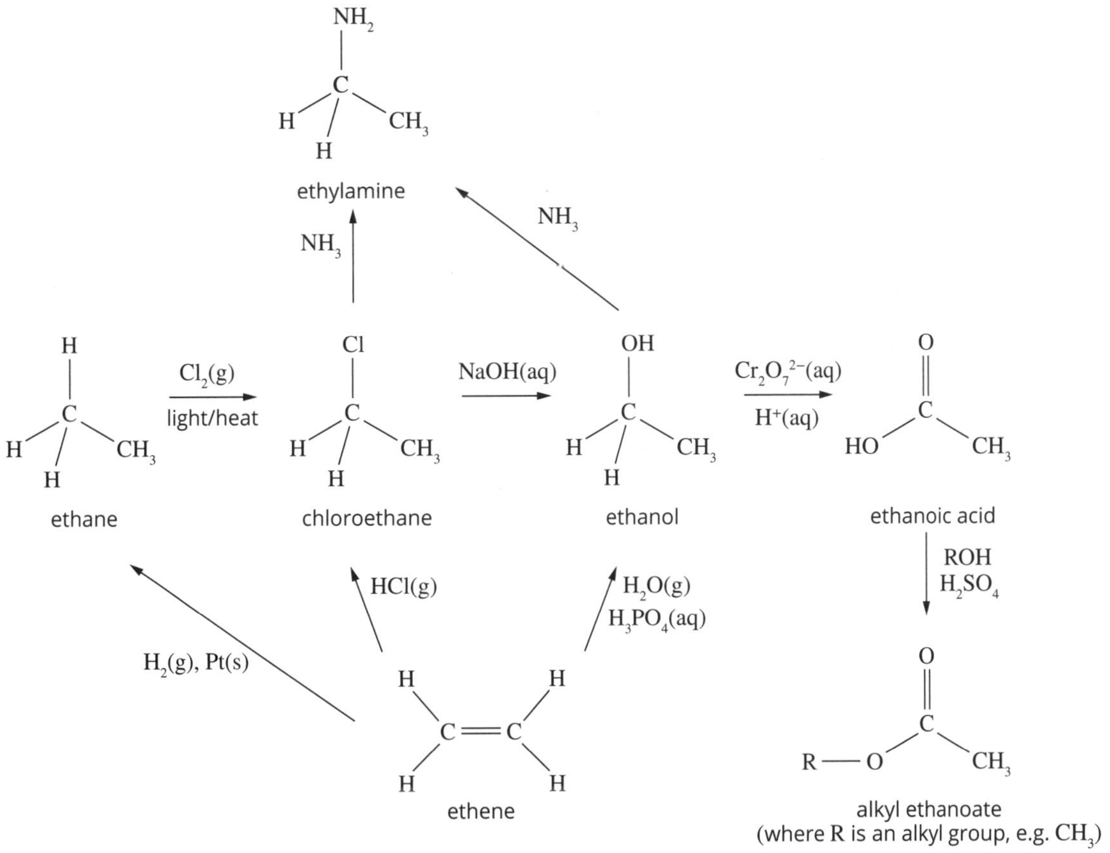

ethylamine
$NH_3$
$NH_3$
ethane
$Cl_2(g)$
light/heat
chloroethane
NaOH(aq)
ethanol
$Cr_2O_7^{2-}(aq)$
$H^+(aq)$
ethanoic acid
ROH
$H_2SO_4$
$H_2(g)$, Pt(s)
HCl(g)
$H_2O(g)$
$H_3PO_4(aq)$
ethene
alkyl ethanoate
(where R is an alkyl group, e.g. $CH_3$)

**Figure 4.1.7** Simple organic reaction pathways involving ethane

- **You will now be able to complete Worksheet 26 and conduct Practical activities 14–18.**

# KEY KNOWLEDGE

## SUSTAINABILITY OF THE PRODUCTION OF ORGANIC CHEMICALS

Industrial chemists and chemical engineers are increasingly trying to develop reaction pathways that reflect **green chemistry** principles, such as minimising waste and energy consumption, use of renewable **feedstocks** and catalysts, and designing safer chemicals. The 12 principles of green chemistry are briefly listed below.

1 *Prevention of waste* is better than cleaning up the waste later.
2 *Maximise the atom economy* of all reactions involved in production of a product.
3 Develop *less hazardous chemical syntheses* to generate safer by-products for all living things and for the sustainability of the planet.
4 Design *safer chemicals* that are purpose-made and non-toxic.
5 Use *safer solvents and reagents* and keep their use to a minimum in reactions.
6 *Design energy-efficient processes* that reduce pollutants and that can be performed at normal temperatures and pressure.
7 *Use renewable feedstocks* wherever possible to minimise consumption of resources.
8 Design processes that *reduce derivatives*, so there are minimum side reactions and by-products, which reduce atom economy.
9 Use *catalysts* to keep the reaction temperatures low and reduce the amounts of reactants consumed while increasing the energy efficiency and the atom economy of the reaction.
10 *Design for degradation*, so that once products are no longer useful, they break down into safe, environmental materials.
11 *Apply real-time analysis for pollution prevention*, so that monitoring of hazards occurs during the reaction process, allowing prevention to occur as the production proceeds.
12 Choose *inherently safer chemistry that prevents accidents*, including fires, accidental release of materials and explosions.

Applying these principles to chemical industries should aid in the energy-efficient production of chemicals that are safe, renewable and recyclable, with minimal waste. This will improve the sustainability of natural resources and reduce pollutants globally, significantly slowing climate change. By implementing a green chemistry approach, chemical industries will move towards a **circular economy**. A circular economy seeks to reuse products and treats waste as a loss of value. For most industries at present, the feedstock, products, catalysts and energy cannot be regarded as being part of a circular process; they are described as being part of a **linear economy**.

- **You will now be able to complete Worksheets 27 and 28.**


ISBN 978 0 6557 0027 2

# Knowledge review—bonding and organic chemistry

1 Identify the correct term from the list below for each definition and complete the following table. This will help you check your knowledge and understanding of the key ideas involved in organic chemistry.

| | | | | | |
|---|---|---|---|---|---|
| lone pair | ethene | tetrahedral | $C_nH_{2n+2}$ | alcohols | propene |
| isomers | –COOH | pentanoic acid | butanoic acid | $C_nH_{2n}$ | –COH |

| Definition | Correct term |
|---|---|
| general formula of an alkene | |
| general formula of an alkane | |
| the smallest alkene | |
| the general name of compounds with a hydroxyl functional group | |
| a pair of outer-shell electrons not involved in bonding | |
| the alkene with three carbon atoms | |
| molecules with the same molecular formula but different structural formulas | |
| the formula of the functional group in carboxylic acids | |
| the shape of a methane molecule | |
| the carboxylic acid with five carbons | |

2 Indicate whether the statements are true or false and, where necessary, make the appropriate corrections.

| Statement | True/False | Correction |
|---|---|---|
| A polar molecule must have polar bonds and be symmetrical. | | |
| The strength of the forces between molecules determines the melting point of the substance. | | |
| Hydrogen bonding is stronger than covalent bonds. | | |
| Dispersion forces are present between all atoms because of the attraction between electrons and protons. | | |
| All carboxylic acids are strong acids. | | |
| 2-Methylbutane and but-2-ene are isomers. | | |

3 **a** Explain the difference between saturated hydrocarbons and unsaturated hydrocarbons, giving an example of each.

**b** What effect does the presence of a halogen have on the physical and chemical properties of haloalkanes when compared to alkanes?

4 Draw the structures of the following compounds and classify them into the homologous series to which they belong.

| Name of compound | Structure | Homologous series |
|---|---|---|
| pentan-2-ol | | |
| 2-methylheptane | | |
| 2-methylhex-2-ene | | |
| methanoic acid | | |

 ISBN 978 0 6557 0027 2

# WORKSHEET 24



# The complexity of carbon—organic compounds

**1** Complete the following table by providing the name, semi-structural formula and structural formula where needed. State whether each statement in the second column is true or false by writing T or F next to it.

| Name of molecule | Statement | T/F | Semi-structural formula | Structural formula |
|---|---|---|---|---|
| ethane | Is soluble in water and has the general formula $C_nH_{2n}$ | | | |
| | Undergoes substitution reactions | | | |
| | Belongs to the homologous series of alkenes | | | |
| 2-chloropentane | Is a structural isomer of 2-chloro-2-methylbutane and is saturated | | | |
| | Has stronger and shorter carbon–carbon bonds than 2-chloropent-2-ene | | | |
| hexanoic acid | Contains a C=C bond and is polar | | | |
| | Is a structural isomer of ethyl butanoate | | | |
| | Reacts with ethanol to produce an ester | | | |
| | Is soluble in water and forms hydrogen bonds | | | |
| | Is a saturated primary alcohol | | $CH_3CH_2CH(OH)CH_3$ | |
| | Does not form hydrogen bonds and has a lower melting point than butane | | | |
| | Belongs to the homologous series of alcohols | | | |
| | Undergoes addition reactions | | $CH_3CH_2CH{=}CHCH_3$ | |
| | Belongs to the homologous series of alkenes | | | |
| | Is less viscous than a long alkane with 10 carbon atoms in the chain | | | |
| | Is a structural isomer of pent-1-ene | | | |
| | Is an unsaturated polar molecule | | $CH_3CH_2CH_2NH_2$ | |
| | Is a structural isomer of $C_3H_9N$ | | | |
| | Is formed by reaction of methanoic acid with propan-1-ol in the presence of sulfuric acid catalyst in an esterification reaction | | | O<br>H ‖ H<br>H—C C C—H<br>H C O H<br>H—<br>H |
| | Has a higher melting point than butane | | | |
| | Contains two functional groups | | | H H<br>N<br>C<br>H—C C—H<br>H H<br>H H H |
| | Can be formed by substitution of a chloroalkane | | | |
| | Forms hydrogen bonds when it dissolves in water | | | |

*(Continued)*

| Name of molecule | Statement | T/F | Semi-structural formula | Structural formula |
|---|---|---|---|---|
| | Will undergo addition reactions | | | $CH_3(H)C{=}C(CH_2CH_3)(CH_3)$ |
| | Has a high melting point because of the strong dipole–dipole forces between the molecules | | | |
| | Reacts with $Br_2$ or $I_2$ to form a saturated compound | | | |

2 Draw the structural formulas of the five structural isomers of hexane and name each isomer.

3 Draw the skeletal structure of 2-hydroxypentanoic acid and circle the bond that is the strongest and shortest.

4 a Draw two-dimensional skeletal structures of benzene and cyclohexane and write their molecular formulas under each.

b Both benzene and cyclohexane have six carbon atoms and are cyclic. State two differences in their structures.

 ISBN 978 0 6557 0027 2

# WORKSHEET 25

Case study

## Focus on functional groups—properties and chemical reactions

Functional groups in organic molecules determine their properties and, therefore, their reactivity. The presence of polar groups in a hydrocarbon chain makes the compound more polar, more soluble in water and often more reactive than the corresponding hydrocarbon. The following table shows the results of several small experiments testing the properties of samples of each of the following types of organic chemicals: saturated and unsaturated hydrocarbons, alcohols, carboxylic acids and esters.

| Compound | Observations |
|---|---|
| A | Sweet-smelling liquid; slightly soluble in water |
| B | White powder; soluble in water; reacted with $NaHCO_3(aq)$ to produce a gas that was shown to be carbon dioxide |
| C | Clear liquid; soluble in water; flammable; reacted with compound B to produce a sweet-smelling liquid similar to A; reacted with a mixture of orange $K_2Cr_2O_7(aq)$ and $H_2SO_4(aq)$ to produce a green $Cr^{3+}$ solution |
| D | Clear liquid; insoluble in water; flammable; changed colour immediately when bromine or iodine solution was added |
| E | Clear liquid; insoluble in water; flammable; did not change colour immediately when bromine or iodine solution was added |

Using the information given, answer the following questions.

**1** State the homologous series to which compound B belongs. Explain your answer.

______

______

**2** Write a general equation for the reaction of compound B with $NaHCO_3$ solution.

______

**3** State the homologous series to which compound C belongs. Explain your answer.

______

______

**4** Write a general equation for the reaction of B and C to produce a compound similar to A.

______

**5** To which homologous series does A belong? ______

**6** **a** Write a general equation for the reaction involving $K_2Cr_2O_7(aq)$, $H_2SO_4(aq)$ and compound C. Show only the organic reactant and product.

______

**b** Has compound C been oxidised or reduced? Explain your answer.

______

______

**7** To which homologous series could the organic product of the reaction in Question 6 belong?

______

8 To which homologous series do you think compound D and compound E belong? Explain your answer.

9 Write an appropriate safety statement for a student performing these experiments.

10 Write a conclusion for these experiments in which you list general properties and reactions that occurred in this experiment for saturated and unsaturated hydrocarbons, alcohols, carboxylic acids and esters.

11 Write semi-structural formulas for the organic reactants and products in the following reactions or sequences of reactions.

**a** An amine is the product when chloroethane reacts with ammonia.

**b** The series of reactions involving butan-1-ol, acidified potassium permanganate solution and concentrated sulfuric acid producing the ester, butyl butanoate

**c** The hydrolysis reaction of pentyl ethanoate


ISBN 978 0 6557 0027 2

# WORKSHEET 26



## Focus on biological molecules—functional group chemistry

**1** Triglycerides are an important class of biomolecules produced by the human body. Use terms from the box to complete the reactions for triglycerides.

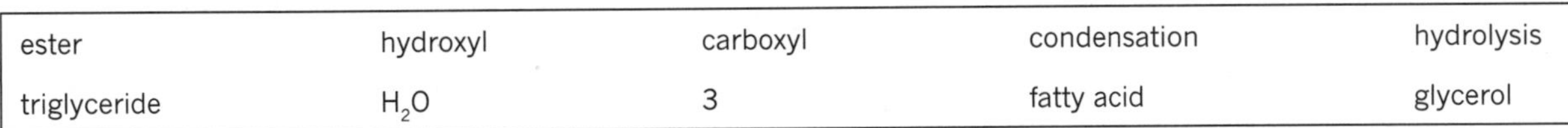

| ester | hydroxyl | carboxyl | condensation | hydrolysis |
|---|---|---|---|---|
| triglyceride | $H_2O$ | 3 | fatty acid | glycerol |

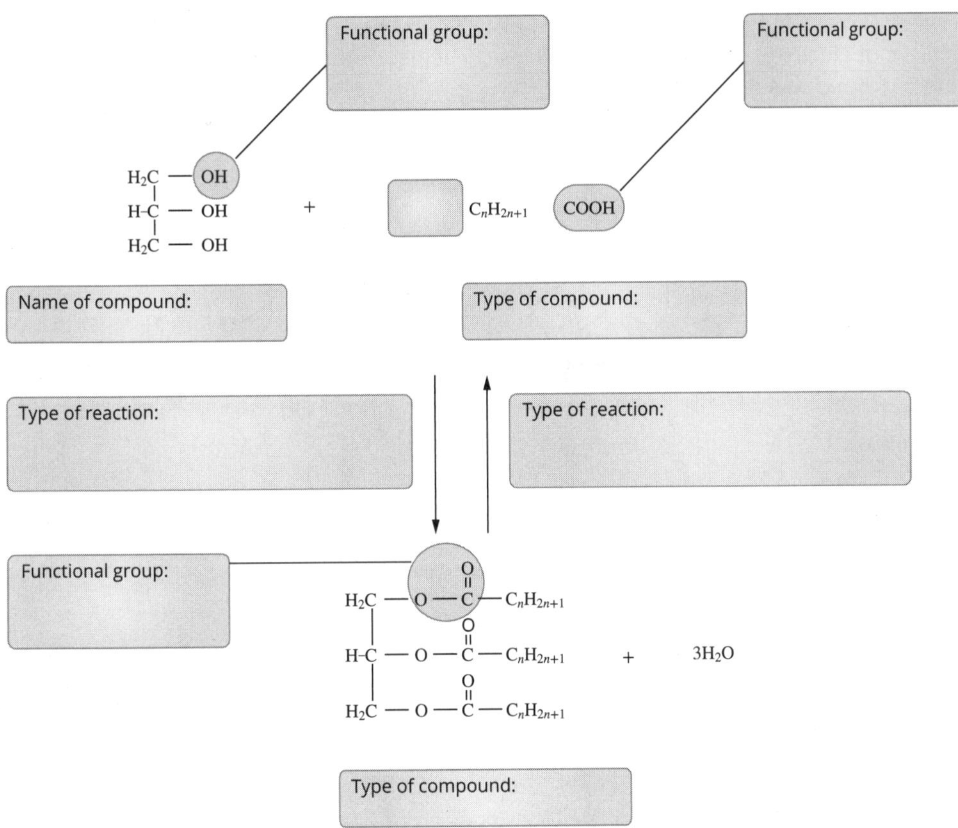


**2** Proteins are polymers that have amide (peptide) links between the monomers.

**a** What is the general name of the monomers that react to form the protein polymer? ____________________

**b** What is the name of the type of reaction when proteins are formed? ____________________

**c** Using the structures given, draw a semi-structural equation (bonding molecule 1 to molecule 2 to molecule 3) to show how a section of a protein could begin to be formed from these three monomers in an enzyme-catalysed reaction. On your diagram, label the functional groups in the protein chain.

$H_2N—CH(CH_3)—COOH$

Molecule 1

$H_2N—CH(CH_2—COOH)—COOH$

Molecule 2

$H_2N—CH(CH_2—C(=O)—NH_2)—COOH$

Molecule 3

Semi-structural equation:

ISBN 978 0 6557 0027 2

**d** Using your answer for the section of protein from part **c**, draw a diagram to show how the functional groups are broken in the polymer, in the presence of the appropriate enzymes, to re-form the monomers.

**3** The structures of glucose and an isomer of glucose, fructose, are shown below. They react to form sucrose in an enzyme-catalysed reaction.

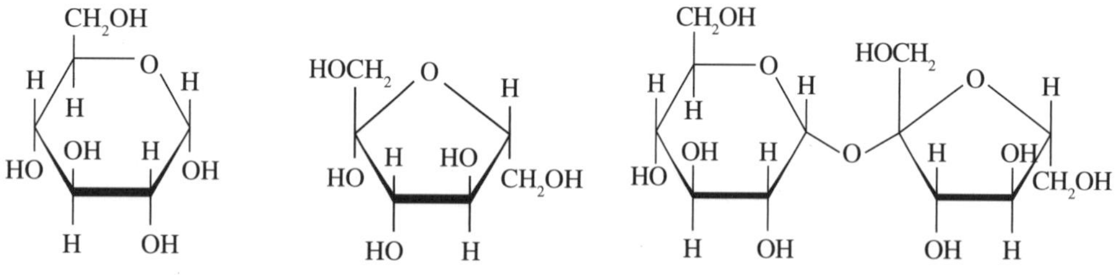

glucose    fructose    sucrose

**a** Circle and name the functional groups in glucose and fructose that react to form sucrose.

______________________________________________

**b** Circle and name the functional group formed in sucrose.

______________________________________________

**c** State the reactant needed (not the enzyme catalyst) to hydrolyse sucrose and reform the monomers.

______________________________________________

**4** Complete the summary with the correct term or expression from the box. Some terms may be used more than once.

| hydroxyl | glycosidic | water | hydrolytic | proteins | esterification | condensation polymerisation reactions | |
|---|---|---|---|---|---|---|---|
| carboxylic acid | amino | amine | amino acids | fatty acids | polymers | concentrated sulfuric acid | condensation |
| triglycerides | carboxyl | alcohol | glycerol | enzyme(s) | catalyst | peptide | |

- The reactions to form esters in the laboratory are similar to those involved in the formation of ________ in the body.
- When a protein is formed in the body, ________ ________ react in the presence of a/an ____________ to form __________ bonds.
- Both the __________ and ____________ functional groups are present in the same monomer molecule in protein formation. The two functional groups are bonded to the same carbon atom.
- The two molecules formed when hydrolysis of a fat occurs are __________, which contains three ________ groups, and _______ ________, which contain three carboxylic acid groups.

 ISBN 978 0 6557 0027 2

**WORKSHEET 26**

- Esters can be made in the laboratory from an ____________ and ________________ ______ in the presence of _________ __________ ___________, which acts as a/an ____________.
- Reactions that form esters can be called either ___________________ reactions or ________________ reactions.
- Carbohydrates are formed by ___________ ______________ ______________. The ether or _________ link is formed by the reaction of two ___________ functional groups.
- A small molecule, ___________, is released in the condensation reactions that form proteins, fats and carbohydrates.
- To undergo the reverse reaction, called a a/an _______________ reaction, these nutrients react with __________ in the presence of an appropriate ____________.

# WORKSHEET 27

Simulation

## Green chemistry—approaches to sustainability

**1 a** Calculations of the atom economy of a reaction can help industrial chemists decide on a suitable synthetic route for a particular chemical. Calculate the atom economy for the first reaction below for the synthesis of ethanol and compare it to the atom economy of the second reaction.

**Reaction 1:** $CH_3CH_2Cl + H_2O \rightarrow CH_3CH_2OH + HCl$

**Reaction 2:** $CH_2{=}CH_2 + H_2O \xrightarrow{H_3PO_4} CH_3CH_2OH$

**b** Calculations of percentage yield are also useful in determining the best pathway to use for a synthesis. In a multi-step process, the yield for each step has an effect on the overall yield. Consider the following reaction sequence for the production of ethyl ethanoate, which is used in perfumes, flavourings and pharmaceuticals.

**Step 1:** A sample of 2000 g of ethene is reacted with water under appropriate conditions to produce 1800 g of ethanol, according to the equation:

$$CH_2{=}CH_2 + H_2O \xrightarrow{H_3PO_4} CH_3CH_2OH$$

**Step 2:** The ethanol produced is reacted with acidified potassium dichromate solution to produce 800.0 g of ethanoic acid, according to the equation:

$$CH_3CH_2OH \xrightarrow{H^+/Cr_2O_7^{2-}} CH_3COOH$$

**Step 3:** The ethanoic acid is then heated with excess ethanol to produce ethyl ethanoate, according to the equation:

$$CH_3CH_2OH + CH_3COOH \xrightarrow{\text{conc. } H_2SO_4} CH_3COOCHCH_2CH_3$$

**i** Calculate the percentage yield for steps 1 and 2.

Step 1:

Step 2:

**ii** If the yield for step 3 is 75%, calculate the percentage yield for the overall reaction pathway.


ISBN 978 0 6557 0027 2

2 When assessing how well a chemical synthesis complies with the principles of green chemistry, there are a number of factors other than percentage yield and atom economy to consider. These include the use of renewable raw materials, how to limit the use of potentially harmful solvents and how to minimise the amount of unwanted products. Indicate at least one advantage of each of the processes listed in the table below.

| Process | Advantage |
|---|---|
| using water as a solvent instead of chloromethane | |
| using a catalyst in a reaction where a catalyst has not been used before | |
| using a lower reaction temperature where possible | |
| using an addition reaction instead of a substitution reaction in an organic process | |
| using a fuel produced from sugar cane instead of one produced from crude oil | |
| using insecticides that are effective against insects, but with a high degree of selectivity | |
| using a biodegradable biopolymer for packaging | |

3 A painkiller, aspirin, is produced by an esterification reaction between salicylic acid and ethanoic anhydride according to the following equation:

salicylic acid + ethanoic anhydride → acetylsalicylic acid (aspirin) + ethanoic acid

A student reacted 160 g of pure salicyclic acid with 103 g of pure ethanoic anhydride. The experiment produced 165 g of aspirin.

**a** Calculate the limiting reactant.

**b** Calculate the theoretical amount, in mol, of aspirin that could be produced and therefore the mass of aspirin.

**c** Calculate the percentage yield of aspirin in this experiment.

**d** Calculate the atom economy of this reaction.

ISBN 978 0 6557 0027 2

**4** Another painkiller, ibuprofen, can be produced by either of two pathways: a green chemistry pathway or a less environmentally friendly ('brown') pathway. Examine these pathways in the diagram.

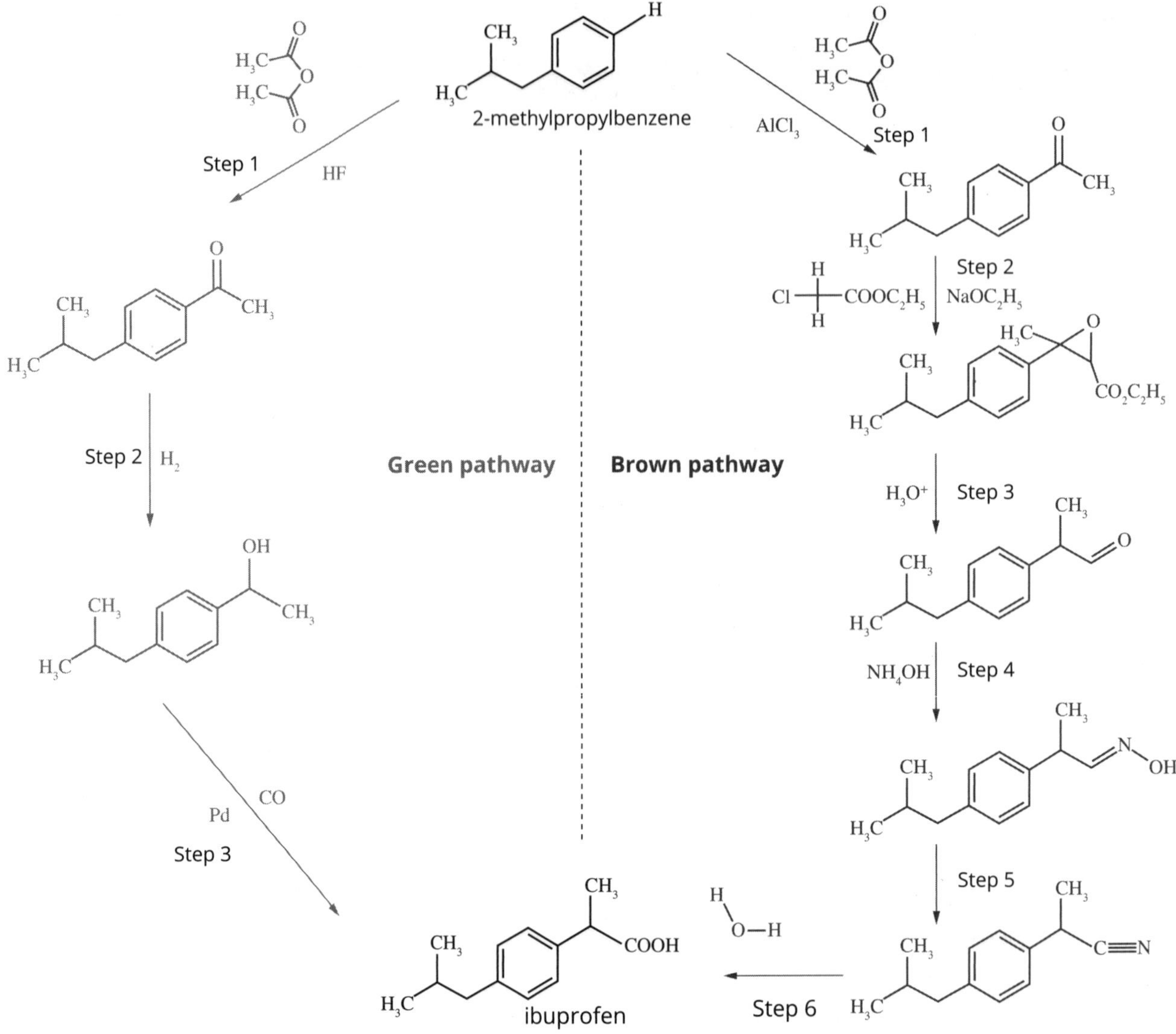


**Figure 4.1.8** Two alternative reaction pathways for the production of ibuprofen: the green pathway and the brown pathway

The left-hand pathway is called the 'green pathway'. Suggest two reasons why the use of fewer steps in a synthesis is considered more environmentally friendly.

 ISBN 978 0 6557 0027 2

5 **a** Using the production and use of ethanol from fossil fuels and bioethanol as examples, and terms from the box, draw two simple flow charts to illustrate the idea of a circular economy and a linear economy. Briefly summarise the meaning of these terms. Use the words in the box as many times as you wish.

| | | | | | |
|---|---|---|---|---|---|
| bioethanol | crude oil | $CO_2$ in the atmosphere | converting ethene to ethanol | used as energy source | fractional distillation |
| ethene | plants | bioethanol + $CO_2$ production | converting plant starch to sugar | photosynthesis | fermentation |

**b** Select two of the green chemistry principles that are in action in the circular economy that are not being followed in the linear economy and briefly explain how they are reflected in the bioethanol manufacturing process.

ISBN 978 0 6557 0027 2

# WORKSHEET 28



## Literacy review—organic chemistry

1 To review your understanding of some of the terms in this topic, state the correct term for each definition in the table below.

| Definition | Term |
|---|---|
| a series of organic compounds with members differing by one $CH_2$ unit | |
| molecules with different arrangements of atoms but the same molecular formula | |
| a group of organic compounds that contain a hydroxyl group | |
| a type of organic compound with a systematic name ending in '-oate' | |
| a type of reaction involving the removal of a C=C bond | |
| an atom or a group of atoms whose properties determine the properties of organic compounds | |
| an organic molecule with a carbonyl functional group at the end of the chain | |
| an organic molecule that contains a –COOH group | |

2 Using the IUPAC rules of nomenclature for organic compounds, help a fellow Year 12 Chemistry student work out the name of the following compound by answering the following questions.

$$CH_3—CH(Cl)—CH_2—C(H)=O$$

**a** How many carbon atoms are there in the longest chain? ____________________

**b** The parent name of the compound is based on the alkane with this number of carbon atoms in it. What alkane is that?

____________________

**c** What are the formulas of the functional groups in this compound?

____________________

**d** What are the names of these functional groups?

____________________

**e** Which functional group determines the name of the compound?

____________________

**f** What is the parent name of the compound with the appropriate ending for this functional group?

____________________

**g** Which carbon is the chloro group bonded to? ____________________

**h** What is the correct name of this compound?

____________________

**i** Why don't you need a number to indicate where the carbonyl group is bonded?

____________________


ISBN 978 0 6557 0027 2

**3** Use the terms and formulas in the box to help you complete the flow chart. (Some may be used more than once and others may not be used at all.) Draw the structural formulas in the boxes above the chemical names in the flow chart where required. Give the chemical required in the boxes beside the arrows.

| | | | | |
|---|---|---|---|---|
| UV light | ethanoic acid | $Cr_2O_7^{2-}/H^+$ | $Cl_2$/UV light | ethene |
| ethyl ethanoate | 1,2-dibromoethane | oxidation | concentrated $H_2SO_4$ | $H_2O/H_3PO_4$ |
| HCl(g) | catalyst | $CO_2(g)$ | $O_2(g)$ | $H_2O(g)$ |
| NaOH(aq) | $Br_2(g)$ | $H_2(g)$ | | |

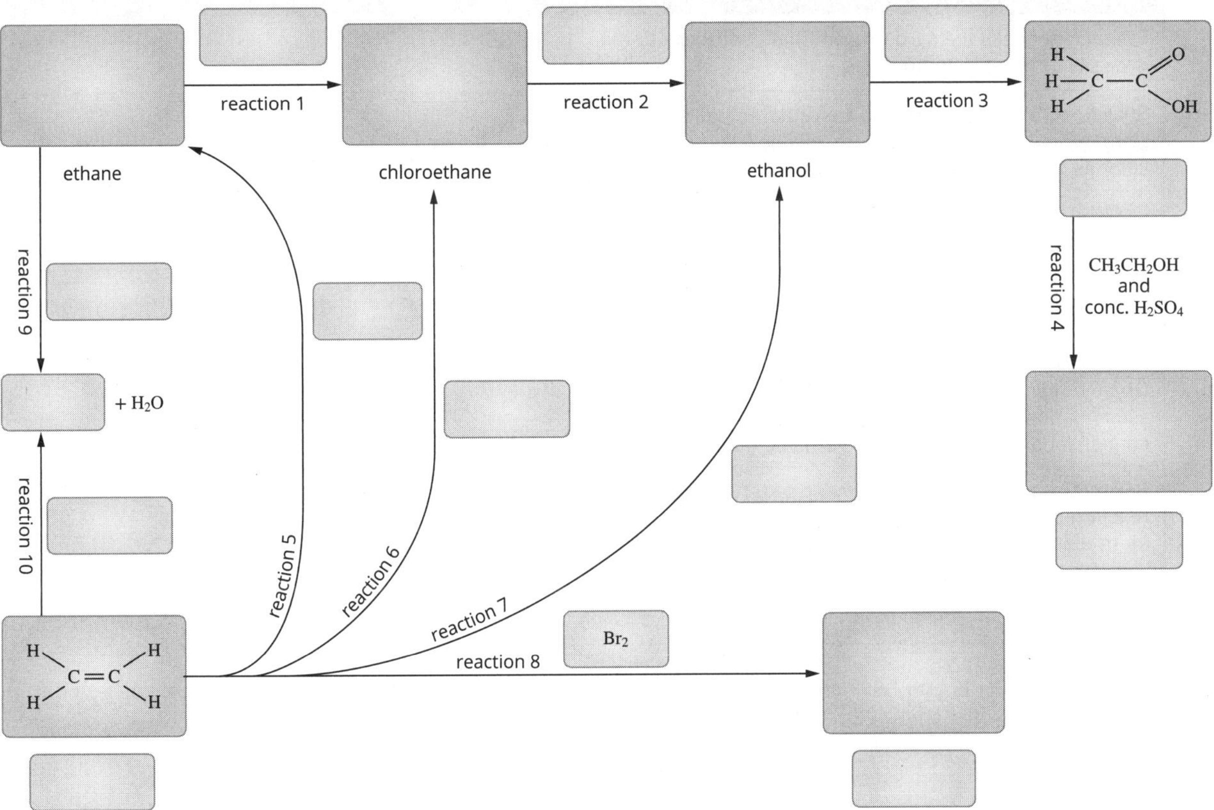


**4** In the table provided, classify the numbered reactions in the flow chart in Question 3, using names from the box below. (There may be more than one classification for a particular reaction.)

| | | | | | | |
|---|---|---|---|---|---|---|
| substitution | combustion | addition | hydrogenation | condensation | esterification | hydrolysis |

| Reaction number | Classification | Reaction number | Classification |
|---|---|---|---|
| 1 | | 6 | |
| 2 | | 7 | |
| 3 | | 8 | |
| 4 | | 9 | |
| 5 | | 10 | |

# WORKSHEET 29

## Reflection—How are organic compounds categorised and synthesised?

The following table summarises the key knowledge covered in this area of study.

1 Reflect on how well you understand the concepts listed. Rate your learning by shading the circle that corresponds to your current level of understanding for each one.

| Key knowledge | Not confident ◄ | | | | ► Very confident |
|---|---|---|---|---|---|
| Characteristics of the carbon atom that contribute to the diversity of organic compounds | ○ | ○ | ○ | ○ | ○ |
| Structures of alkanes, alkenes, benzene, haloalkanes, primary amines, primary amides, alcohols, aldehydes, ketones, carboxylic acids and esters | ○ | ○ | ○ | ○ | ○ |
| IUPAC naming of organic compounds | ○ | ○ | ○ | ○ | ○ |
| Trends in boiling points, melting points and viscosity within homologous series with reference to structure and bonding | ○ | ○ | ○ | ○ | ○ |
| Organic reactions and pathways (substitution, addition, esterification, condensation, transesterification, condensation polymerisation, hydrolytic reactions) | ○ | ○ | ○ | ○ | ○ |
| Calculations of percentage yield and atom economy of reaction pathways; advantages of processes with high atom economies | ○ | ○ | ○ | ○ | ○ |
| The sustainability of the production of chemicals with reference to green chemistry principles | ○ | ○ | ○ | ○ | ○ |

2 Consider the points you have shaded from 'Not confident' to 'Very confident'. List specific ideas you can identify that were challenging.

_______________

_______________

_______________

3 Write down two different strategies that you will apply to help further your understanding of these ideas.

_______________

_______________

_______________

 ISBN 978 0 6557 0027 2

# PRACTICAL ACTIVITY 14

Controlled experiment

## Preparing artificial fragrances and flavours

### SUGGESTED DURATION

- 50 minutes (depending how many esters made per student)

### INTRODUCTION

Esters are commonly used as artificial flavourings in foods such as ice cream and sweets. They are partially responsible for many familiar odours, including those of coffee, perfumes and fruit.

### AIM

To prepare several esters that are widely used as artificial fragrances and flavours

As directed by your teacher, complete the pre-lab safety information by referring to the safety data sheets (SDSs) or your teacher's risk assessment for the activity.

**PRE-LAB SAFETY INFORMATION**

| Material used | Hazard | Control |
|---|---|---|
| concentrated sulfuric acid, $H_2SO_4$ | | |
| glacial ethanoic acid, $CH_3COOH(l)$ | | |
| salicylic acid, $HOC_6H_4COOH(s)$ | | |
| decanoic acid, $CH_3(CH_2)_8COOH(l)$ | | |
| pentan-1-ol, $CH_3CH_2CH_2CH_2CH_2OH(l)$ | | |
| ethanol, $CH_3CH_2OH(l)$ | | |
| esters | • irritating to eyes, respiratory system and skin | Smell esters produced for a very short time (1–2 seconds), cautiously wafting vapour towards the nose. |

Complete and indicate that you have understood the information in the safety table.

Name (print): __________

I understand the safety information (signature): __________

### MATERIALS

- small volumes of the alcohols pentan-1-ol, $CH_3CH_2CH_2CH_2CH_2OH(l)$, and ethanol, $CH_3CH_2OH(l)$
- small amounts (in small plastic dropper bottles) of the carboxylic acids: glacial ethanoic (acetic) acid, $CH_3COOH(l)$, decanoic acid, $CH_3(CH_2)_8COOH(l)$ and salicylic acid, $HOC_6H_4COOH(s)$
- dropping bottle of concentrated sulfuric acid, $H_2SO_4$
- 6 semi-micro test tubes
- semi-micro test-tube rack
- small tongs
- 250 mL beaker containing 100 mL boiling water
- 2 × 250 mL beakers
- marking pen
- safety glasses
- safety gloves

### METHOD

1. Label two semi-micro test tubes 'A' and 'B'. Place 10 drops of pentan-1-ol in each tube.
2. Wearing gloves, add 10 drops of glacial ethanoic acid to test tube A and a similar volume of salicylic acid to test tube B. Then add 2 drops of concentrated sulfuric acid to each tube.
3. Heat the mixtures for 10 minutes in a beaker of boiling water and then pour each one into a 250 mL beaker containing 200 mL of cold water.
4. Try to identify the odour of the esters produced by cautiously wafting the vapour from the ester towards you. Note the name of the carboxylic acids and alcohol used in this test and describe the smell of the esters in the results table.
5. Wash out the beakers thoroughly and repeat steps 1–4 using clean semi-micro test tubes and other combinations of alcohols and carboxylic acids. Record the results in Table 1.

## PRACTICAL ACTIVITY 14

### RESULTS

Table 1 Experimental results

| Combination | Alcohol | Carboxylic acid | Ester | Smell |
|---|---|---|---|---|
| 1 | | | | |
| 2 | | | | |
| 3 | | | | |
| 4 | | | | |
| 5 | | | | |

### DISCUSSION

1 Salicylic acid has the formula $C_6H_4(OH)COOH$. Complete Table 2 by writing equations for each of the reactions you performed in this experiment and naming the esters formed. (The names of esters made from salicylic acid end with salicylate.)

Table 2 Esters produced

| Combination | Equation | Name of ester formed |
|---|---|---|
| 1 | | |
| 2 | | |
| 3 | | |
| 4 | | |
| 5 | | |

2 What is the role of the concentrated sulfuric acid in these reactions?

3 Discuss the validity of this practical activity.

### CONCLUSION

Write a general equation for the formation of an ester from an alcohol and a carboxylic acid.


ISBN 978 0 6557 0027 2

# PRACTICAL ACTIVITY 15

# Reactions, properties and functional group tests of some organic compounds

## SUGGESTED DURATION

- 80 minutes

## INTRODUCTION

The organic chemicals used in this experiment undergo reactions typical of the different classes of organic compounds to which they belong. The chemicals and the classes they represent are cyclohexane (saturated hydrocarbon), cyclohexene (unsaturated hydrocarbon), 2-chloro-2-methylpropane (chloroalkane), ethanol (alcohol) and ethanoic acid (carboxylic acid). Cyclohexane and cyclohexene are shown in Figure 4.1.9.

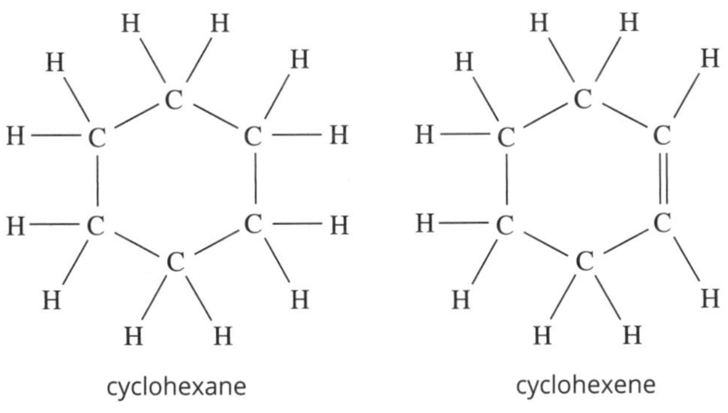


**Figure 4.1.9** The structural formulas of cyclohexane and cyclohexene

## AIM

To investigate the reactions and properties of saturated and unsaturated hydrocarbons, chloroalkanes, alcohols and carboxylic acids

**PRE-LAB SAFETY INFORMATION**

| Material used | Hazard | Control |
|---|---|---|
| concentrated sulfuric acid, $H_2SO_4$ | • extremely corrosive; causes severe burns<br>• harmful by inhalation, ingestion and skin contact | Use small quantities. Handle with extreme care and avoid contact. Dilute small spills with water. Wear gloves, safety glasses and a laboratory coat. |
| glacial ethanoic acid, $CH_3COOH(l)$ | • strongly corrosive; causes serious burns<br>• very harmful if swallowed | Use small quantities. Handle with extreme care and avoid contact. Dilute small spills with water. Wear gloves, safety glasses and a laboratory coat. |
| iodine solution in hexane, $I_2$(soln) | • harmful by inhalation and skin contact | Use small quantities. Avoid contact with eyes or skin. Work in a fume hood. Avoid breathing vapour. Wear gloves, safety glasses and a laboratory coat. |
| 2 M nitric acid, $HNO_3(aq)$ | • contact with combustible materials may cause fire<br>• causes burns | Use small quantities. Wear gloves, safety glasses and a laboratory coat. |
| potassium hydroxide, KOH(s) | • corrosive and can cause severe burns<br>• harmful if swallowed | Use small quantities. Wear gloves, safety glasses and a laboratory coat. |

*(Continued)*

## MATERIALS

- small dropping bottles of:
  - cyclohexane, $C_6H_{12}(l)$
  - cyclohexene, $C_6H_{10}(l)$
  - iodine in hexane, $I_2$(soln)
  - 2-chloro-2-methylpropane (tert-butyl chloride), $(CH_3)_3CCl(l)$
  - ethanol, $CH_3CH_2OH(l)$
  - 0.1 M silver nitrate solution, $AgNO_3(aq)$
  - concentrated sulfuric acid, $H_2SO_4$
  - glacial ethanoic (acetic) acid, $CH_3COOH(l)$
  - 0.02 M potassium permanganate solution, $KMnO_4(aq)$
  - 2 M nitric acid, $HNO_3(aq)$
  - 1 M sulfuric acid, $H_2SO_4(aq)$
- potassium hydroxide, KOH(s)
- sodium hydrogen carbonate, $NaHCO_3(s)$
- deionised water
- 3 strips of blue litmus paper
- 2 strips of red litmus paper
- 250 mL beaker
- 8 semi-micro test tubes
- 2 semi-micro test-tube stoppers
- semi-micro test-tube rack
- semi-micro test-tube holder
- semi-micro spatula
- tweezers
- 2 dropping pipettes
- 2 small watch glasses
- matches
- beaker of hot water (almost boiling)
- disposable plastic gloves
- safety glasses

| PRE-LAB SAFETY INFORMATION *(Continued)* | | |
|---|---|---|
| Material used | Hazard | Control |
| organic chemicals used in this experiment | • highly flammable<br>• harmful irritant to skin, eyes and respiratory system | Keep bottles firmly stoppered and away from flames. Wear gloves, safety glasses and a laboratory coat. |
| 0.1 M silver nitrate solution, $AgNO_3(aq)$ | • stains skin, clothes and benchtops | Handle with care. Wear a laboratory coat and gloves. |
| Please indicate that you have understood the information in the safety table.<br><br>Name (print): ______<br><br>I understand the safety information (signature): ______ | | |

## METHOD

## PART A • REACTIONS OF SATURATED AND UNSATURATED HYDROCARBONS

1 ▪ Shake 5 drops of cyclohexane with 10 drops of water in a test tube. Note in the results table whether the saturated hydrocarbon is soluble in water. Repeat with the unsaturated hydrocarbon cyclohexene.

2 ▪ Place 2 drops of cyclohexane on one watch glass and 2 drops of cyclohexene on another. Light the two liquids with a match and record the appearance of each flame (colour, smoke, soot etc.).

3 ▪ In a fume hood, place cyclohexane to a depth of 1 cm in one test tube and cyclohexene to the same depth in another tube. Add 2 drops of a solution of iodine in hexane to each liquid. If there is no immediate reaction, stopper the tube and place it in sunlight for 5 minutes. Record your observations in Table 1.

## PART B • REACTIONS OF CHLOROALKANES

Use a fume hood during all of Part B.

1 ▪ Shake 5 drops of 2-chloro-2-methylpropane with 10 drops of water in a test tube. Note in the results table whether the chloroalkane is soluble in water.

2 ▪ Mix 2 drops of 2-chloro-2-methylpropane and 5 drops of ethanol with half a small pellet of solid potassium hydroxide. (The ethanol helps to dissolve the potassium hydroxide.) Warm the mixture in a beaker of hot water for 1 minute.

Test for the presence of $Cl^-$ ions by acidifying the mixture with 2 M nitric acid, using litmus paper to test the acidity, and adding 3 drops of silver nitrate solution. Note whether a white precipitate of silver chloride forms.

## PART C • REACTIONS OF ALCOHOLS

1 ▪ Add 5 drops of ethanol to 10 drops of deionised water in a test tube and shake the tube. Note whether the alcohol is soluble in water. Test the solution with red and blue litmus paper.

2 ▪ Add 4 drops of glacial ethanoic (acetic) acid to 8 drops of ethanol in a dry test tube. Carefully add 1 drop of concentrated sulfuric acid. Heat the mixture in a beaker of hot water for 2 minutes and then pour it into a beaker of cold water. Note the fruity odour of the chemical formed.

3 ▪ Place 5 drops of ethanol, 10 drops of 1 M sulfuric acid and 2 drops of potassium permanganate solution in a test tube. Heat the mixture in the beaker of hot water for 2 minutes and note the colour change due to the formation of colourless $Mn^{2+}$ ions.

## PART D • REACTIONS OF CARBOXYLIC ACIDS

1 ▪ Shake 5 drops of glacial ethanoic acid with 10 drops of deionised water in a test tube. Note whether ethanoic acid is soluble in water. Test the solution with red and blue litmus paper.

2 ▪ Add a small quantity of sodium hydrogen carbonate to the solution of ethanoic acid from step 1. Record your observations in Table 1.


ISBN 978 0 6557 0027 2

## RESULTS AND DISCUSSION

1 For each test, note the inferences (reasoned conclusions) that can be drawn from the observations and write equations, where appropriate.

**Table 1 Test results and equations**

| Test | Observations | Inferences | Equations (where appropriate) |
|---|---|---|---|
| **Part A** Solubility of cyclohexane and cyclohexene | | | |
| Burning cyclohexane and cyclohexene | | | |
| Action of iodine on cyclohexane and cyclohexene | | | |
| **Part B** Solubility of the chloroalkane | | | |
| Action of KOH on the chloroalkane | | | |
| **Part C** Solubility of ethanol in water; litmus test | | | |
| Reaction of ethanol with ethanoic acid | | | |
| Action of $MnO_4^-/H^+$ on ethanol | | | |
| **Part D** Solubility of ethanoic acid in water; litmus test | | | |
| Reaction of ethanoic acid with $NaHCO_3$ | | | |

2 Comment on the reproducibility of your results.

## CONCLUSION

Summarise the reactions and properties of alkanes, alkenes, chloroalkanes, alcohols and carboxylic acids in Table 2.

**Table 2 Summary of reactions and properties**

| Homologous series | Reactions and properties |
|---|---|
| alkanes (saturated hydrocarbons) | |
| alkenes (unsaturated hydrocarbons) | |
| chloroalkanes | |
| alcohols | |
| carboxylic acids | |

ISBN 978 0 6557 0027 2

# PRACTICAL ACTIVITY 16

Modelling

## Modelling functional groups and organic reactions

### SUGGESTED DURATION

- 80 minutes, depending on the number of molecules constructed

### INTRODUCTION

This activity will help you to visualise and appreciate the three-dimensional structures of organic compounds, common functional groups and structural isomers. Upon completion of this task, you will be able to identify and explain the role of functional groups, discuss the effect that they have on the bonding and reactions of organic compounds and construct reaction pathways, including writing equations.

### AIM

- To examine the bonding, shape and nomenclature of a number of organic molecules with common functional groups
- To investigate the concept of structural isomers
- To model reactions involving common functional groups

### METHOD

### PART A • FUNCTIONAL GROUPS

For each of the molecules in the Part A materials list, complete Table 1.

**a** Write the semi-structural formula.

**b** Draw the structural formula.

**c** State the strongest type(s) of bonding between molecules.

### PART B • STRUCTURAL ISOMERS

For the following molecules, draw the structural isomers in Table 2 and construct models of them. Write the systematic name of each isomer where possible.

**a** dichloroethene

**b** pentane

**c** butanol

**d** hydroxypropanoic acid

### PART C • REACTIONS

For each of the pairs of reactants in the Part C materials list:

**a** construct a three-dimensional model of each organic reactant

**b** rearrange the atoms in the reactants to form models of the products

**c** complete your answers in Table 3 by writing equations for the organic reactions.

### MATERIALS

- molecular model building kit for constructing some of the following molecules:

**Part A**

- alkanes: methane, butane
- alkenes: ethene, propene, but-1-ene, but-2-ene
- chloroalkane: 1,1-dichloroethane
- alcohols: ethanol, butan-2-ol
- carboxylic acids: methanoic acid, propanoic acid
- ester: ethyl propanoate

**Part B**

- dichloroethene
- pentane
- butanol
- hydroxypropanoic acid

**Part C**

- ethane and chlorine
- ethene and hydrogen chloride
- ethanol and ethanoic acid


ISBN 978 0 6557 0027 2

## PRACTICAL ACTIVITY 16

## RESULTS

**Table 1 Part A results: Functional groups**

| Name | Semi-structural formula | Structural formula | Strongest type of bonding between molecules |
|---|---|---|---|
| methane | | | |
| butane | | | |
| ethene | | | |
| propene | | | |
| but-1-ene | | | |
| but-2-ene | | | |
| 1,1-dichloroethane | | | |
| ethanol | | | |
| butan-2-ol | | | |
| methanoic acid | | | |
| propanoic acid | | | |
| ethyl propanoate | | | |

| Table 2 Part B results: Structural isomers |
|---|
| dichloroethane |
| pentane |
| butanol |
| hydroxypropanoic acid |

| Table 3 Part C results: Equations for the reactions | |
|---|---|
| ethane and chlorine | |
| ethene and hydrogen chloride | |
| ethanol and ethanoic acid | |

## DISCUSSION

1 What type of bonding is present within the molecules?

2 Complete Table 4 by naming the functional groups present in the molecules constructed in Part A.

Table 4 Names of functional groups in Part A

| Homologous series | Functional group present | Homologous series | Functional group present |
|---|---|---|---|
| alkanes | | alcohols | |
| alkenes | | carboxylic acid | |
| chloroalkanes | | esters | |

3 Construct a flow chart that shows how ethyl ethanoate can be prepared from ethane and from ethene.

4 Write equations using structural formulas for each of the reactions in the flow charts in Question 3.

## CONCLUSION

How do functional groups affect the bonding and reactions of organic compounds?

 ISBN 978 0 6557 0027 2

# PRACTICAL ACTIVITY 17

Controlled experiment

# Performing a condensation polymerisation reaction (teacher demonstration only)

This practical activity should only be done as a teacher demonstration, in a fume hood, for a class practical activity.

## SUGGESTED DURATION

- 50 minutes

## INTRODUCTION

Two immiscible organic liquids are poured into a beaker. A strand of nylon is pulled out from the interface of the liquids and wound onto a glass rod. Figure 4.1.10 shows the process involved.

The reaction between sebacoyl chloride and hexamethylene diamine (hexane-1,6-diamine) can be represented as follows:

$$ClCO(CH_2)_8COCl + NH_2(CH_2)_6NH_2 \rightarrow -CO(CH_2)_8-CONH-(CH_2)_6NH- + HCl$$

In this reaction, the two monomers react together to form a long polymer chain. The link formed is called an amide link. Nylon is the trade name for the amide that is formed. A small molecule, HCl, is produced in this condensation polymerisation reaction.

A similar condensation polymerisation reaction occurs when amino acids react to form proteins, where the amide bonds are specifically called peptide bonds.

Hydrogen bonding occurs between the –CONH– amide links in neighbouring chains. As a consequence, nylon can be drawn out into fibres.

## MATERIALS

- 2 × 100 mL beakers
- 2 glass rods
- 20 mL hexane, $C_6H_{14}(l)$
- 30 mL aqueous solution containing 3 g hexamethylene diamine (hexane-1,6-diamine), $NH_2(CH_2)_6NH_2$, 0.3 g sodium carbonate, $Na_2CO_3$, and a few drops of phenolphthalein solution
- 30 mL of 10% sebacoyl chloride, $ClCO(CH_2)_8COCl$, in hexane
- tweezers

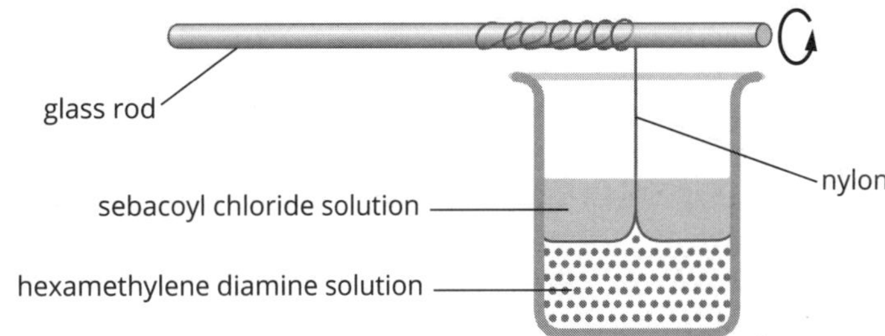

**Figure 4.1.10** The experimental set-up

## AIM

To prepare the amide nylon by the condensation polymerisation reaction of an acid chloride and an amine

| PRE-LAB SAFETY INFORMATION | | |
|---|---|---|
| **Material used** | **Hazard** | **Control** |
| hexane, $C_6H_{14}(l)$ | • extremely flammable and a skin irritant | Wear safety glasses, gloves and a laboratory coat. Perform in a fume hood with excellent ventilation. |
| hexamethylene diamine, $NH_2(CH_2)_6NH_2$ | • strong base and highly corrosive<br>• reacts violently with acid and can be readily absorbed into the skin, causing burns and blisters<br>• if inhaled, can cause shortness of breath, sore throat and laboured breathing | Wear safety glasses, gloves and a laboratory coat. Perform in a fume hood with excellent ventilation. No naked flames or contact with hot surfaces should occur. |
| sebacoyl chloride, $ClCO(CH_2)_8COCl$ | • highly corrosive and reacts violently with water, liberating toxic gases<br>• is an eye and skin irritant, and extremely destructive to the respiratory tract if inhaled | Wear safety glasses, gloves and a laboratory coat. Perform in a fume hood with excellent ventilation. No naked flames or contact with hot surfaces should occur. |

Complete and indicate that you have understood the information in the safety table.

Name (print) ______________________

I understand the safety information (signature): ______________________

## PRACTICAL ACTIVITY 17

### METHOD

1. Mix a small sample of hexane and water in a beaker. Do they mix? If not, which substance is on top? Which substance is more dense?
2. In a fume hood, place about 30 mL of the hexamethylene diamine solution in a beaker. Carefully add 30 mL of sebacoyl chloride solution by running it down the side of the beaker, so that the two liquids do not mix.
3. Use tweezers to draw the polymer formed at the liquid interface from the beaker. Keep the thread away from the sides of the beaker, or it will adhere and spoil the effect.
4. Wind the thread onto a glass rod as shown in Figure 4.1.10. It should be possible to draw out a long thread.
5. After some of the thread has been obtained, use another glass rod to mix the remaining liquid in the beaker. A ball of the amide polymer, nylon, forms.

### RESULTS

Record your observations, clearly describing the substance formed.

### DISCUSSION

1. Why are hexane and water chosen as solvents for sebacoyl chloride and hexamethyl diamine respectively? Would ethanol and water be suitable?

2. Draw the structural formulas of the reactants and products for the equation for the reaction. Label the amide link and circle the functional groups from which the amide link was formed.

3. Using the diagram below, draw the structural formula of the reactants and products for the equations when the following amino acids react. Label and circle the peptide link and the functional groups from which the peptide link was formed.

$$\begin{array}{ccccc} & & CH_2 - COOH & & \\ & & | & & \\ H_2N & - & CH & - & COOH \end{array}$$

$$\begin{array}{ccccc} & & CH_3 & & \\ & & | & & \\ H_2N & - & CH & - & COOH \end{array}$$

### CONCLUSION


ISBN 978 0 6557 0027 2

# PRACTICAL ACTIVITY 18

Modelling

# Modelling proteins, fats and carbohydrates

## SUGGESTED DURATION

- 80 minutes

## MATERIALS

- molecular model kits
- sticky tape
- scissors

**Part A**

- 50 cm each of three different coloured paper streamers (not crepe)

**Part B**

- 3 × 40 cm paper streamers

**Part C**

- 20 square basic Lego® blocks

## INTRODUCTION

Proteins are condensation polymers of amino acids. Each amino acid has the same basic structure of a carboxyl group, amino group, hydrogen atom and side chain (R) all bonded to a central carbon atom, as shown in Figure 4.1.11. The R group gives each amino acid, and therefore the protein, its distinctive properties.

The primary structure of a protein is the sequence of amino acids held together by covalent bonds. The protein chain is coiled, folded and pleated into a secondary structure, held together by hydrogen bonds. Interactions between the R groups are responsible for the formation of a three-dimensional tertiary structure.

Fats and oils contain large molecules, called triglycerides, which are formed by a condensation reaction between a glycerol molecule and three fatty acid molecules.

Carbohydrates have the formula $C_x(H_2O)_y$. The polysaccharides starch, cellulose and glycogen are condensation polymers of the monosaccharide glucose. Starch is considered to be an energy food because the body easily digests the carbohydrate into glucose molecules, which can then undergo cellular respiration to produce energy for the various body processes.

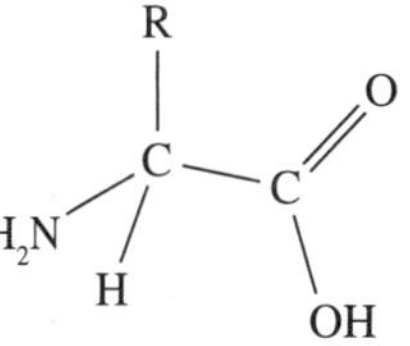


**Figure 4.1.11** The general structure of an amino acid

## AIM

- To make models of the structures of macronutrients in food:
  - for proteins: representing the primary and secondary structures
  - for fats: showing glycerol and fatty acids
  - for carbohydrates: the structure of monosaccharides and polysaccharides
- To draw and label the basic units of these nutrients and label the functional groups in them

## METHOD

### PART A • PROTEINS

1. Use the molecular model kit to build the general structure of an amino acid with a hydrogen atom as the R group. Draw this structure and label it 'glycine'.
2. Remove the R group and replace it with a $-CH_3$ group. Draw this structure and label it 'alanine'.
3. Make another amino acid replacing the R group with $(CH_3)_2CH-$. Draw this amino acid and label it 'valine'.
4. Join alanine and valine together, removing a $H_2O$ molecule as you do. This represents a condensation reaction to form a dipeptide. Draw the structures for the condensation reaction you have modelled.
5. On your diagram, circle the peptide (amide) link and label it.

*You will learn about the secondary and tertiary structures of enzymes in the next area of study. The next three steps model the secondary level of structure.*

6. To make the secondary structure, cut four 10 cm strips of each coloured streamer and label each with the code for three different amino acids such as Gly on one, Tyr on the second and Leu on the third.
7. Maintaining the same order, stick these together to make a long chain.
8. Hydrogen bonds can form between an N–H group in one amino acid and the C=O group in an amino acid in another part of the chain. Twist the streamer to show the effect these bonds can have on the shape of the chain. This represents the secondary structure of the protein.

PRACTICAL ACTIVITY 18

## PART B • FATS

1 ▪ Use the molecular model kit to build a glycerol molecule and three separate –COOH units.

2 ▪ Bend each 40 cm length of streamer into a zig-zag length. To make stearic acid, bend each streamer 17 times to represent the fatty acid chain $C_{17}H_{35}$.

3 ▪ Use sticky tape to stick the streamer to the carbon atom of a –COOH unit. Repeat for each –COOH unit and streamer. Now you have three fatty acid molecules.

4 ▪ To represent the condensation reaction, join each fatty acid molecule to the glycerol, removing a $H_2O$ molecule each time.

5 ▪ Draw the structural formula for each reactant and the product in this condensation reaction. Circle and label the functional groups in the reactants and products.

## PART C • CARBOHYDRATES

1 ▪ Use the molecular model kit to build the structural isomers glucose and fructose.

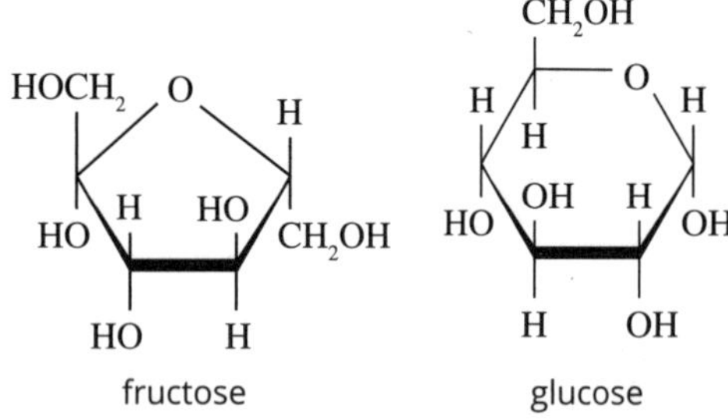


2 ▪ Join them together to represent the condensation reaction that occurs to form the disaccharide sucrose. Note that an $H_2O$ molecule is also formed in this reaction.

3 ▪ Draw the structural formula for each reactant and the product in this condensation reaction. Circle and label the functional groups in the reactants and products.

4 ▪ To represent part of the polysaccharide cellulose, join at least 20 Lego blocks together. Starch is a branched polymer. Add more Lego blocks to create branches every four or five blocks. To represent the similar structure of glycogen, place branches every two or three blocks apart.

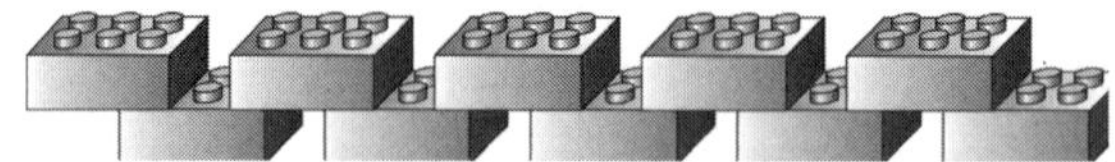

### RESULTS

Draw structural formulas of the reactants and products in condensation reactions involving the formation of:

proteins


ISBN 978 0 6557 0027 2

fats

carbohydrates

## DISCUSSION

**1** What are the major functional groups in an amino acid?

**2** What is meant by the peptide link in a protein?

**3** Which functional group causes a fatty acid molecule to be soluble in water?

**4** Which small molecule is produced during the condensation reaction between the glucose and fructose molecules to form sucrose?

## CONCLUSION

# EXAM QUESTIONS

## Multiple-choice questions

**Question 1** VCE Chemistry 2016 (A) 2

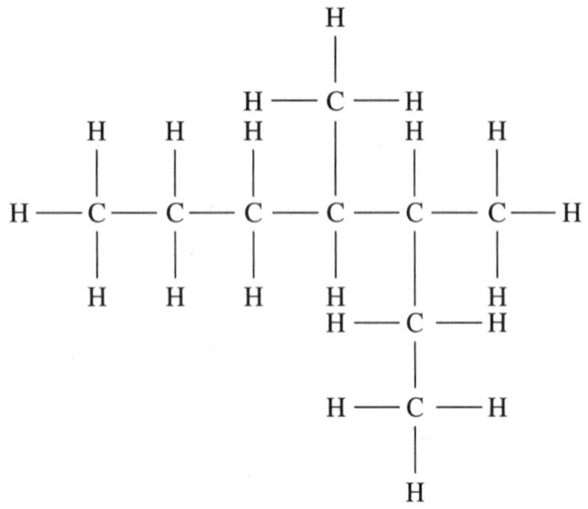

What is the correct systematic name for the compound shown above?

**A.** 4-methyl-5-ethylhexane

**B.** 2-ethyl-3-methylhexane

**C.** 4,5-dimethylheptane

**D.** 3,4-dimethylheptane

**Question 2** VCE Chemistry 2016 (A) 18

The molecule with the structural formula shown below reacts with hydrogen bromide, HBr, to form $C_5H_{11}Br$.

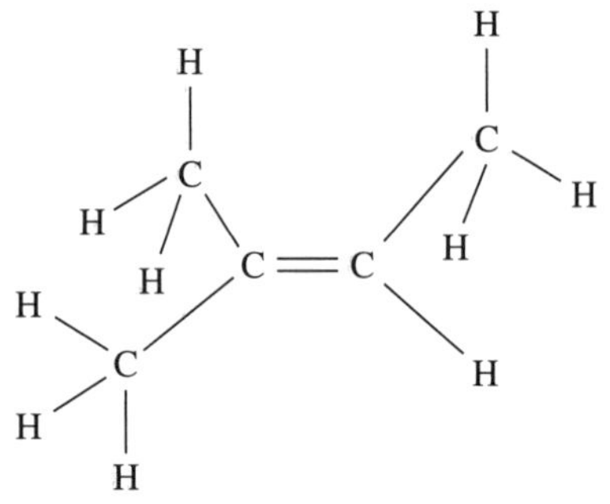

The number of different structural isomers theoretically possible to be produced by this reaction is

**A.** 1

**B.** 2

**C.** 3

**D.** 4

**Question 3** VCE Chemistry 2017 (A) 3

A hydrolytic reaction occurs when

**A.** a dipeptide is formed.

**B.** a triglyceride is formed.

**C.** water is a reaction product.

**D.** glucose is formed from maltose.


ISBN 978 0 6557 0027 2

# EXAM QUESTIONS

**Question 4** VCE Chemistry 2019 (A) 3

A compound has the following skeletal formula.

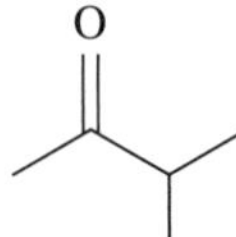


The molar mass of the compound is

**A.** 71 g $mol^{-1}$

**B.** 74 g $mol^{-1}$

**C.** 85 g $mol^{-1}$

**D.** 86 g $mol^{-1}$

**Question 5** VCE Chemistry 2018 (A) 21

A student wants to use a physical property to distinguish between two alcohols, octan-1-ol and propan-1-ol. Both alcohols are colourless liquids at standard laboratory conditions (SLC).

The student should use

**A.** density because propan-1-ol has a much higher density than octan-1-ol.

**B.** boiling point because octan-1-ol has a higher boiling point than propan-1-ol.

**C.** electrical conductivity because octan-1-ol has a higher conductivity than propan-1-ol.

**D.** spectroscopy because it is not possible to distinguish between the alcohols using their physical properties.

**Question 6** VCE Chemistry 2020 (A) 4

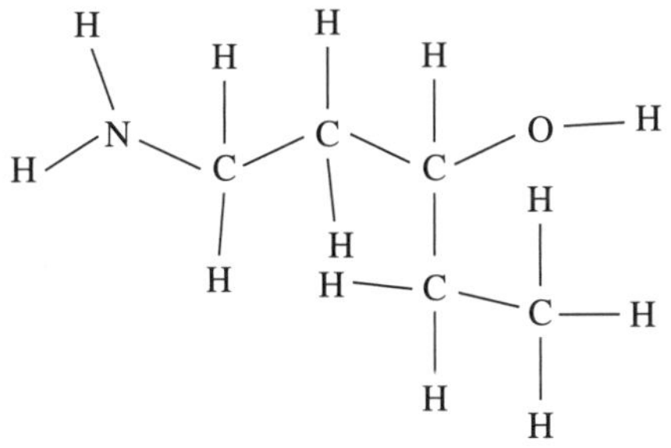

What is the IUPAC name of the molecule shown above?

**A.** 3-hydroxy-3-ethyl-propan-1-amine

**B.** 3-amino-1-methylpropan-1-ol

**C.** 3-hydroxypentan-1-amine

**D.** 1-aminopentan-3-ol

**Question 7** VCE Chemistry 2017 (A) 5

Which one of the following is a biofuel?

**A.** ethanol produced from crude oil

**B.** ethanol produced from cellulose

**C.** propane produced from natural gas

**D.** electricity produced by hydropower

# EXAM QUESTIONS

**Question 8** VCE Chemistry 2020 (A) 7

How many structural isomers have the molecular formula $C_3H_6BrCl$?

**A.** 4

**B.** 5

**C.** 6

**D.** 7

**Question 9** VCE Chemistry 2020 (A) 16

The following table provides information about three organic compounds, X, Y and Z.

| Compound | Structural formula | Molar mass (g mol$^{-1}$) | Boiling point (°C) |
|---|---|---|---|
| X | $CH_3CH_2CH_2OH$ (H–C(H)(H)–C(H)(H)–C(H)(H)–O–H) | 60 | 97 |
| Y | $CH_3COOH$ (H–C(H)(H)–C(=O)–O–H) | 60 | 118 |
| Z | $HCOOCH_3$ (H–C(H)(H)–O–C(=O)–H) | 60 | ? |

Which one of the following is the best estimate for the boiling point of compound Z?

**A.** 31°C

**B.** 101°C

**C.** 114°C

**D.** 156°C

**Question 10** VCE Chemistry 2019 (A) 16

The number of carbon-to-carbon double bonds (C=C) in a molecule can be identified by reacting the molecule with hydrogen gas, $H_2$. The type of reaction involved is an addition reaction.

10.0 g of a fatty acid becomes fully saturated after reacting with 0.263 g of $H_2$.

The fatty acid is

**A.** oleic acid ($M$ = 282 g mol$^{-1}$).

**B.** linolenic acid ($M$ = 278 g mol$^{-1}$).

**C.** arachidic acid ($M$ = 312 g mol$^{-1}$).

**D.** arachidonic acid ($M$ = 304 g mol$^{-1}$).


ISBN 978 0 6557 0027 2

# EXAM QUESTIONS

## Short-answer questions

**Question 1** (6 marks) VCE Chemistry 2015 (B) 5

**a.** A reaction pathway is designed for the synthesis of the compound that has the structural formula shown below.

```
              H
              |
          H — C — H
    H     H   |            O     H
    |     |   |            ||    |
H — C  —  C — C — O  —  C  — C — H
    |     |   |                  |
    H     H   H                  H
```

The table below gives a list of available organic reactants and reagents.

| Letter | Available organic reactants and reagents |
|---|---|
| A | acidified $KMnO_4$ |
| B | concentrated $H_2SO_4$ |
| C | $H_2O$ and $H_3PO_4$ |
| D | H–C(H)(H)–C(H)=C(H)–C(H)(H)–H |
| E | H₂C=CH₂ (H and H on each C) |
| F | H–C(H)(H)–C(H)(H)–C(H)(H)–C(H)(H)–O–H |
| G | H–C(H)(H)–C(H)(H)–O–H |

# EXAM QUESTIONS

Complete the reaction pathway design flow chart below. Write the corresponding letter for the structural formula of all organic reactants in each of the boxes provided. The corresponding letter for the formula of other necessary reagents should be shown in the boxes next to the arrows. 5 marks

**Reaction pathway design flow chart**

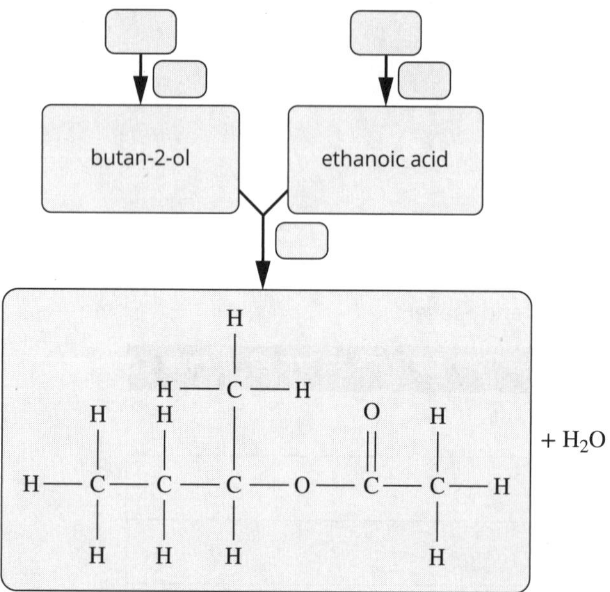


**b.** In the space below, draw the full structural formula of an isomer of butan-2-ol. 1 mark

 ISBN 978 0 6557 0027 2

# EXAM QUESTIONS

**Question 2** (8 marks) VCE Chemistry 2014 (B) 2

Compounds B and F may be synthesised as follows.

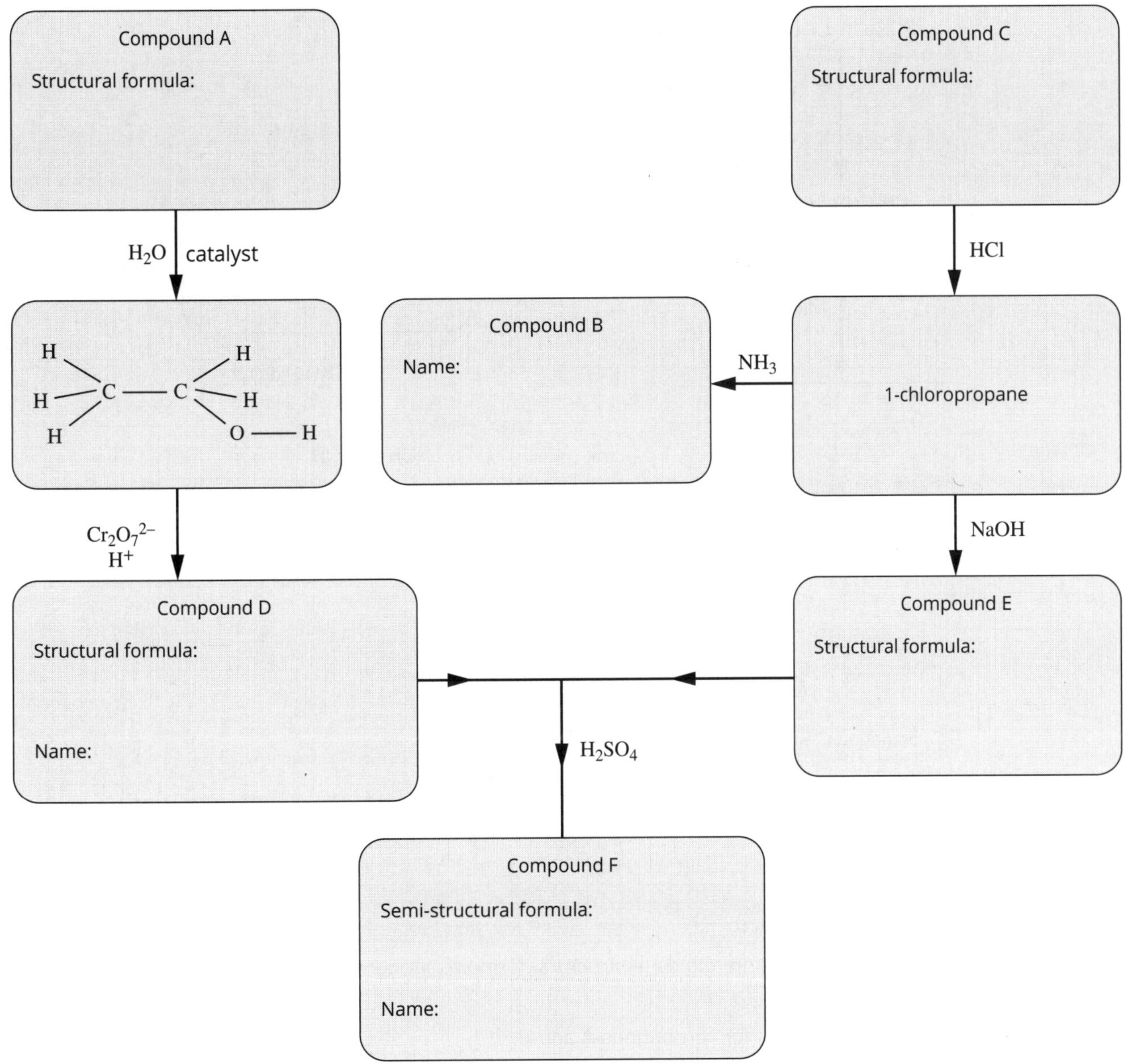


**a.** Draw the structural formulas of compounds A, C, D and E in the boxes provided. 4 marks

**b.** Write the systematic **names** of compounds B and D in the appropriate boxes. 2 marks

**c.** Insert the semi-structural formula and systematic name of compound F in the box provided. 2 marks

**Question 3** (7 marks) VCE Chemistry 2013 (B) 6

The reaction pathway below represents the synthesis of compound C.

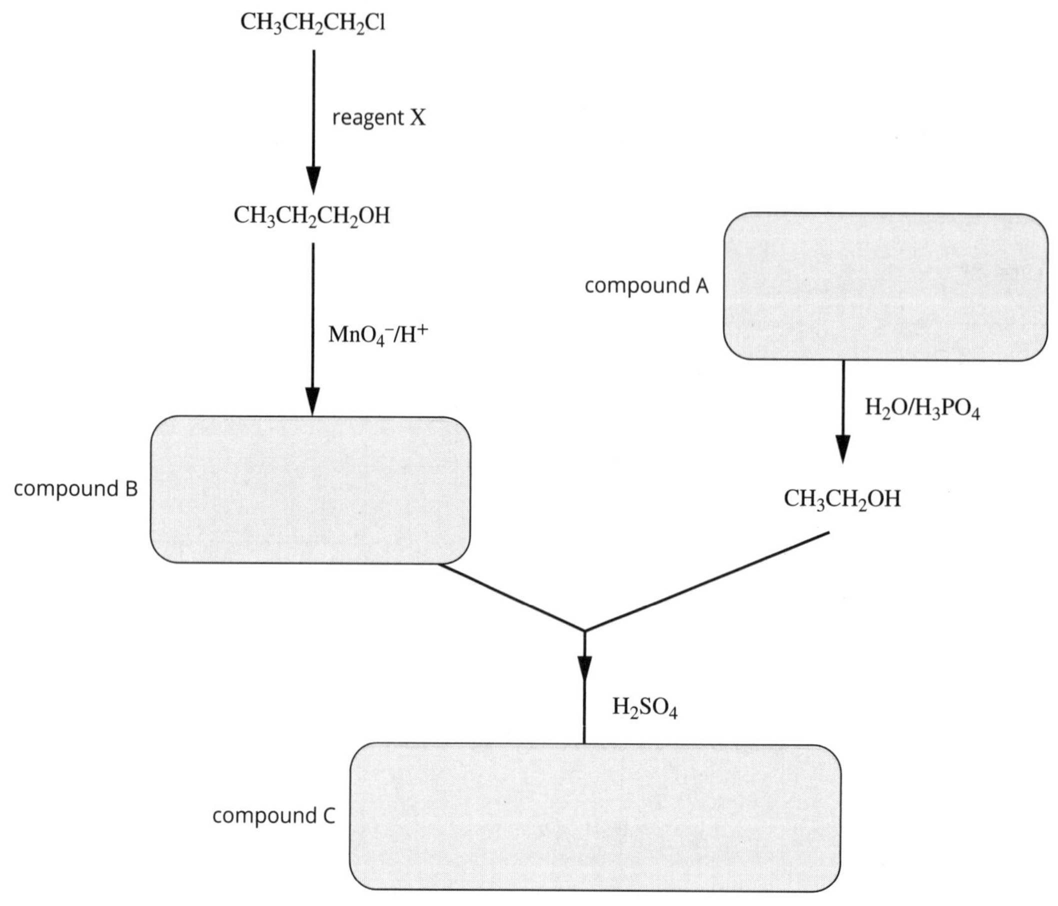


**a.** Identify reagent X. 1 mark

________________________________

**b.** In the appropriate boxes above, write the semi-structural formulas for compounds A, B and C. 3 marks

**c.** Give the systematic IUPAC names for compounds A and B. 2 marks

compound A ________________________________

compound B ________________________________

**d.** Sketch the energy profile for the complete combustion of compound C using the axis below, labelling the energy of the reactants, the products and the activation energy. 1 mark

Energy

 ISBN 978 0 6557 0027 2

# EXAM QUESTIONS

**Question 4** (8 marks) VCE Chemistry 2017 (B) 1

Industrially, ethanol, $C_2H_5OH$, is made by either of two methods.

One method uses ethene, $C_2H_4$, which is derived from crude oil. The other method uses a sugar, such as sucrose, $C_{12}H_{22}O_{11}$, and yeast, in aqueous solution. The production of $C_2H_5OH$ from $C_{12}H_{22}O_{11}$ and yeast proceeds according to the equation

$$C_{12}H_{22}O_{11}(aq) + H_2O(l) \rightarrow 4C_2H_5OH(aq) + 4CO_2(g)$$

**a.** Determine the mass, in grams, of pure $C_2H_5OH$ that would be produced from 1.250 kg of $C_{12}H_{22}O_{11}$ dissolved in water.

$M(C_{12}H_{22}O_{11}) = 342 \text{ g mol}^{-1}$ 2 marks

______________________________________________

______________________________________________

______________________________________________

______________________________________________

**b. i.** Complete the reaction by writing the formula for the reactant in the box provided below. 1 mark

$$C_2H_4(g) + \boxed{\phantom{XXXXX}} \xrightarrow{\text{catalyst}} C_2H_5OH(g)$$

**ii.** Classify this type of reaction. 1 mark

______________________________________________

**c.** $C_2H_5OH$ can be converted into ethanoic acid, $CH_3COOH$, in the presence of Reagent X. Write the formula for Reagent X in the box provided below. 1 mark

Reagent X

$$C_2H_5OH \longrightarrow CH_3COOH$$

**d.** $CH_3COOH$ can be used in the production of esters.

**i.** Write a balanced chemical equation for the reaction of $CH_3COOH$ with propan-1-ol using semi-structural formulas for all organic compounds. 2 marks

______________________________________________

**ii.** Write the IUPAC name for the ester product of the equation written in part **d.i.** 1 mark

______________________________________________

# EXAM QUESTIONS

**Question 5** (7 marks) VCE Chemistry 2020 (B) 3

Below is a reaction pathway beginning with hex-3-ene.

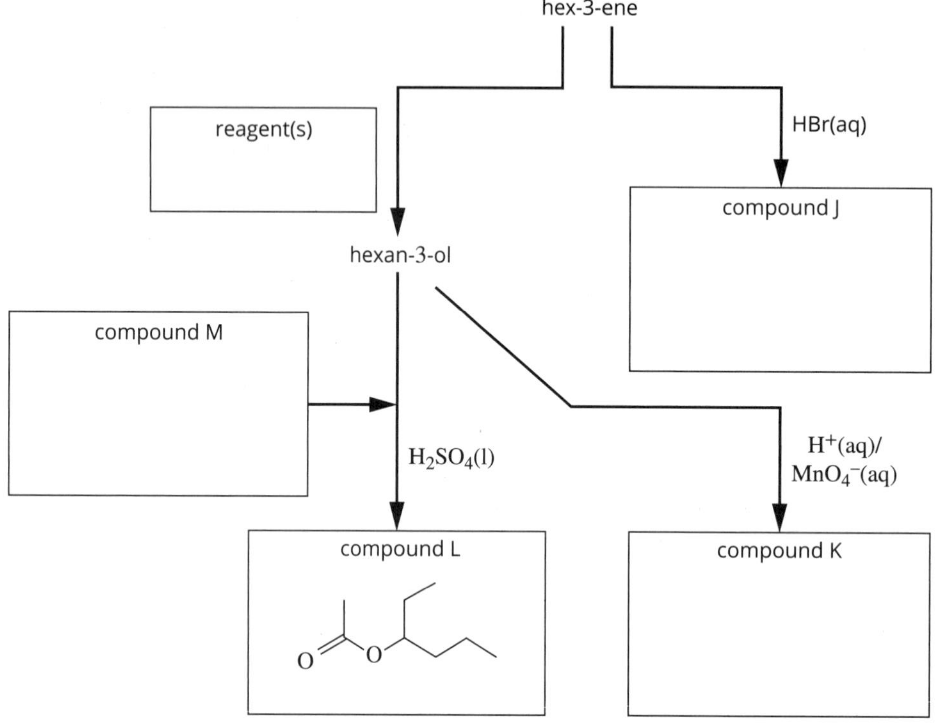


a. Write the IUPAC name of compound J in the box provided. 1 mark

b. State the reagent(s) required to convert hex-3-ene to hexan-3-ol in the box provided. 1 mark

c. Draw the structural formula for a tertiary alcohol that is an isomer of hexan-3-ol. 1 mark

d. Hexan-3-ol is reacted with compound M under acidic conditions to produce compound L.

Draw the semi-structural formula for compound M in the box provided. 1 mark

e. i. Draw the semi-structural formula for compound K in the box provided. 1 mark

ii. Name the class of organic compound (homologous series) to which compound K belongs. 1 mark

f. What type of reaction produces compound K from hexan-3-ol? 1 mark


ISBN 978 0 6557 0027 2

UNIT 4

# How are carbon-based compounds designed for purpose?

## AREA OF STUDY 2

## How are organic compounds analysed and used?

### Outcome 2

On completion of this unit the student should be able to apply qualitative and quantitative tests to analyse organic compounds and their structural characteristics, deduce structures of organic compounds using instrumental analysis data, explain how some medicines function, and experimentally analyse how some natural medicines can be extracted and purified.

### Key knowledge

#### Laboratory analysis of organic compounds

- qualitative tests for the presence of carbon–carbon double bonds, hydroxyl and carboxyl functional groups
- applications and principles of laboratory analysis techniques in verifying components and purity of consumer products, including melting point determination and distillation (simple and fractional)
- measurement of the degree of unsaturation of compounds using iodine
- volumetric analysis, including calculations of excess and limiting reactants using redox titrations (excluding back titrations)

#### Instrumental analysis of organic compounds

- applications of mass spectrometry (excluding features of instrumentation and operation) and interpretation of qualitative and quantitative data, including identification of molecular ion peak, determination of molecular mass and identification of simple fragments
- identification of bond types by qualitative infrared spectroscopy (IR) data analysis using characteristic absorption bands
- structural determination of organic compounds by low resolution carbon-13 nuclear magnetic resonance ($^{13}$C-NMR) spectral analysis, using chemical shift values to deduce the number and nature of different carbon environments
- structural determination of organic compounds by low and high resolution proton nuclear magnetic resonance ($^{1}$H-NMR) spectral analysis, using chemical shift values, integration curves (where the height is proportional to the area underneath a peak) and peak splitting patterns (excluding coupling constants), and application of the $n + 1$ rule (where $n$ is the number of neighbouring protons) to deduce the number and nature of different proton environments
- the principles of chromatography, including high performance liquid chromatography (HPLC) and the use of retention times and the construction of a calibration curve to determine the concentration of an organic compound in a solution (excluding features of instrumentation and operation)
- deduction of the structures of simple organic compounds using a combination of mass spectrometry (MS), infrared spectroscopy (IR), proton nuclear magnetic resonance ($^{1}$H-NMR) and carbon-13 nuclear magnetic resonance ($^{13}$C-NMR) (limited to data analysis)
- the roles and applications of laboratory and instrumental analysis, with reference to product purity and the identification of organic compounds or functional groups in isolation or within a mixture

**Medicinal chemistry**

- extraction and purification of natural plant compounds as possible active ingredients for medicines, using solvent extraction and distillation
- identification of the structure and functional groups of organic molecules that are medicines
- significance of isomers and the identification of chiral centres (carbon atom surrounded by four different groups) in the effectiveness of medicines
- enzymes as protein-based catalysts in living systems: primary, secondary, tertiary and quaternary structures and changes in enzyme function in terms of structure and bonding as a result of increased temperature (denaturation), decreased temperature (lowered activity), or changes in pH (formation of zwitterions and denaturation)
- medicines that function as competitive enzyme inhibitors: organic molecules that bind through lock-and-key mechanism to an active site preventing binding of the actual substrate

VCE Chemistry Study Design extracts © VCAA (2022); reproduced by permission.

# Laboratory analysis of organic compounds

This topic describes how qualitative and quantitative tests can be used in the laboratory to analyse organic compounds, and how relatively simple techniques such as melting point determination and volumetric analysis can be used to verify the components and purity of consumer products.

- **As revision, you will now be able to complete Worksheet 30.**

## ANALYSIS TECHNIQUES—DISTILLATION AND MELTING POINT DETERMINATION

Organic chemists create new compounds for many purposes, including for use as medicines, fuels, plastics and detergents. From the organic chemistry you have studied previously, you will not be surprised to learn that when organic products are formed, by-products and isomers are often present in the final mixture. It is important to be able to separate the desired product from other materials and to test its identity and purity.

### Distillation

One technique used by chemists is **distillation**, which separates two or more liquids with different boiling points. **Simple distillation** of a liquid mixture can be performed in the laboratory using the equipment shown in Figure 4.2.1.

The mixture is heated to the lower boiling point of the two liquids in the round-bottomed flask and the vapour rises up the column. It passes down the condensing tube, where it cools below the boiling point and is collected as a liquid (**distillate**) in the collecting vessel.

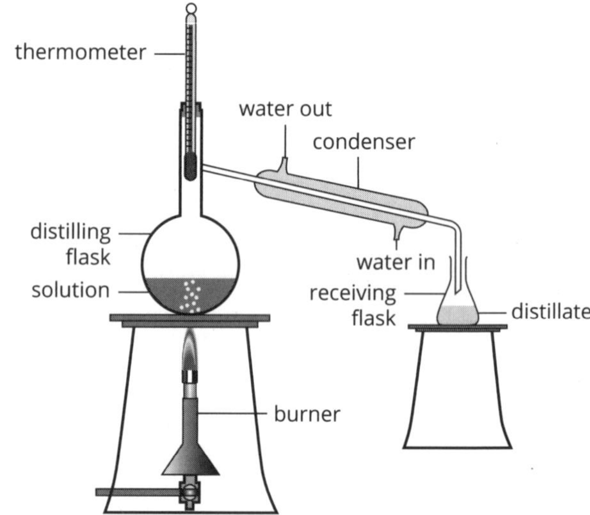


**Figure 4.2.1** Simple distillation equipment, which can be used to separate two or more liquids

When a mixture consists of different liquids, the technique of **fractional distillation** is employed (Figure 4.2.2 on the following page). Fractional distillation columns are packed with glass beads, on which the vapours can easily condense. The temperature is lower near the top and gases rise and condense, running back down the column.

 ISBN 978 0 6557 0027 2

## KEY KNOWLEDGE

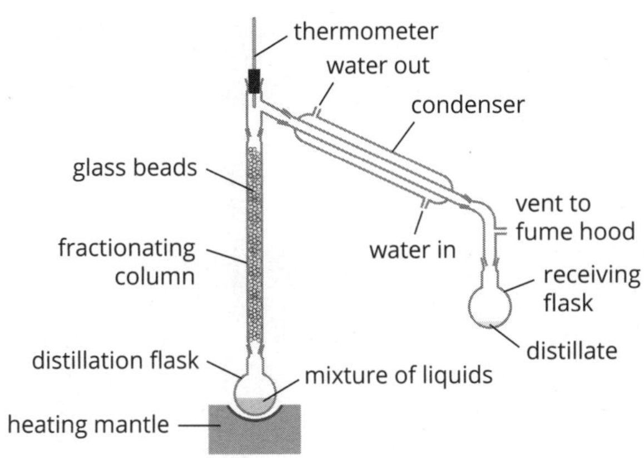


**Figure 4.2.2** Fractional distillation equipment, which is used in the laboratory

The gas that rises to the top of the fractionating column, where the temperature is close to the boiling point of the desired material, will pass into the condensing tube. It is collected as a liquid (distillate) in a receiving flask. In some instances, the distillate itself may not be pure and may be a mixture of several compounds that have similar boiling points.

As you learnt in Unit 3, crude oil is composed mainly of alkanes that are separated by fractional distillation using a large fractionating tower with collecting vessels at different levels of the tower. As the gases rise through the tower, the temperature decreases. The gases condense at different temperatures and are collected as fractions containing molecules with a similar number of carbon atoms. The compounds collected from the top of the tower are small alkanes with low boiling points, while long-chain alkanes remain at the bottom of the tower.

### Purity

**Melting point determination** provides information about the identity and purity of a sample. One type of apparatus consists of a thermometer, an eyepiece, a melting point tube, tube channels and a heating control knob, as seen in Figure 4.2.3. To measure a melting point (MP), a small quantity of the sample is placed in a thin glass capillary tube and the tube is inserted in the tube channel, where its temperature is raised. By watching the sample through the eyepiece, the temperatures at which the sample begins to melt and is completely molten can be measured.

The melting point of a compound depends on its structure and the intermolecular forces present. These forces vary depending on the type of compound. If a sample is pure, the forces between the molecules are equal and the sample will melt over a small range (e.g. 0.5–2°C). The MP of a sample can be compared to a known MP to identify a compound, and the MP range indicates its purity.

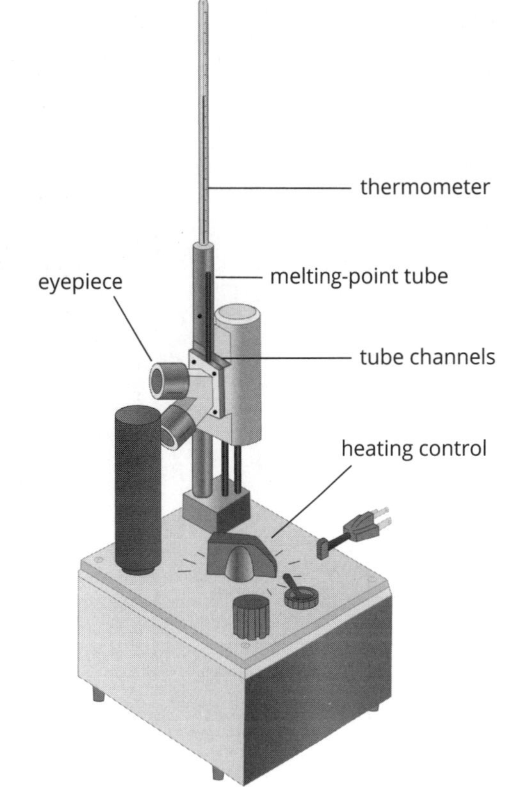


**Figure 4.2.3** Equipment to determine melting points

In summary, if:

- a compound is pure, it has a sharp MP
- there are two or more compounds in a sample, the intermolecular forces will vary. The MP will be lower than that of the pure compound and the MP range will be larger (e.g. more than 5°C)
- the MP of a sample is higher than that of a pure compound, the sample is not the same as the pure compound.

## QUALITATIVE TESTS FOR FUNCTIONAL GROUPS

Once the purity of an organic sample has been determined by melting point determination, chemical tests can enable us to determine the functional groups present in an organic compound and help find its structure. The tests for three important functional groups are listed in Table 4.2.1 on the following page.

## KEY KNOWLEDGE

**Table 4.2.1** Tests for functional groups

| Functional group | Test | Explanation |
|---|---|---|
| carbon–carbon double bond | $Br_2$ or $I_2$ test | 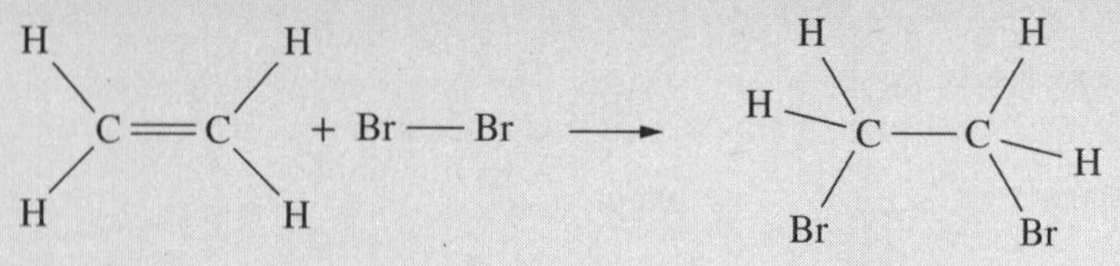 $H_2C{=}CH_2 + Br{-}Br \rightarrow H_2CBr{-}CBrH_2$ <br> $CH_2{=}CH_2 + Br_2 \longrightarrow BrCH_2CH_2Br$ <br> 1,2 -dibromoethane <br> Only C=C double bonds react instantly with the $Br_2$ molecule, causing a colour change from orange to colourless. ($I_2$ can be used instead of $Br_2$.) |
| hydroxyl group | • Na metal test <br> • ester test <br> • test with acidified potassium dichromate or potassium permanganate (for primary and secondary alcohols) | • Produces $H_2$ gas with Na metal: e.g. $2CH_3OH + 2Na \rightarrow 2CH_3ONa + H_2$ <br> • Produces an ester when reacted with a carboxylic acid. Esters are sweet smelling ('fruity'). For example, ethanol reacts with ethanonic acid as shown: 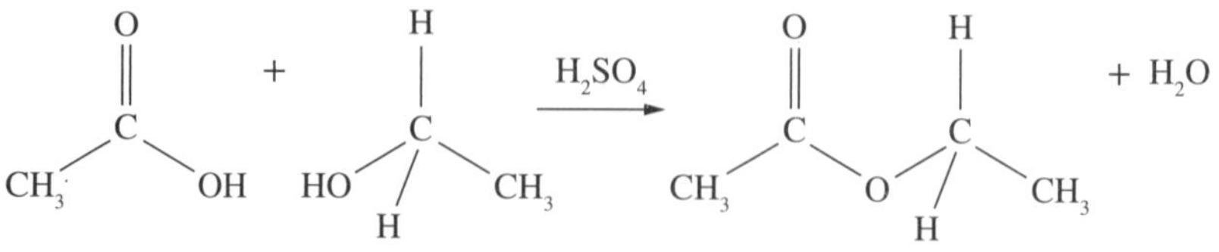 <br> ethanoic acid + ethanol $\xrightarrow{H_2SO_4}$ ethyl ethanoate + water <br> $CH_3COOH(l) + CH_3CH_2OH(l) \xrightarrow{\text{conc. } H_2SO_4} CH_3COOCH_2CH_3(l) + H_2O(l)$ <br> • If a reaction occurs with acidified $K_2Cr_2O_7$, the orange $Cr_2O_7^{2-}$ solution turns green as $Cr^{3+}$ ions are formed (or a purple $MnO_4^-$ solution turns colourless as $Mn^{2+}$ ions are formed). |
| carboxyl group | • acidity test <br> • Na metal test <br> • ester test <br> • test with carbonate produces $CO_2$ gas | • Turns litmus red <br> • Produces $H_2$ gas with Na metal: e.g. $2CH_3COOH + 2Na \rightarrow 2CH_3COONa + H_2$ <br> • Produces an ester when reacted with an alcohol (see hydroxyl group test above). Esters smell fruity. <br> • Produces $CO_2$ gas with carbonates, e.g. <br> $2CH_3COOH(aq) + Na_2CO_3(aq) \rightarrow 2CH_3COONa(aq) + CO_2(g) + H_2O(l)$ <br> Limewater test for $CO_2$: turns limewater, $Ca(OH)_2$, milky <br> $CO_2(g) + Ca(OH)_2(aq) \rightarrow CaCO_3(s) + H_2O(l)$ |

- **You will now be able to revise Practical activity 15 (in Unit 4 Area of Study 1).**

## DEGREE OF UNSATURATION USING IODINE NUMBER

Earlier in Unit 4, you learnt a formula to calculate the degree of unsaturation of a pure compound. Another way chemists can measure the degree of unsaturation of a substance is by determining its **iodine number**.

The iodine number is usually used to measure the degree of unsaturation of fats and oils, which are composed of mixtures of different molecules. It is defined as the mass of iodine in grams that will react with 100 g of the substance, as shown in the unbalanced equation in Figure 4.2.4.

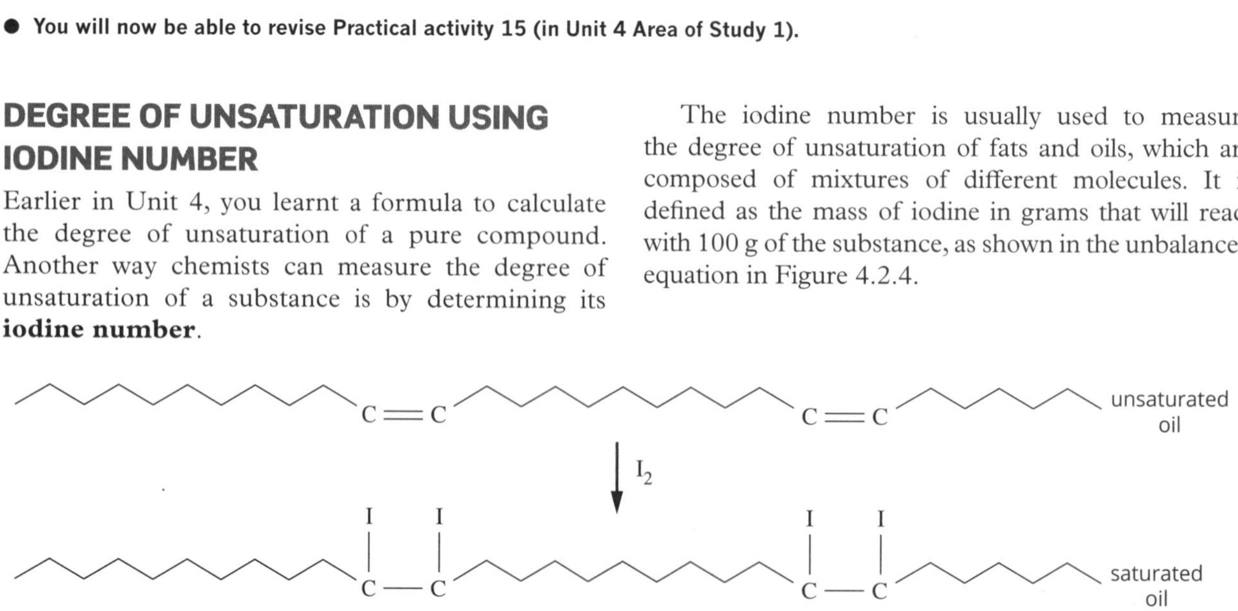

unsaturated oil (C=C ... C=C) $\xrightarrow{I_2}$ saturated oil (CI–CI ... CI–CI)

**Figure 4.2.4** An equation showing the reaction between an unsaturated oil with iodine to produce a saturated oil

 ISBN 978 0 6557 0027 2

The higher the iodine number, the more unsaturated the fat or oil, and the more reactive and less stable the oil is. A fat or oil with a high iodine number is likely to be oxidised and become rancid more quickly. Table 4.2.2 gives the iodine numbers of some common fats and oils. The data allows chemists to determine possible uses for the substances, and indicates the most suitable fats or oils to consume in the diet. Although controversial, research suggests that consuming unsaturated fats may have health benefits, including lowering the risk for cardiovascular disease and early death.

**Table 4.2.2** The iodine number ranges of common fats and oils

| Fat or oil | Iodine number (g per 100 g oil) |
|---|---|
| beeswax | 7–16 |
| butter | 25–42 |
| coconut oil | 6–11 |
| fish oil | 190–205 |
| linseed oil | 170–204 |
| olive oil | 75–94 |
| peanut oil | 82–107 |
| sesame oil | 100–120 |

Coconut oil is highly saturated, making it excellent for making soap (since saturated oils tend to be solids at room temperature), but you may be wise to avoid eating too much of it. Linseed oil is highly unsaturated, which means it would be a good drying oil (an oil that thickens and hardens when exposed to air), and is excellent for use as a base for paint.

- **You will now be able to conduct Practical activity 19.**

## VOLUMETRIC ANALYSIS

**Volumetric analysis** enables the concentration of a substance being analysed to be calculated. Acids and bases or substances that can act as oxidising agents or reducing agents can be analysed by volumetric titrations. You are specifically required to learn about redox titrations, which are similar to the acid–base titrations that you learnt about in Year 11.

### Terminology

Some of the terms used in volumetric analysis are listed in Table 4.2.3.

**Table 4.2.3** Terms used in redox volumetric analysis

| Term | Description |
|---|---|
| primary redox standard | A substance so pure that the amount can be accurately calculated from its mass. For a chemical to be a primary standard it should be easily obtained, have a known formula and be easily stored without reacting with water or gases in the atmosphere; for example, $K_2Cr_2O_7$. |
| standard solution | A solution with an accurately known concentration. Figure 4.2.5 shows a standard solution being made from a primary standard. |
| burette | A calibrated glass tube that delivers variable volumes accurately |
| titre | The volume delivered by a burette |
| pipette | A calibrated glass tube that delivers a fixed volume accurately |
| aliquot | The volume delivered by a pipette |
| end point | The point at which the indicator changes colour. Redox titrations often do not require an indicator because one of the reactants usually changes colour at the equivalence point. |
| equivalence point | The point in a titration at which the reactants have been mixed in stoichiometric proportions |

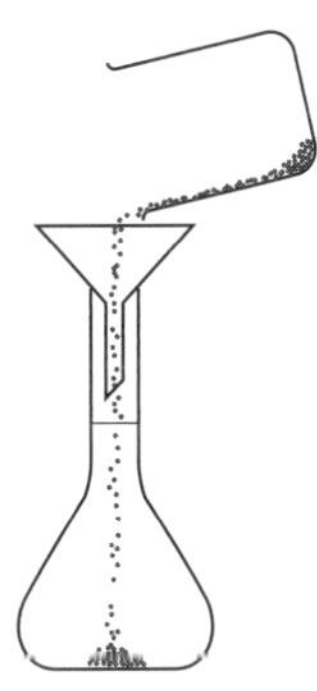


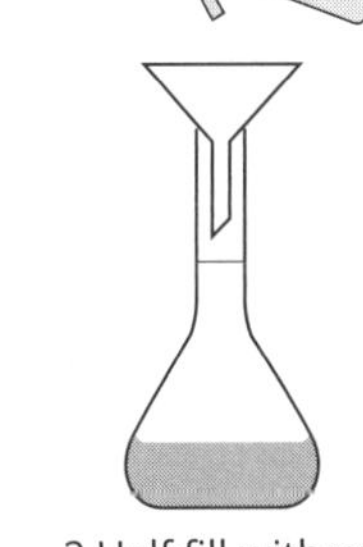

**Figure 4.2.5** The steps in making a standard solution from a primary standard

### Redox titrations

In a simple redox titration (Figure 4.2.6 on the following page), one solution (the titrant) is dispensed from a burette into a conical flask that contains a known volume of the solution being analysed. The volume of the solution in the flask is precisely measured by pipette. The conjugate redox pairs of at least one of the reactants usually have different colours, so the **equivalence point** can be detected by the colour change that occurs and an indicator is not required.

## KEY KNOWLEDGE

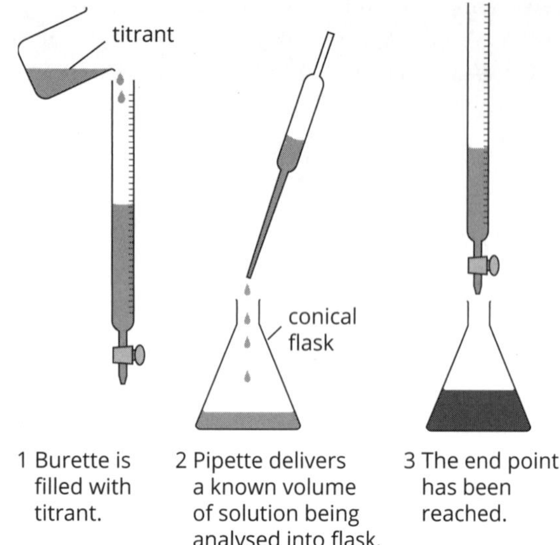


**Figure 4.2.6** The steps in a direct redox titration

The concentration of the solution may then be calculated from the concentration and aliquot of the titrant and the titre of the solution under analysis by stoichiometry.

### Excess and limiting reactants

In some chemical reactions the reactants are not mixed together in the ratio given by the equation. One reactant is therefore not completely consumed; this reactant is called the **excess reactant**. The reactant that is completely consumed is called the limiting reactant.

When performing a stoichiometric calculation for such reactions, it is necessary to determine the limiting reactant. This follows the same steps described in the Key knowledge in Unit 3, Area of Study 1. The steps involved in the calculation are as follows.

1 Write a balanced equation.
2 Determine the number of moles of each reactant.
3 Determine the number of moles of the limiting 'known reactant'.
4 Determine the mole ratio from the balanced chemical equation and use it to calculate the amount, in mol, of the 'unknown substance':

$$\frac{n(\text{unknown})}{n(\text{known})} = \frac{\text{coefficient of unknown substance}}{\text{coefficient of known substance}}$$

$$n(\text{unknown}) = n(\text{known}) \times (\text{mole ratio})$$

These steps are shown in the following worked example.

**Example:** 0.10 g of liquid cyclohexene and 20.0 mL of 0.10 M bromine solution are mixed in a test for unsaturation. Calculate which reactant is in excess and by how much.

| Thinking | Working | |
|---|---|---|
| 1 Write a balanced equation. | $C_6H_{10} + Br_2 \rightarrow C_6H_{10}Br_2$ | |
| 2 Determine the number of moles of each reactant. | $n(C_6H_{10})$<br>$n = \frac{m}{M}$<br>$= \frac{0.10}{82.14}$<br>$= 0.0012$ mol | $n(Br_2)$<br>$n = c \times V$<br>$= 0.10 \times 0.0200$<br>$= 0.0020$ mol |
| 3 Compare the values for the moles of reactants to determine which is in excess and which is the limiting reactant. | The equation shows that 1 mol of $C_6H_{10}$ reacts with 1 mol of $Br_2$.<br>0.0012 mol of $Br_2$ is needed for all of the $C_6H_{10}$ to react.<br>0.0020 mol of $Br_2$ is present, so $Br_2$ is in excess.<br>$C_6H_{10}$ is the limiting reactant. | |
| 4 Use the mole ratio from the balanced chemical equation to calculate the amount, in mol, of the unknown substance, i.e. the amount of excess reactant that reacted:<br>$\frac{n\text{ (unknown substance)}}{n\text{ (known substance)}} = \frac{\text{coefficient (unknown substance)}}{\text{coefficient (known substance)}}$ | $\frac{n(Br_2)}{n(C_6H_{10})} = \frac{1}{1}$<br>$n(Br_2) = n(C_6H_{10})$<br>$= 0.0012$ | |
| 5 Calculate the quantity required for the 'unknown substance', i.e. amount in excess. | $n(Br_2)$ in excess = $n(Br_2)$ initially – $n(Br_2)$ reacted<br>= 0.0020 – 0.0012 = 0.0008 mol<br>$n(Br_2)$ is in excess by 0.0008 mol | |

- **You will now be able to complete Worksheet 31 and conduct Practical activity 20.**


ISBN 978 0 6557 0027 2

# Instrumental analysis of organic compounds

Modern instrumental techniques for the analysis of chemical systems have largely replaced more traditional methods such as volumetric analysis. In this topic, you will study how these techniques can be used to deduce structures of organic compounds and ensure the quality of consumer items.

## SPECTROSCOPIC TECHNIQUES

Atoms and molecules absorb energies of various wavelengths. By irradiating a sample with energy of different wavelengths, an absorption spectrum is produced.

The absorbed radiation may cause:

- bonds to stretch or bend more vigorously
- electrons to jump to higher energy levels
- nuclei with magnetic properties to flip when in a magnetic field.

The type of instrumental analysis used depends on the solution being analysed and the information required.

The electromagnetic spectrum covers a continuous range of energy and wavelengths from radio waves at the low-energy (high wavelength) end to gamma (γ) rays at the high-energy (low wavelength) end. Energy from different parts of the spectrum can be used for different spectroscopic techniques, for both qualitative and quantitative analysis (Table 4.2.4).

**Table 4.2.4** Spectroscopic techniques

| Mass spectrometry (MS) | |
|---|---|
| **Description** | The interaction of a gaseous sample with a high-energy electron beam forms positive ions.<br>These cations are accelerated by an electric field and then deflected by a magnetic field. |
| **Basis of technique** | The electron beam removes electrons, creating cations.<br>As a result of the presence of electric and magnetic fields, the cations move in curved paths depending on their mass–charge (*m/z*) ratio.<br>**Fragmentation** patterns: the cation formed when a molecule loses an electron can be unstable. Some of them can break up into smaller fragments, where one fragment is a cation and the other is an uncharged free radical (atom or group of atoms with unpaired electrons).<br>The tallest (most intense) peak is called the **base peak** and is due to the most abundant fragment ion. It is given a relative intensity of 100% and all other peaks are measured against it. |
| **Major uses** | Used to:<br>• determine molecular mass (found from the mass of the **molecular** or **parent molecular ion**, which is the ion produced when an electron is lost from a molecule)<br>• identify the isotopes of an element<br>• assist in identifying structures of molecules |
| **Interpretation of data** | The peak with the highest *m/z* ratio is usually due to the molecular (or parent) ion (see spectrum below). Only positively charged ions will produce peaks in the mass spectrum.<br>Characteristic fragmentation patterns result from the break-up of the unstable molecular ion and are useful for determining the structure of molecules.<br>The mass spectrum of a compound can be complex, because of:<br>• different isotopes<br>• fragmentation of the molecule.<br>For example, the mass spectrum of ethanol is shown below. The highest value peak is *m/z* 46. This is the parent ion, $CH_3CH_2OH^+$, and its *m/z* value is equal to the molar mass of ethanol. The other peaks result from atoms or groups of atoms breaking apart from the ethanol molecule ion, e.g. *m/z* 15 is due to a $CH_3^+$ fragment, and *m/z* 31 is due to $CH_2OH^+$. 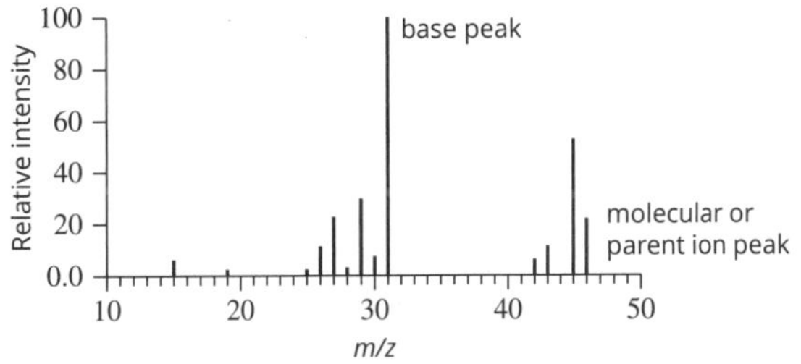<br> |

*(Continued)*

ISBN 978 0 6557 0027 2

# KEY KNOWLEDGE

| Table 4.2.4 Spectroscopic techniques | |
|---|---|
| **Infrared (IR) spectroscopy** | |
| **Description** | Polar bonds in molecules interact with IR light.<br>The frequency absorbed depends on the nature of the bond:<br>• single bonds absorb lower energy than a double bond between the same atoms<br>• single bonds between heavy atoms absorb lower energy than those between lighter atoms. |
| **Basis of technique** | IR light causes bonds in molecules to stretch (change distance between atoms) and bend (change angle between bonds). 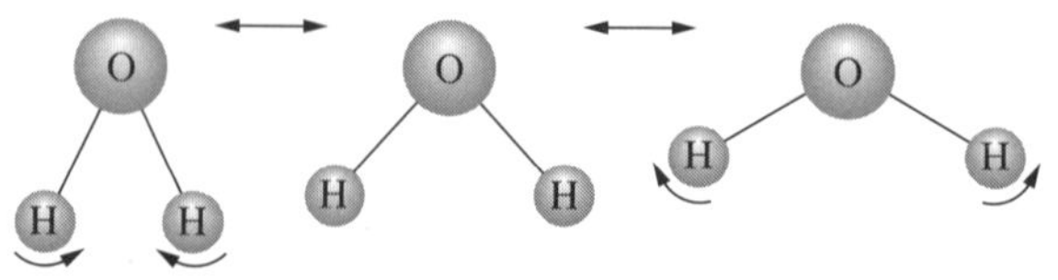<br>O H H O H H O H H<br>Bending in water molecules |
| **Major uses** | Used to identify types of bonds and functional groups present, e.g.<br>• C–O absorbs at 1050–1410 $cm^{-1}$<br>• O–H absorbs at 3200–3650 $cm^{-1}$<br>• C=O absorbs at 1670–1750 $cm^{-1}$. |
| **Interpretation of data** | The IR spectrum of gaseous ethanol shows the frequencies absorbed by the molecule. These are compared to those listed in databases to identify functional groups and multiple bonds. 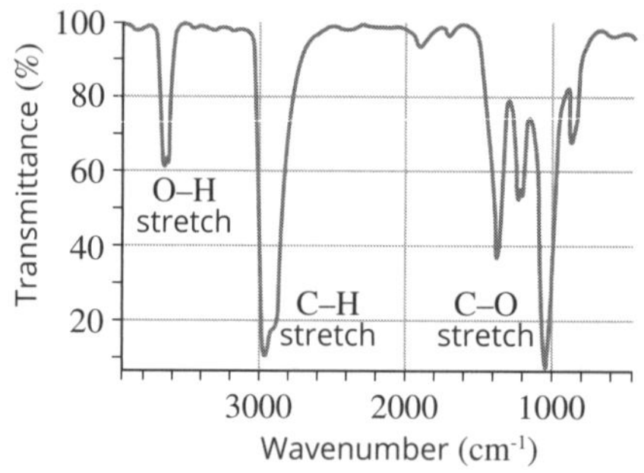<br>Transmittance (%) 100 80 60 40 20<br>O–H stretch C–H stretch C–O stretch<br>3000 2000 1000<br>Wavenumber ($cm^{-1}$)<br>Note that in IR spectra, by convention, frequency is plotted on the horizontal axis and is measured in wavenumbers (unit of $cm^{-1}$), whereas transmittance is plotted on the vertical axis (so peaks are inverted). |
| **Nuclear magnetic resonance (NMR) spectroscopy** | |
| **Description** | The nuclei of atoms in a strong magnetic field interact with radio wave energy. |
| **Basis of technique** | Energy is absorbed by nuclei with an odd number of protons or neutrons such as $^{13}C$ and $^{1}H$. When this occurs, the nuclei are said to be in resonance.<br>Atoms in the same chemical environment are **equivalent**. Equivalent atoms have the same **chemical shift** and form one signal.<br>Atoms in different chemical environments absorb different amounts of energy. The chemical shift (δ) of a signal in a spectrum indicates how far the signal is from the signal of a reference compound and is measured in ppm.<br>Tetramethylsilane (TMS) is widely used as the reference (0 ppm). |
| **Major uses** | Used to provide information about the C–H backbone of organic compounds. |

*(Continued)*


ISBN 978 0 6557 0027 2

# KEY KNOWLEDGE

**Table 4.2.4** Spectroscopic techniques

| | |
|---|---|
| **Interpretation of data** | For a $^{1}H$ or $^{13}C$ NMR spectrum, the:<br>• number of signals indicates the number of different H or C chemical environments<br>• chemical shift for each signal allows the type of chemical environments of each H or C atom to be identified.<br>The area under each peak in a $^{1}H$ spectrum is proportional to the number of different H atoms in that chemical environment.<br>In a $^{1}H$ NMR spectrum, signals can show **splitting patterns** that can provide further information about the molecule's structure. The number of signals in the splitting pattern is given by $n + 1$, where $n$ is the number of non-equivalent hydrogen atoms on adjacent carbon atoms.<br>For example, the spectrum of $CH_3CHCl_2$ has two major signals due to the methyl protons and the proton in the –CH group. As a result of splitting, the:<br>• $–CH_3$ proton signal is a doublet<br>• –CH proton signal is a quartet. 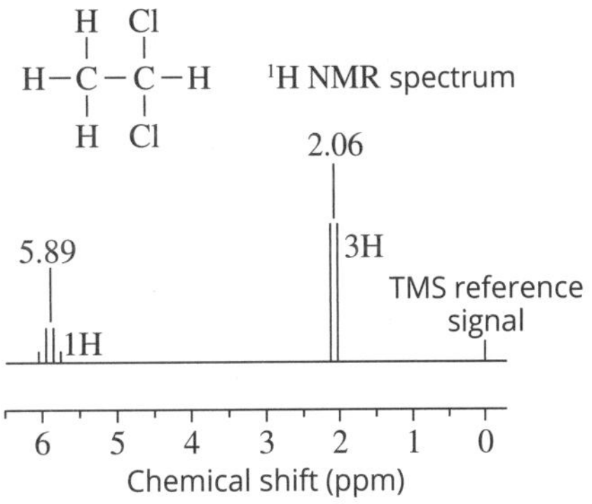<br>$^{1}H$ NMR spectrum<br>2.06<br>5.89<br>3H<br>1H<br>TMS reference signal<br>6 5 4 3 2 1 0<br>Chemical shift (ppm) |

● **You will now be able to complete Worksheets 32 and 33.**

## HIGH-PERFORMANCE LIQUID CHROMATOGRAPHY

**High-performance liquid chromatography (HPLC)** can be regarded as an advanced version of the simple techniques of chromatography that you studied in Unit 2—paper chromatography and thin-layer chromatography. HPLC can be used for both qualitative and quantitative analysis of numerous organic substances; for example, contaminants in water and drugs present in blood. HPLC is a method that involves the separation of the components of a mixture, depending on their polar or non-polar nature (Table 4.2.5 on the following page).

# KEY KNOWLEDGE

**Table 4.2.5** Analysis using HPLC

| Essential features of the instrument | Process and sensitivity | Use |
|---|---|---|
| High-performance liquid chromatography (HPLC)<br>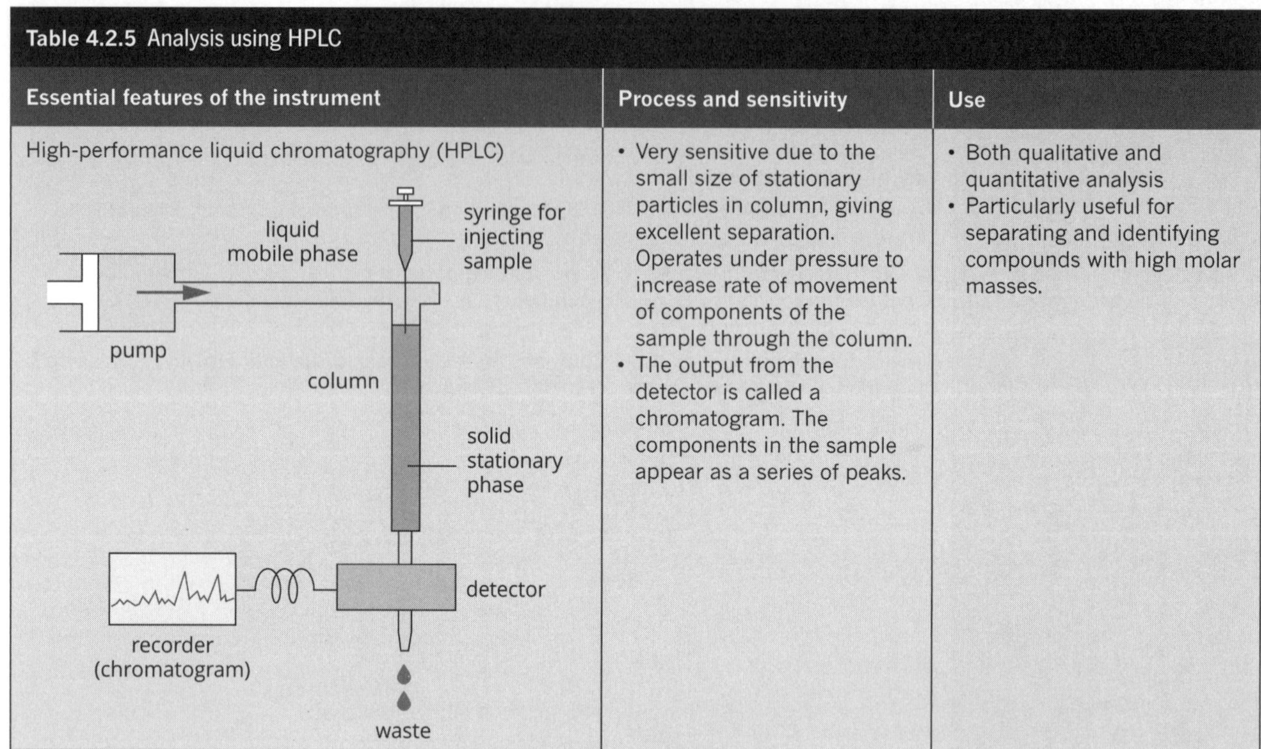<br> | • Very sensitive due to the small size of stationary particles in column, giving excellent separation. Operates under pressure to increase rate of movement of components of the sample through the column.<br>• The output from the detector is called a chromatogram. The components in the sample appear as a series of peaks. | • Both qualitative and quantitative analysis<br>• Particularly useful for separating and identifying compounds with high molar masses. |

## Principles of HPLC

To separate the components in a mixture by HPLC, a sample of the mixture is dissolved in a solvent. This solution is called the **mobile phase** and it is passed through a column containing a solid material, called the **stationary phase**. Separation occurs as a result of the different abilities of the components to adsorb (adhere) onto the stationary phase and **desorb** (dissolve back) into the mobile phase. The more strongly a component bonds to the stationary phase, the slower the movement of the component over the stationary phase.

## Qualitative analysis

In Year 11 you learnt that components in a mixture can be identified by paper or thin-layer chromatography by measuring their individual $R_f$ (retardation factor) values and comparing them to values for known compounds. The $R_f$ value is defined as the distance the component moved from the origin (where the sample was initially placed) divided by the distance the solvent moved from the origin.

In the case of HPLC, the time taken for each component to pass through the HPLC column is called the component's **retention time, $R_t$**. The identity of the components can be found by comparing the $R_t$ values to values for known compounds.

The values of $R_t$ are characteristic for the component for the conditions under which the chromatogram was obtained. Compounds that adsorb strongly to the stationary phase move slowly and have high $R_t$ values.

## Quantitative analysis using HPLC

The concentration of a component in a sample can be determined by HPLC by comparing the area under its peak in the chromatogram with that of standard solutions. Calibration involves four main steps.

1. Standard solutions of the component are prepared and a fixed volume of each is injected into the instrument. The area under the peak for each solution is measured.
2. A calibration curve is plotted using the results obtained for the standard solutions (Figure 4.2.7).
3. The sample is injected and its peak area is measured.
4. The sample's concentration is determined using the calibration curve.

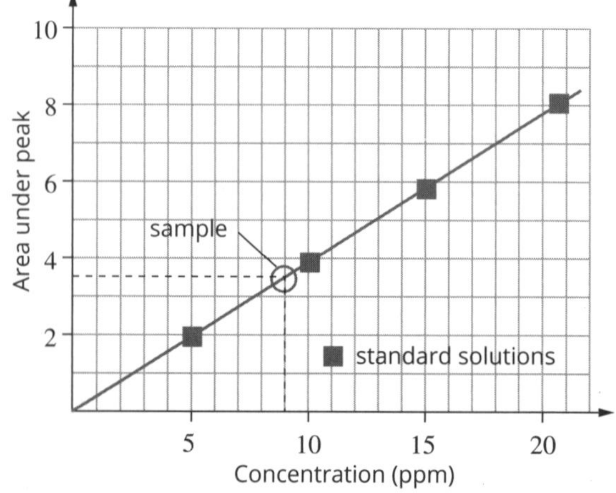


**Figure 4.2.7** A calibration curve used in quantitative analysis by HPLC

● **You will now be able to conduct Practical activity 21.**

 ISBN 978 0 6557 0027 2

# KEY KNOWLEDGE

## STRUCTURE DETERMINATION BY SPECTROSCOPY

The structure of organic molecules can be determined with the aid of all of the types of spectroscopy described above. The steps involved are shown in Figure 4.2.8.

**Find molecular formula**

- Use percentages of each element present in a compound obtained from elemental analysis to determine the empirical formula.
- The molar mass of the molecule can be determined from the mass spectrum.
- Use the relative mass of the empirical formula to calculate the molecular formula.

↓

**Find functional groups**

- Use IR spectroscopy to determine the functional groups present in the molecule.
- Results from simple chemical tests (e.g. reaction with $Br_2$) may also be helpful.

↓

**Find the structural formula**

- Use $^{13}C$ and $^{1}H$ NMR spectra to find the position of the functional groups and the backbone of the molecule.

**Figure 4.2.8** Finding a structural formula by using spectroscopic techniques

## COMPOUND IDENTIFICATION BY USING LABORATORY AND INSTRUMENTAL ANALYSIS

The laboratory and instrumental techniques described above allow chemists to identify organic compounds and test their purity.

HPLC is often combined with MS and the various spectroscopic methods to separate components of mixtures and to identify and quantify these components. Databases of standards allow components to be rapidly identified. If a substance is impure, there will be additional peaks in its chromatogram or spectrum, indicating the presence of other compounds.

The selection of which of the various laboratory and instrumental analyses to use depends on the solution to be investigated and the resources available. For example, IR will help determine the functional groups present in a sample and NMR provides information about the C–C or C–H backbone of the compound. An organic chemist will usually use a combination of different techniques.

- **You will now be able to complete Worksheet 34.**

# Medicinal chemistry

Medicinal chemistry is a field of study in which synthetic organic chemistry overlaps with molecular biology in the areas of drug development. A drug is a substance that affects how the body functions by changing particular chemical processes in the body (Table 4.2.6).

**Table 4.2.6** Some examples of drugs and their uses

| Types of drugs | Uses |
|---|---|
| painkillers, antibiotics, sedatives | medical purposes |
| caffeine, peppermint, vanilla | in food and drinks |
| nicotine and alcohol | legal (but harmful) recreational use |
| heroin and hallucinogens | illegal recreational use |

The use of herbal medicine has its origins in ancient cultures, including those of indigenous peoples. Materials from plants have long been used as remedies to prevent or cure infections. Very often, the active ingredients in modern drugs are extracted from the natural materials in plants. Two examples are listed in Table 4.2.7.

**Table 4.2.7** Some medicines and their traditional uses

| Modern medication | History of use |
|---|---|
| aspirin | In 400 BCE, the Greeks recommended an infusion of willow leaves and bark to relieve fever, pain and inflammation. The active ingredient, salicin, in willow bark was isolated in 1829. It converts to salicyclic acid in the body. |
| terpinen-4-ol | The Bundjalung people of eastern Australia have used tea tree oil in a variety of ways, including inhaling to treat coughs and applying to the body to maintain healthy skin. One of the components, terpinen-4-ol, has been isolated and found to have antibacterial properties. It is used to stop infection and to treat bacterial and fungal nail infections. |

# KEY KNOWLEDGE

## EXTRACTION AND PURIFICATION

Separation methods, such as **solvent extraction** and distillation, are often used in association with the combined techniques of high-performance liquid chromatography and mass spectrometry (HPLC-MS) in the design and production of new medicines from natural materials.

Table 4.2.8 describes how these processes are used for processing plant material.

### Purification techniques

After chemicals are extracted from plant material, the sample can be injected into an HPLC instrument. HPLC can separate even a complex mixture into its individual components, including the required active ingredient. The components can then be directly fed into a mass spectrometer or NMR instrument to identify them. These combined techniques are essential in drug design and are used in many other types of analyses, including the drug testing of athletes. Use of these techniques accelerates the purification and identification of biologically active molecules.

## IDENTIFICATION OF FUNCTIONAL GROUPS AND MOLECULAR STRUCTURE

The functional groups in organic molecules determine the properties and chemistry of the compounds. They are the reactive sites in the molecule. If we know which functional groups are present, and where these sites are located in the molecular structure, then we can predict how organic molecules and medicines will react. The laboratory tests and spectroscopic techniques discussed earlier are used to identify the functional groups and structures of the organic molecules that are medicines.

The major functional groups in organic chemistry, and in medicines in particular, are hydroxyl, carbonyl, carboxyl, amino, phosphate, and groups containing sulfur. These groups are important in biological molecules such as proteins, lipids, carbohydrates and DNA, and therefore in the medicines that interact with them.

The chemical structure of a medicine determines both its physical and chemical properties, as well as how it will be absorbed and processed by the body.

Although you aren't required to be familiar with it for Unit 4, X-ray crystallography is the ultimate technique for determining the molecular structure of a compound because it allows us to know the position of the atoms in the compound precisely. High-powered X-rays are aimed at a crystal of the compound, causing the X-rays to diffract in many different directions. The intensity and angles of the scattered rays can be analysed to generate a three-dimensional picture of the electron density in the crystal, hence showing the position, identity and bonding of the atoms.

- **You will now be able to complete Worksheet 35.**

**Table 4.2.8** Extraction and isolation techniques

| Technique | Description of process |
|---|---|
| solvent extraction | • A solvent is used to dissolve the required substance in a plant material.<br>• The solution is filtered from the undissolved plant material.<br>• The solute and solvent are separated by distillation and the solute is concentrated.<br>• Methanol is considered an effective extraction solvent for many situations, giving high extraction yields. Ideally, the solvent should have low polarity, high volatility (evaporate easily) and low toxicity. |
| distillation | • Separates the components extracted from the plant material, using differences in their boiling points.<br>• Steam distillation uses steam to break up the cells in plants and extract the volatile oil components, which readily mix with the steam.<br>– The oils and steam condense and separate due to the non-polar nature of the oils.<br>– This technique is suitable when the required components are volatile, stable and non-polar, allowing easy separation.<br>– Steam distillation is commonly used to produce aromatic oils, e.g. tea tree oil. |


ISBN 978 0 6557 0027 2

# KEY KNOWLEDGE

## STEREOISOMERS

Different isomers of a compound can interact differently with the various environments of the body. For this reason, it is essential that we can distinguish between isomers in order to maximise their effectiveness as medicines. As discussed in Unit 4 Area of Study 1, there are two types of isomers: structural isomers and stereoisomers.

Two molecules are stereoisomers if their atoms are connected in the same order but oriented in space differently. Optical isomers have a different three-dimensional arrangement of atoms around a carbon atom. This carbon atom is called a **chiral centre**.

A chiral centre is present when a carbon atom has four different groups bonded to it. There may be four carbon atoms bonded to this chiral centre, but each of the carbon atoms must be part of a different group. When a molecule contains a chiral centre, its mirror image is not superimposable (Figure 4.2.9), and the two molecules are called **enantiomers**.

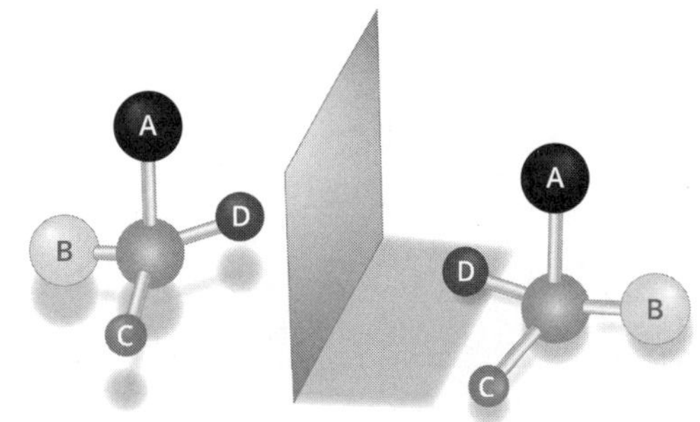


**Figure 4.2.9** A representation of a chiral molecule. Its mirror image cannot be superimposed on the original molecule.

Many properties of enantiomers, such as solubility, melting and boiling points, are the same. However, enantiomers interact differently with other chiral molecules, such as the enzymes present in the body.

The drug thalidomide, which has one chiral centre (Figure 4.2.10), was used in the 1950s and 1960s to reduce morning sickness during pregnancy. However, although one enantiomer was useful, the other caused birth defects. These days thalidomide is used in the treatment of some cancers because it stops the growth of cancer cells.

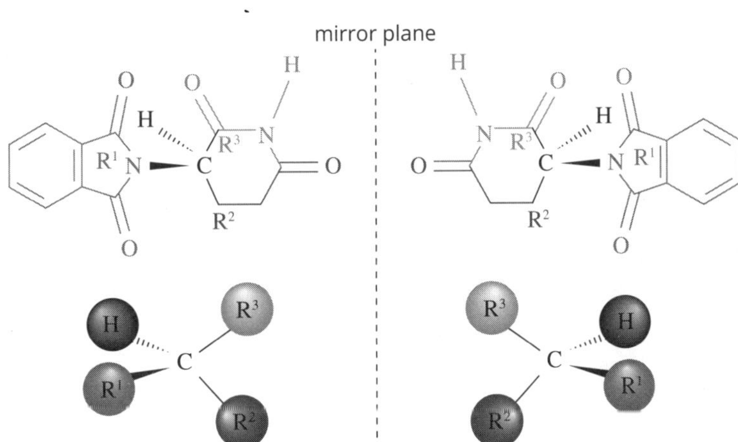


**Figure 4.2.10** Enantiomers of thalidomide (some hydrogen and carbon atoms are omitted for clarity). The chiral centres are redrawn below each structure to show the four different groups bonded to each one. Bonds going in and out of the plane of the paper are shown as dashed and solid wedges.

Iboprofen (Figure 4.2.11) is another drug that is chiral. One optical isomer reduces pain and swelling, while the other optical isomer has no effect on the body.

**Figure 4.2.11** One isomer of iboprofen reduces back pain. The carbon atom (marked *), which is bonded to the four groups (–COOH, –H, $–CH_3$ and the rest of the molecule) is the chiral centre.

## ENZYMES

Enzymes are proteins that act as biological catalysts in biochemical reactions in living things. Proteins are polymers made from **2-amino acids** (α-amino acids). These small monomer molecules contain a **carboxyl functional group** and an **amino functional group** bonded to the same carbon atom (Figure 4.2.12 on the following page). Different amino acids have different R groups, or side chains, that may be polar or non-polar. Except for the simplest amino acid, glycine, all amino acids are chiral because there are four different groups bonded to carbon 2.

Because enzymes are proteins, which are made from amino acids, enzymes are also chiral. This is most significant when they interact with medicines and catalyse reactions.

# KEY KNOWLEDGE

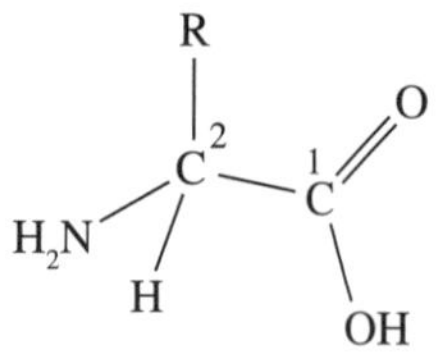

Figure 4.2.12 The general structure of a 2-amino acid: R represents a number of possible functional groups.

## Enzyme structure and function

The structure of enzymes differs in the number, type and sequence of amino acids from which they are made. The chemical composition, amino acid sequence and three-dimensional shape are unique for each enzyme and determine its function in an organism. The four levels of structure in enzymes are shown in Table 4.2.9.

**Table 4.2.9** The structure of enzymes

| Structure | Description |
|---|---|
| primary structure | The order in which the amino acids making up the enzyme are joined together 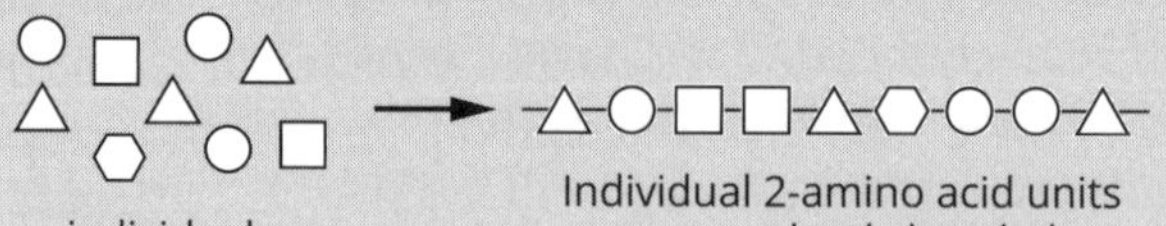<br>individual 2-amino acids<br>Individual 2-amino acid units are covalently bonded by peptide links in protein. |
| secondary structure | Folding and twisting of the chain, held in place by hydrogen bonds between the CO and NH groups in adjacent parts of the chain. Some enzymes form the three-dimensional $\alpha$-helical shape shown below. Other sections can form a shape called a $\beta$-pleated sheet.<br>C=O--------H-N C=O--------H-N C=O--------H-N<br>------- Hydrogen bonds between sections of keratin chain in wool and hair give these proteins a helical structure. 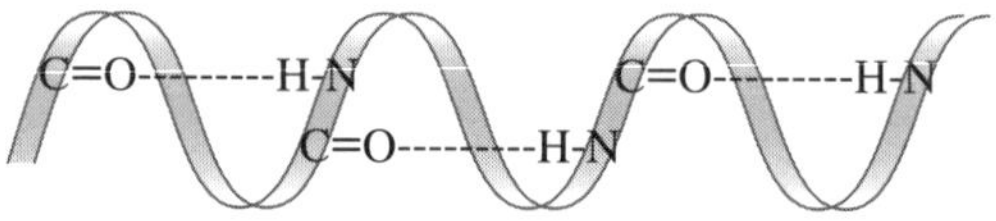 |
| tertiary structure | The overall three-dimensional structure of the enzyme, which is formed by the presence of different functional groups in the R (side chain) part of the amino acids, creates shapes that are held in place by disulfide (S–S) covalent bonds, ionic bonds between $-NH_3^+$ and $-COO^-$ groups, dipole–dipole forces, hydrogen bonds and dispersion forces. The different types of bonding are a consequence of the different polarities of the R groups. <br>various bonds that hold the three-dimensional structure together |
| quaternary structure | Some enzymes have a quaternary structure, where two or more polypeptide chains bond to each other to create a unique configuration.<br>For example, haemoglobin has four polypeptide chains bonded together.  |


ISBN 978 0 6557 0027 2

As catalysts, enzymes:

- can produce significantly faster rates than inorganic catalysts
- are unchanged after the reaction
- provide an alternative pathway for a reaction to occur, which decreases the activation energy, thereby increasing the reaction rate
- operate under mild conditions
- are often selective in the reactions they catalyse (many enzymes catalyse one specific reaction or type of reaction).

Each enzyme has a unique three-dimensional shape. They catalyse a reaction by interacting with a reactant molecule, called the **substrate**, which fits into an **active site** in the enzyme. The active site is usually a flexible hollow on the enzyme's surface. The substrate forms specific bonds with the enzyme, which results in the weakening or breaking of bonds within the substrate, allowing reaction to occur.

## Model for the action of enzymes

One model for the catalytic action of an enzyme is described as a lock-and-key mechanism. In this model, shown in Figure 4.2.13, the active site is rigid and complementary to the shape of the substrate, which explains the specificity of enzymes. A substrate molecule needs to have the right shape and configuration of functional groups to fit the active site. As a result, only one optical isomer of a chiral substrate is able to bind to a chiral enzyme and therefore able to react.

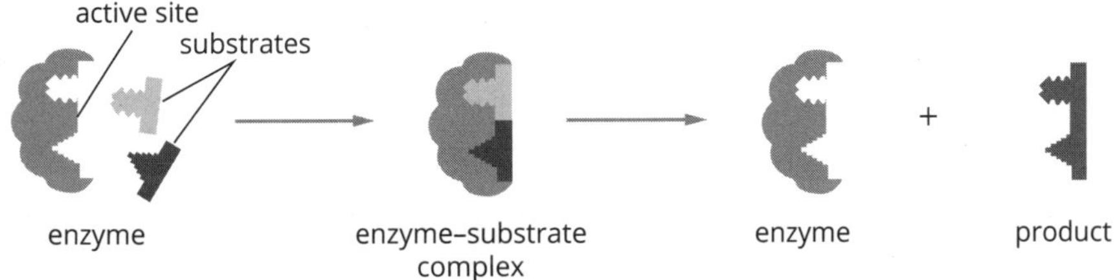


**Figure 4.2.13** The lock-and-key model for the action of an enzyme in a reaction

## Denaturation

**Denaturation** is a process that destroys the biological activity of proteins, including enzymes. During denaturation, the three-dimensional shape of the enzyme is changed, resulting in a change in the shape of the active site. Changes of conditions cause unfolding of the enzyme chains, due to the breaking of the bonds between the side chains (R groups), which results in the destruction of the tertiary structure. Depending on the nature of the R groups, the bonds broken may be hydrogen bonds, dispersion forces, covalent bonds, ionic bonds or dipole–dipole forces.

Although the hydrogen bonding in the secondary structure may be affected by denaturation, the primary structure is not affected at all. Only reaction with water (**hydrolysis**) can break the covalent bonds (peptide links) in the primary structure.

Coagulation and clumping result when denaturation occurs (Figure 4.2.14). This prevents the enzyme returning to its original shape, with the result that it can no longer function as intended. Because the shape of the active site is changed, substrate molecules can no longer bind to the enzyme.

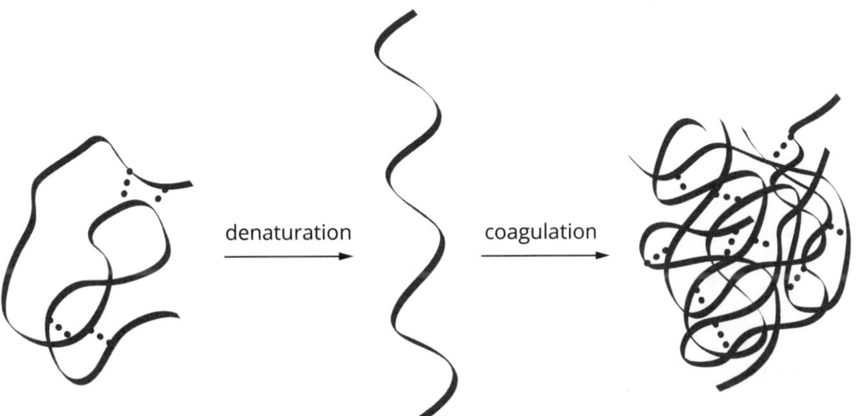


**Figure 4.2.14** Denaturation destroys the tertiary structure of the enzyme and causes the chains to unravel and then clump together.

ISBN 978 0 6557 0027 2

The changes in conditions that can cause denaturation are listed and explained in Table 4.2.10 on the following page.

**Table 4.2.10** The conditions that cause denaturation of an enzyme or a change in the rate of reaction

| | Condition | Explanation |
|---|---|---|
| Causes denaturation | Addition of certain chemicals, e.g. some metal ions | Heavy metals (e.g. Ag, Hg and Pb) form bonds with sulfur atoms in the R (side chain) part of the amino acids and prevent disulfide bonds forming, therefore changing the shape of the active site. This makes the enzyme ineffective and therefore denatures it. |
| | Change in pH | Depending on the pH, amino acids can be positively charged, negatively charged or uncharged. Some polar functional R groups of amino acids can also become charged in solutions of different pH. Hence, a change in pH can alter the charge of an enzyme, which in turn can alter its overall shape and cause denaturation.<br>The structures of amino acids in acidic, basic and neutral solutions are shown in Table 4.2.11. |
| | Increased temperature | Heat causes the bonds in the secondary and tertiary structure to break, creating a new tertiary structure and changing the shape of the active site. This denatures the enzyme. (Heat does not break the primary structure, i.e. the peptide links.) As seen in Figure 4.2.15, enzymes are only effective over a small range of temperature. |
| | **Condition** | **Explanation** |
| Decreases rate; does not cause denaturation | Decreased temperature | A decrease in temperature lowers the kinetic energies of the enzyme and substrate molecules, resulting in less frequent and less energetic collisions between the molecules. As a result, the rate of reaction between the enzyme and substrate decreases.<br>However this does not denature the enzyme because there is no change to the active site and if the temperature increases the enzyme's rate of reaction will increase, as will its effectiveness. |

**Table 4.2.11** The structure of a 2-amino acid in solutions of different pH

| Highly acidic conditions—low pH | Approximately neutral conditions—pH ≈7 | Highly basic conditions—high pH |
|---|---|---|
| $R-CH(NH_3^+)-C(=O)-OH$ | $R-CH(NH_3^+)-C(=O)-O^-$ | $R-CH(NH_2)-C(=O)-O^-$ |
| The basic amino group acts as a base and accepts a proton. | **Zwitterions** form at approximately neutral pH. Zwitterions are neutral species that have one positive and one negative charge in the molecule ion. | The acidic carboxyl group acts as an acid and donates a proton. |

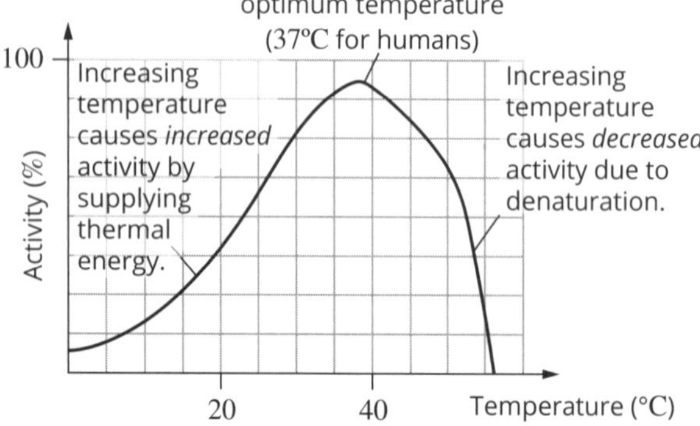


**Figure 4.2.15** The effect of temperature on the rate of reaction of an enzyme-catalysed reaction

● **You will now be able to conduct Practical activity 22.**

## COMPETITIVE ENZYME INHIBITORS

Some medicines act as **competitive enzyme inhibitors**. These medicines bind to the active site of an enzyme, preventing a substrate molecule from binding to the same site at the same time and causing an undesirable reaction. This process is described as competitive because the substrate and the drug molecules are both trying to bind to the active site at the same time. At one instant the inhibitor will bind and then the substrate will bind at another moment.

By increasing the concentration of the inhibitor, the inhibitor is more likely than the substrate to come into contact with the enzyme's active site. Equally, decreasing the inhibitor's concentration will increase the rate of the enzyme–substrate reaction, by allowing the substrate to bind to the enzyme's active site more frequently (Figure 4.2.16 on the following page).

 ISBN 978 0 6557 0027 2

# KEY KNOWLEDGE

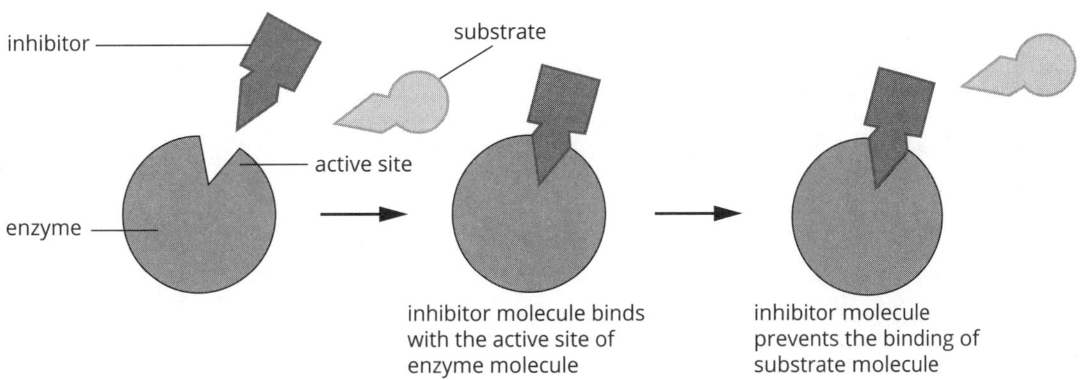


**Figure 4.2.16** An example of how a medicine can act as a competitive enzyme inhibitor

Although these medicines usually work by binding to the active site, some bind to another part of the enzyme, causing a change in the structure and therefore the shape of the active site. Table 4.2.12 gives examples of medicines that are competitive enzyme inhibitors.

**Table 4.2.12** Examples of competitive enzyme inhibitors

| Medicine | How it works |
|---|---|
| penicillin | • used as an antibiotic<br>• used to treat bacterial infections<br>• stops the active site of an enzyme, which many bacteria use in the last step as they synthesise a new cell wall after they divide |
| methatrexate | • used as a chemotherapy drug<br>• binds to the enzyme, preventing a coenzyme (a molecule that assists the catalysis reaction) from binding to the enzyme before the substrate<br>• restricts synthesis of DNA and RNA<br>• prevents cancer cells from growing and dividing. |

- **You will now be able to complete Worksheets 36 and 37.**

# WORKSHEET 30



## Knowledge review—chromatography, catalysts and stoichiometry

1 Explain the effect a catalyst has on a reaction. Include the term 'activation energy' in your explanation.

2 Figure 4.2.17 shows a thin-layer chromatography experiment.

a Why do the two components separate?

b Calculate the $R_f$ value for the component of the sample that moved the faster.

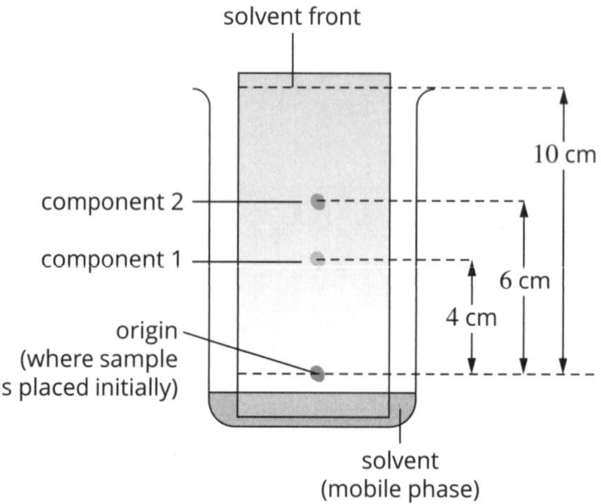


**Figure 4.2.17** A TLC experiment

3 Concentrated sulfuric acid and a sodium chloride solution react according to the equation:

$$H_2SO_4(l) + NaCl(aq) \rightarrow HCl(g) + NaHSO_4(aq)$$

Calculate the maximum possible mass of HCl gas produced when 234.0 g of sodium chloride reacts completely with sulfuric acid.



ISBN 978 0 6557 0027 2

# WORKSHEET 31

Case study

## Calculations—stoichiometry involving limiting reactants and redox titrations

1 A 0.345 g sample of magnesium reacted with 40.00 mL of 1.00 mol $L^{-1}$ HCl solution. The reaction is represented by the equation:

$$Mg(s) + 2HCl(aq) \rightarrow MgCl_2(aq) + H_2(g)$$

Calculate the maximum volume of hydrogen gas produced at standard laboratory conditions (SLC).

2 A fuel cell used in a police breathalyser contains $K_2Cr_2O_7$. The $Cr_2O_7^{2-}$ ion is a strong oxidising agent that converts ethanol, $CH_3CH_2OH$, to ethanoic acid, $CH_3COOH$, under acidic conditions, according to the balanced overall equation:

$$3CH_3CH_2OH(aq) + 2Cr_2O_7^{2-}(aq) + 16H^+(aq) \rightarrow 4Cr^{3+}(aq) + 11H_2O(l) + 3CH_3COOH$$

In a laboratory where this reaction was tested, 20.00 mL of 0.100 M ethanol reacted with 20.00 mL of 0.100 M $Cr_2O_7^{2-}$ solution.

**a** Determine the reactant that is in excess.

**b** Calculate how many grams of ethanoic acid are formed.

3 A pipette is used to deliver a 25.00 mL aliquot of hydrogen peroxide solution to a conical flask. This aliquot is titrated with 0.0525 M potassium permanganate solution delivered from a burette. The $H_2O_2$ and the $MnO_4^-$ ions react in the conical flask. The products of the overall redox reaction are $O_2$ and $Mn^{2+}$ ions. The average titre is 31.12 mL.

**a** Write balanced half-equations and a balanced overall equation for this redox reaction.

______________________________

______________________________

______________________________

**b** State the oxidising agent and the reducing agent in the reaction.

______________________________

**c** Explain the effect of rinsing the:

**i** pipette with distilled water only

**ii** conical flask with distilled water only

**d** Calculate the concentration of the hydrogen peroxide solution, in M.

 ISBN 978 0 6557 0027 2

# WORKSHEET 32

Simulation

## Analysis—identification of organic compounds and functional groups

**1** Three bottles of clear liquids labelled 'Organic compound 1', 'Organic compound 2' and 'Organic compound 3' need to be relabelled with the correct name of each compound. A student knows that the compounds are listed in Table 4.2.13 and that one is an alcohol, one a carboxylic acid and the third an ester made from the alcohol and the acid. However the ester had been contaminated with at least one other ester.

The student plans to use the data from Table 4.2.13, together with the results of laboratory tests for the functional groups in carboxylic acids and alcohols (see Table 4.2.1, page 174) and information from IR spectroscopy, to determine the identity of the compound in each bottle.

**Table 4.2.13** Data

| Compound | Data | Compound | Data |
|---|---|---|---|
| ethanoic acid | $C_2H_4O_2$; MP 16.5°C | propyl ethanoate | fruity smell; $C_5H_{10}O_2$ |
| propanoic acid | $C_3H_6O_2$; MP –20.7°C | ethyl propanoate | pineapple smell; $C_5H_{10}O_2$ |
| decanoic acid | $C_{10}H_{20}O_2$; MP 31.5°C | ethanol | $C_2H_6O$; MP –114°C |
| decyl ethanoate | waxy smell; $C_{12}H_{24}O_2$ | propan-1-ol | $C_3H_8O$; MP –126°C |
| ethyl decanoate | waxy smell; $C_{12}H_{24}O_2$ | decan-1-ol | $C_{10}H_{22}O$; MP 6.4°C |

The student recorded the following observations when the solutions in each bottle were tested.

| | Organic compound 1 | Organic compound 2 | Organic compound 3 |
|---|---|---|---|
| laboratory tests | Bubbles were produced when reacted with sodium metal. No bubbles were produced when reacted with sodium carbonate.<br>A waxy smelling compound was produced when reacted with a carboxylic acid.<br>MP recorded as 6.0–6.5°C. | Bubbles were produced when reacted with sodium metal. Bubbles were produced when reacted with sodium carbonate.<br>A waxy smelling compound was produced when reacted with an alcohol.<br>MP recorded as 15.5–16.5°C. | A waxy, mildly fruity smell was recorded. |
| spectroscopic results | IR spectrum shows a broad band around 3300 $cm^{-1}$. | IR spectrum shows two bands: around 3300 $cm^{-1}$ and one around 1700 $cm^{-1}$. | IR spectrum shows a band around 1700 $cm^{-1}$. |

**a** Name the type of each compound and its identity, explaining why you made your decision.

Organic compound 1:

________________________________________

________________________________________

________________________________________

Organic compound 2:

________________________________________

________________________________________

________________________________________

Organic compound 3:

**b** Suggest what the contaminating ester could be and explain why.

**c** Explain what the melting point indicates about the purity of compounds 1 and 2.

**d** Should you say the melting points are *accurate* or *precise*? Explain why in terms of the meanings of these words.

**2** At another time, two bottles were found to be labelled 'pentane or pentene?'. Describe how you could test to determine the identity of the chemicals.

**3** In an experiment to determine the iodine number of an unsaturated hydrocarbon with molar mass of 82.14 g $mol^{-1}$, 645.16 g of iodine reacted with 100 g of the unsaturated hydrocarbon. Calculate the iodine number of the compound and determine how many double bonds are present in the molecule.

 ISBN 978 0 6557 0027 2

# WORKSHEET 33



## Spectroscopy—nuclear magnetic resonance

**1** Five $^1H$ NMR spectra and five $^{13}C$ NMR spectra are shown. Using the labels A–J, correctly match each chemical listed in the table below with its corresponding spectrum. The $^1H$ NMR spectra shown are low-resolution spectra (some of the signals would be split in high-resolution spectra).

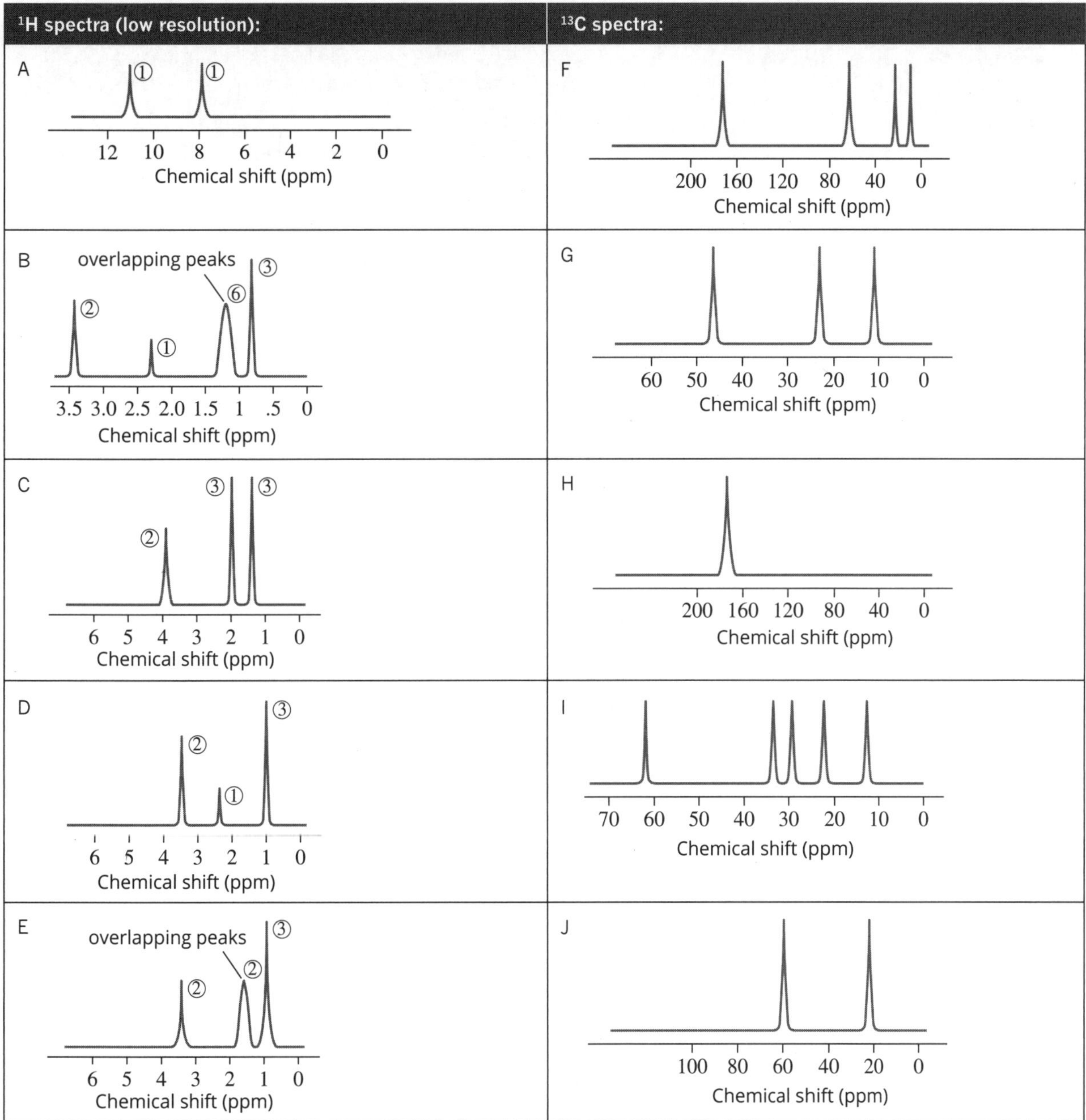


| NMR spectra | | |
|---|---|---|
| **Chemical** | **Corresponding $^1H$ spectrum (A–E)** | **Corresponding $^{13}C$ spectrum (F–J)** |
| $CH_3CH_2OH$ | | |
| $CH_3COOCH_2CH_3$ | | |
| HCOOH | | |
| $CH_3CH_2CH_2Cl$ | | |
| $CH_3CH_2CH_2CH_2CH_2OH$ | | |

**2** NMR spectroscopy is widely used by analytical chemists. It provides detailed information about the nuclei of certain atoms and the carbon and hydrogen backbones of organic molecules. Examine the structures of propane and ethanoic acid and complete the table below.

$$\overset{1}{\mathbf{CH_3}} - \overset{2}{\mathbf{CH_2}} - CH_3 \qquad \overset{3}{\mathbf{CH_3}} - \overset{4}{\mathbf{COOH}}$$

propane ethanoic acid

| Propane and ethanoic acid data | | | | |
|---|---|---|---|---|
| **Compound** | **Groups in which protons are found** | **Number of different types of protons** | **Number of different types of carbon atoms** | **Expected splitting of the signals of the bolded H atoms, numbered 1–4** |
| propane | | | | |
| ethanoic acid | | | | |

 ISBN 978 0 6557 0027 2

# WORKSHEET 34

Simulation

# Using spectroscopy to determine structure

It has been known from early times that the bark of willow trees contains a chemical that assists in the relief of pain. This chemical itself has no direct medicinal effect, but in the human body it is converted into an active chemical, salicylic acid. Chemists can use different spectroscopic techniques to discover the identity and structure of such compounds.

## Identifying salicylic acid

1 Mass spectrometry can be used to confirm the identity of a sample of a white powder thought to be salicylic acid. The fragmentation pattern observed in a mass spectrum is characteristic of the compound being investigated. By analysing this pattern and comparing it with a database of known chemicals, the chemist can decide if the sample is salicylic acid.

a Label the molecular ion in the mass spectrum of the sample with an asterisk (Figure 4.2.18).

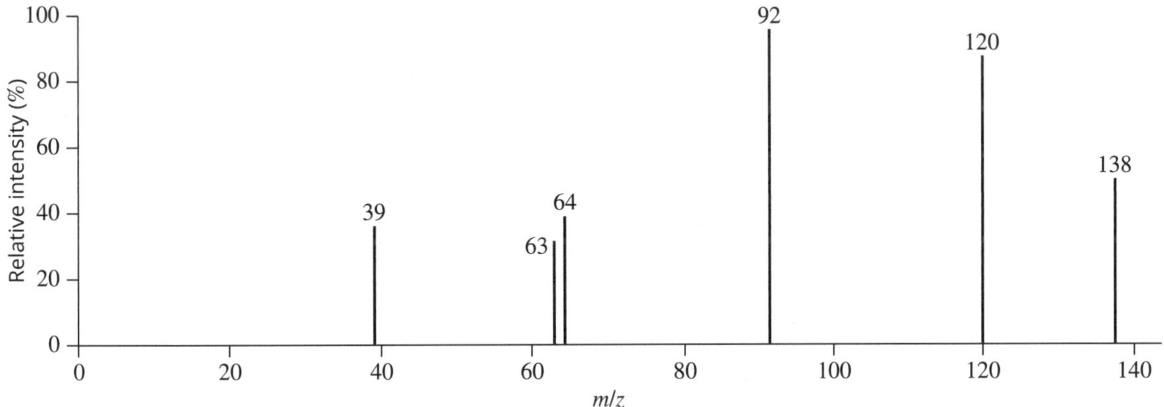


**Figure 4.2.18** The mass spectrum of the white powder

b Likely molar mass: ___________ g mol$^{-1}$

c Knowing that the molecular formula of salicylic acid is $C_7H_6O_3$, is the white powder likely to be salicylic acid? Explain.

d What are possible formulas of the cations that have the following masses corresponding to peaks in the mass spectrum?

92: ___________

120: ___________

## Determining the structure

2 The functional groups present in a molecule can be determined by IR spectroscopy. The IR spectrum of salicylic acid is shown in Figure 4.2.19.

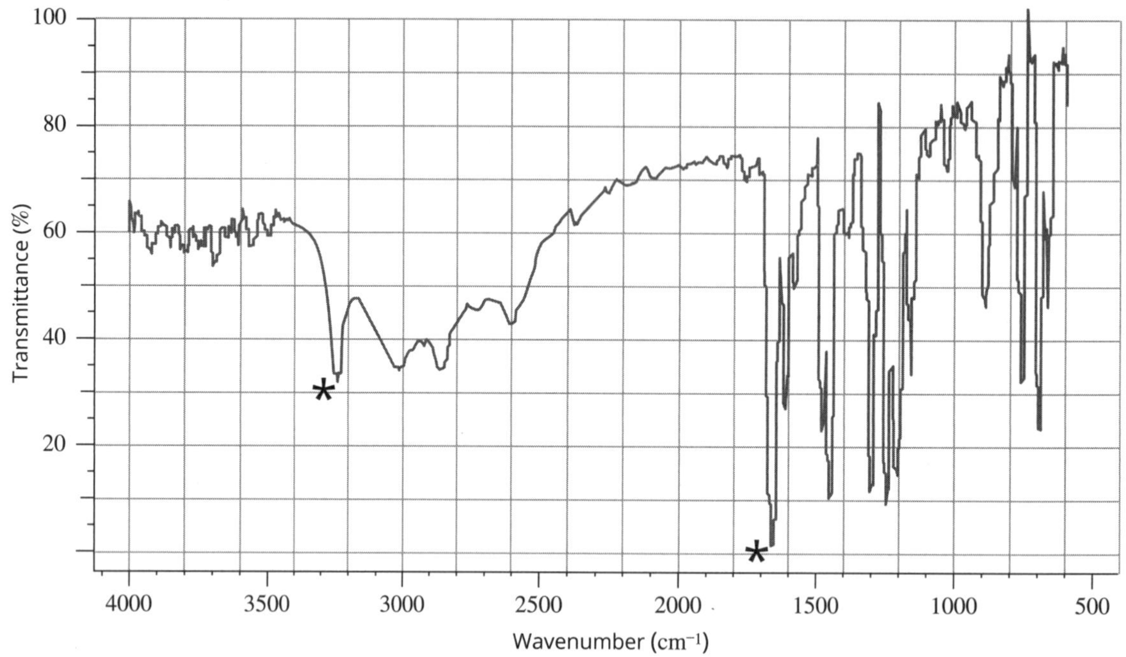


**Figure 4.2.19** The IR spectrum of salicylic acid

Table 4.2.14 shows the characteristic absorption bands of different functional groups. The peaks in the IR spectrum of salicylic acid between 1400 and 1600 cm$^{-1}$ are found in compounds called arenes, which are based on benzene. Benzene ($C_6H_6$) can be represented by the structure shown in Figure 4.2.20. In arenes, a hydrogen atom bonded to a carbon atom in the benzene ring can be replaced by other groups. Phenol is an example of an arene.

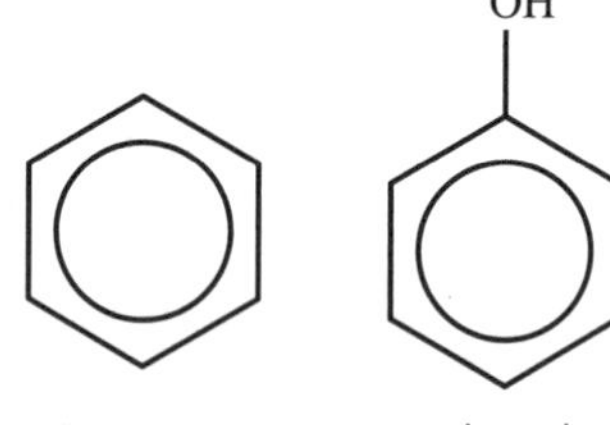


**Figure 4.2.20** A benzene ring forms the basis of the arene phenol.

**Table 4.2.14** Characteristic absorption bands of functional groups

| Band | Location | Frequency (cm$^{-1}$) and intensity |
|---|---|---|
| C=O | in carboxylic acids | 1680–1740, strong |
| C=O | in esters | 1720–1840, strong |
| C–H | in alkanes | 2850–2950, medium–strong |
| C=O | in ketones | 1680–1850, strong |
| C–H | in alkenes, arenes | 3000–3250, strong |
| C–O | in alcohols, esters | 1050–1410, strong |
| O–H | in alcohols, phenols | 3200–3600, strong |
| O–H | in carboxylic acids | 2500–3300, strong |

Use Table 4.2.14 to identify the functional groups in salicylic acid that correspond to the graph peaks marked with asterisks in Figure 4.2.19.

1700 cm$^{-1}$: ______________________________

3300 cm$^{-1}$: ______________________________

 ISBN 978 0 6557 0027 2

**3** In a $^{13}C$ NMR spectrum, the number of peaks indicates the number of carbon atoms in different chemical environments.

Examine the simplified $^{13}C$ NMR spectrum shown in Figure 4.2.21.

How many different carbon atoms are present in salicylic acid? ______________________

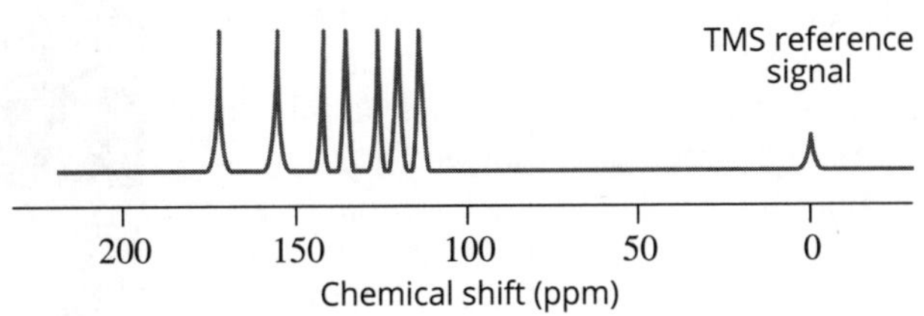


**Figure 4.2.21** A simplified $^{13}C$ NMR spectrum of salicylic acid

**4** Complete the following summary of the information you have found about the structure of salicylic acid.

Molar mass: ______________________

Molecular formula: ______________________

Functional groups present: ______________________

Number of carbon atoms in different environments in the molecule: ______________________

**5** Draw a possible molecular structure for salicylic acid.

ISBN 978 0 6557 0027 2

# WORKSHEET 35

Case study

## Chromatography and medicines

**Figure 4.2.22** *The Incredulity of St Thomas* (circa 1504)

1 The paint used by artists must be viscous enough to stop it running everywhere as it is applied, but it must not be too sticky. Although the paint must dry hard to be durable, it cannot dry so fast that it reduces the artist's freedom. Different media have been used through the ages. Chemists can use analytical techniques on the paint from a picture to determine the medium employed and restore damaged paintings appropriately. A painting by Cima da Conegliano called *The Incredulity of St Thomas*, painted in 1504, was fully restored by the National Gallery in London (Figure 4.2.22).

Cima may have used either walnut oil or linseed oil. Both oils contain significant amounts of esters based on palmitic and stearic acids. The oils contain these esters in different ratios, so by working out the palmitate-to-stearate ester ratio using gas chromatography, we can tell which oil Cima used to bind his pigments. (Gas chromatography is a similar technique to HPLC but uses a gas as the mobile phase.) The ratio for linseed oil is between 1 : 1.4 and 1 : 1.9 and for walnut oil is 1 : 0.3.

Examine the chromatogram shown in Figure 4.2.23 from a sample of paint from Cima's painting.

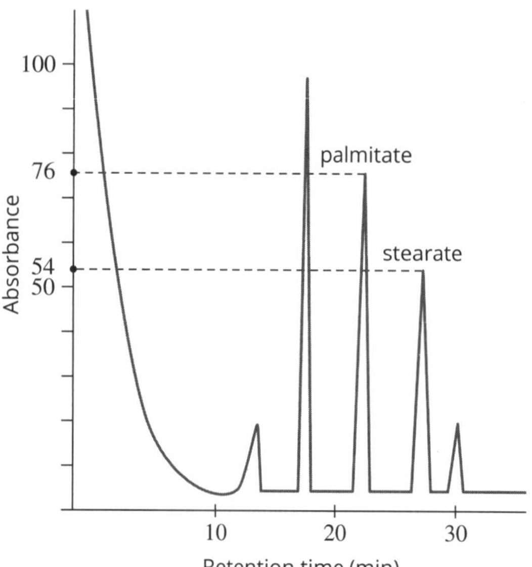


**Figure 4.2.23** A chromatogram of a sample of paint from Cima's painting

a What does the area under each peak in a chromatogram represent?

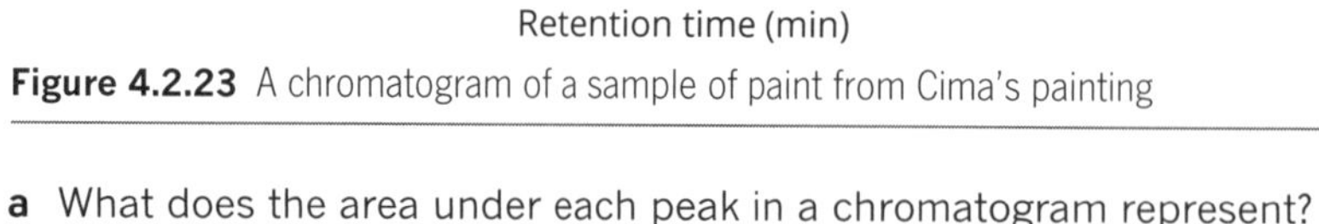

b How would chemists have known which peak was due to a palmitate ester and which was due to a stearate ester?

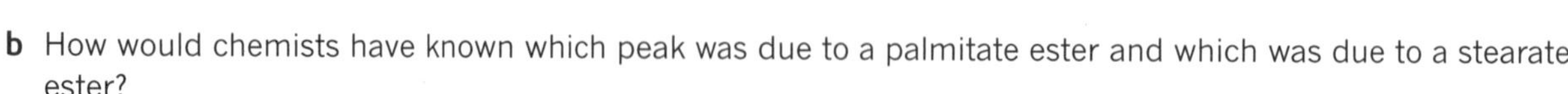

c Assume that the peak areas of one ester can be compared with peak areas of the other and that peak height is proportional to peak area. Estimate the palmitate-to-stearate ester ratio using the graph. Determine the oil Cima used.

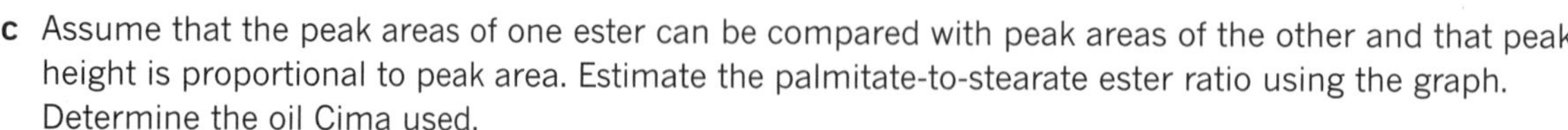

 ISBN 978 0 6557 0027 2

**2** Traditional medicine, including the medicine of Indigenous Australians, has used plant material for centuries to treat a variety of diseases and conditions, such as insect bites and stings, nausea, pain and swelling. Such medicinal plants provide the basic active ingredient for many medicines in use today. HPLC can be used to separate and extract the components of medicinal plants accurately and precisely. Without the aid of HPLC, the extraction and purification of these plant components could be a long and difficult process.

**a** Briefly explain how tea tree oil can be extracted from the plant by steam distillation.

**b** Before determining the concentration of the active component in the plant, the pure extracted component must be identified. Explain how the $R_t$ value of a component can help identification.

**c** The following tests could be used in the identification process of an unknown component extracted from a plant. For each test, indicate what useful information could be found.

| Test | Information to be obtained from the test |
|---|---|
| melting point | |
| IR spectroscopy | |
| mass spectroscopy | |
| NMR spectroscopy | |

**d** Once the component has been identified, HPLC can be used to analyse the concentration of the component in the sample. Explain the steps involved in the analysis, using the terms 'standard solutions' and 'calibration curve'.

**e** Plot a calibration curve below using the following results from standard solutions of an active ingredient from a plant.

| Calibration results | |
|---|---|
| **Concentration of the standard solutions (ppm)** | **Area under the peak** |
| 5 | 2.0 |
| 10 | 4.0 |
| 15 | 6.0 |
| 20 | 8.0 |
| Sample | 5.3 |

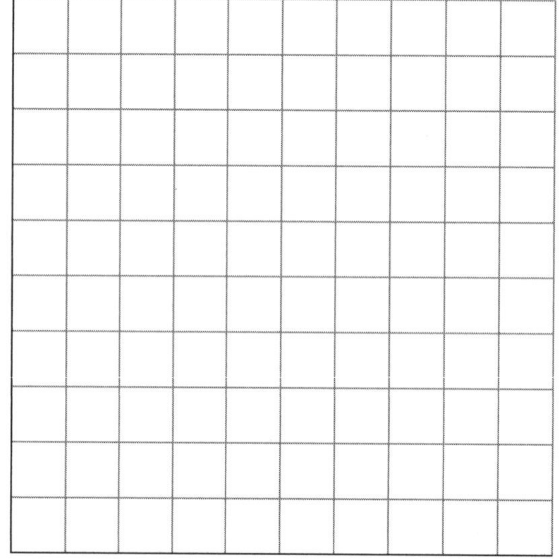

**f** Determine the concentration of the active component in the plant material.

 ISBN 978 0 6557 0027 2

# WORKSHEET 36



## Enzymes—formation, structure and function

Enzymes are polymers that play a key role in living things.

**1 a** Enzymes are made of repeating units (monomers), called ____________ ____________, which consist of a central carbon atom bonded to four different groups.

**b** Draw a general structural formula for an enzyme monomer.

**c** What property is displayed by a molecule in which a carbon atom is bonded to four different groups?

______________________________________________________________________

**d** Draw a three-dimensional representation of the formula of the enzyme monomer in part b, using wedge-shaped bonds, and draw its mirror image.

**e** Is this mirror image superimposable on the original molecule? Explain your answer.

______________________________________________________________________

**f** What is the general name of such a pair of molecules?

______________________________________________________________________

**2 a** What type of polymerisation reaction occurs when monomers react to form enzymes?

______________________________________________________________________

**b** Using the formula of the monomer from Question 1, write an equation for the reaction involving two monomers only. Show all reactants and products as structural formulas. Label and name three different functional groups in the reactants and products involved in this reaction.

**c** Complete the table, using your knowledge of the structure of enzymes.

| Level of structure | Description of structure and bonding type involved |
|---|---|
| primary | |
| secondary | |
| tertiary | |

ISBN 978 0 6557 0027 2

**3** **a** Briefly explain the process known as denaturing and describe three ways an enzyme can be denatured.

**b** In an experiment to determine the effect of temperature on the rate of an enzyme-catalysed reaction, the following graph was plotted.

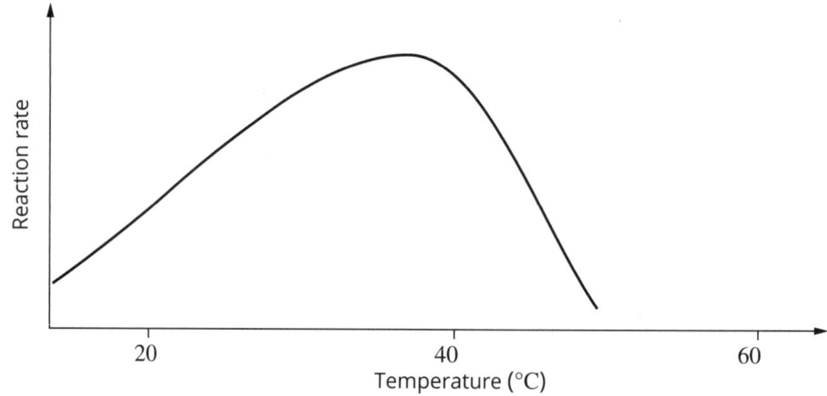


**i** Explain the significance of the peak at approximately 37°C.

**ii** Explain the effect of the temperature change on the reaction rate to the left of this peak.

**iii** Explain the effect of the temperature change on the reaction rate to the right of this peak.

**c** What is the difference between denaturing an enzyme and hydrolysing an enzyme?

**4** Draw structural formulas to show a 2-amino acid in acidic, neutral and basic solutions. Label the zwitterion.

| Highly acidic conditions—low pH | Approximately neutral conditions—pH ≈ 7 | Highly basic conditions—high pH |
|---|---|---|
| | | |

 ISBN 978 0 6557 0027 2

5 a Complete the diagram and explain the mechanism by which an enzyme catalyses chemical reactions in the body.

1

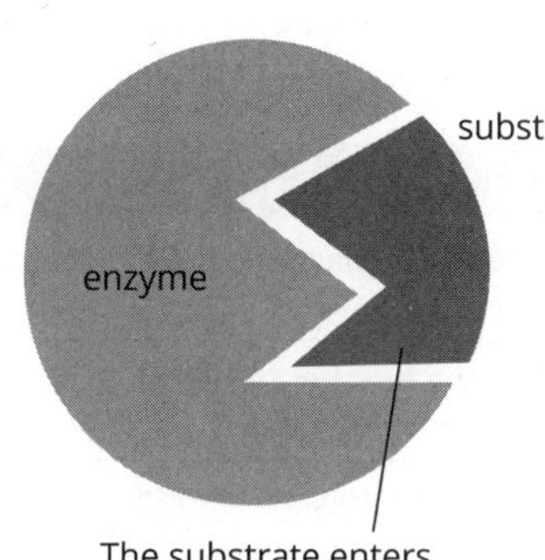


2

3

Explanation:

______________________________________________________________________

______________________________________________________________________

b Explain how this model for enzyme action can account for the difference in behaviour of a pair of enantiomers in an enzyme-catalysed reaction.

______________________________________________________________________

______________________________________________________________________

______________________________________________________________________

c Viagra is the brand-name of a competitive enzyme inhibitor that is used to treat erectile dysfunction.

i What is a competitive enzyme inhibitor?

______________________________________________________________________

______________________________________________________________________

______________________________________________________________________

ii How does the inhibitor decrease the enzyme–substrate reaction rate?

______________________________________________________________________

______________________________________________________________________

ISBN 978 0 6557 0027 2

# Literacy review—laboratory and instrumental analysis

**1** For the phrases in the left-hand column concerning NMR spectra, select the correct phrase from the right-hand column to complete the sentence and write the letter in the middle column. (Some phrases in the right-hand column will not be used.)

| Phrase | Correct letter | Phrase |
|---|---|---|
| 1 The number of peaks in a $^{13}C$ spectrum indicates … | | A … when nuclei held in a magnetic field interact with radio waves. |
| 2 An NMR spectrum can be obtained for nuclei with … | | B … proportional to the number of H atoms in a particular chemical environment. |
| 3 The chemical shift is … | | C … an odd number of protons or an odd number of neutrons. |
| 4 The chemical shift for each peak gives information about … | | D … an even number of protons and an even number of neutrons. |
| 5 The area under a peak in a $^{1}H$ spectrum is … | | E … the number of different C chemical environments. |
| 6 NMR spectra are produced … | | F … the number of C atoms in the molecule. |
| | | G … measured in ppm. |
| | | H … the type of environment of the H or C atom. |
| | | I … when electrons held in a magnetic field interact with radio waves. |

**2** Choose from the terms listed to complete the summary statements outlining key points about chromatography. Some terms may be used more than once or not at all.

| | | | | | | |
|---|---|---|---|---|---|---|
| qualitative | quantitative | column | separation | standard | enzymes | desorb |
| concentration | stationary | $R_t$ | adsorb(s) | calibration | pressure | binds |

High-performance liquid chromatography (HPLC) can be used for both ____________ and ____________ analysis. Components are separated when they ____________ onto the stationary phase and ____________ and dissolve into the mobile phase. In HPLC, the time taken for a component to pass through the column is measured by the ____________ value. If a component ____________ strongly to the ____________ phase it will have a high ____________ value. HPLC can be used to separate high molecular mass organic compounds such as ____________. To increase the ____________ of components in a sample, the sample is passed through a ____________ packed with fine particles under high ____________.


ISBN 978 0 6557 0027 2

**3** Complete the following sentences using terms from the box below. Some terms may be used more than once or not at all.

| hydrogen bonds | S–S covalent links | helical | covalent bonds | ionic bonds |
|---|---|---|---|---|
| three-dimensional | dipole–dipole forces | amino acids | dispersion forces | |

In the primary structure of enzymes, the number and sequence of ____________ ____________ is unique for each enzyme. They are bonded to each other by ____________ ____________. The secondary structure is held in place by ____________ ____________ between the CO and NH groups in adjacent parts of the chain. The side chains of the amino acids protrude from the enzyme chain and have different properties that determine the ____________ ____________ shape. The bond types that hold the enzyme chain together in particular shapes may be ____________ ____________, ____________ ____________, ____________ ____________, ____________ ____________, ____________ ____________ and ____________ ____________. Some enzymes have a quaternary structure.

**4** Match the definitions in the left-hand column with the terms in the right-hand column. Put the letter for the correct term in the middle column beside the appropriate definition.

| Definition | Correct letter | Term |
|---|---|---|
| 1 Carbon atom with four different groups bonded to it | | A competitive enzyme inhibitor |
| 2 Isomers with the same connectivity, but different orientation | | B fractional distillation |
| 3 The separation of the liquid components in a mixture on the basis of their boiling temperatures | | C iodine number |
| 4 A measure of the degree of unsaturation of a fat or oil | | D chemical shift |
| 5 The multi-peaked signal that a proton gives in a $^1$H NMR spectrum when there are hydrogen atoms in different neighbouring chemical environments | | E splitting pattern |
| 6 A model involving an active site with a complementary shape to that of a substrate, accounting for the specificity of enzymes | | F chiral centre |
| 7 Same molecular formula, different structure | | G isomers |
| 8 It binds to an active site, preventing substrate molecules from binding to the site at the same time | | H stereoisomers |
| 9 The position of a signal in an NMR spectrum relative to tetramethylsilane | | I lock-and-key mechanism |

# WORKSHEET 38

## Reflection—How are organic compounds analysed and used?

The following table summarises the key knowledge covered in this area of study.

1 Reflect on how well you understand the concepts listed. Rate your learning by shading the circle that corresponds to your current level of understanding for each one.

| Key knowledge | Not confident ⟵ | | | ⟶ Very confident |
|---|---|---|---|---|
| Tests for the presence of C=C, –OH and –COOH functional groups | ○ | ○ | ○ | ○ |
| MP determination and distillation for purity and isolation | ○ | ○ | ○ | ○ |
| Measurement of the degree of unsaturation of compounds using iodine | ○ | ○ | ○ | ○ |
| Volumetric analysis using redox titrations | ○ | ○ | ○ | ○ |
| Applications of MS | ○ | ○ | ○ | ○ |
| Identification of bond types with IR spectroscopy | ○ | ○ | ○ | ○ |
| Structural determination using $^{13}C$ NMR and $^{1}H$ NMR | ○ | ○ | ○ | ○ |
| The principles of chromatography, including use of HPLC | ○ | ○ | ○ | ○ |
| Deduction of the structures of simple organic compounds using a combination of MS, IR, $^{1}H$ NMR and $^{13}C$ NMR | ○ | ○ | ○ | ○ |
| Use of laboratory and instrumental analysis for identifying organic compounds or functional groups and product purity | ○ | ○ | ○ | ○ |
| Extraction and purification of natural plant compounds for medicines, using solvent extraction and distillation | ○ | ○ | ○ | ○ |
| Functional groups in medicines; importance of isomers and chiral centres | ○ | ○ | ○ | ○ |
| Enzymes as catalysts in living systems: their structures and effect on their function of denaturation, decreased temperature and changes in pH | ○ | ○ | ○ | ○ |
| Competitive enzyme inhibitors | ○ | ○ | ○ | ○ |

2 Consider the points you have shaded from 'Not confident' to 'Very confident'. List specific ideas you can identify that were challenging.

________________________________________

________________________________________

3 Write down two different strategies that you will apply to help further your understanding of these ideas.

________________________________________

________________________________________

 ISBN 978 0 6557 0027 2

# PRACTICAL ACTIVITY 19

Case study

## Determining the iodine number of an oil—a secondary data exercise

### SUGGESTED DURATION

- 80 minutes

### INTRODUCTION

Fats and oils contain mixtures of triglycerides with varying levels of unsaturation. Saturated fats are composed of triglycerides made from fatty acids that contain only C–C single bonds. They are mostly found in animal products, including full-cream milk, butter, cheese and meat. They are also found in some vegetable products, such as coconut oil.

Monounsaturated oils are composed of triglycerides made from fatty acids that contain one C=C bond. These oils come from seeds or nuts such as olives, peanuts and canola. Polyunsaturated oils, made from fatty acids containing multiple C=C bonds, come from plant products such as sesame seed, corn, safflower seed and sunflower seed.

Although controversial, the general consensus among experts is that saturated fat is less healthy than unsaturated fat, potentially increasing the risk of heart disease, and there is continuing debate about the benefits of monounsaturated oils over polyunsaturated oils.

As a consequence, there is considerable interest in the degree of unsaturation of the fats and oils in commercial products as a guide to their nutritional value. Furthermore, the degree of unsaturation provides an indication of the oil's resistance to oxidation by oxygen in the air: the more double bonds that are present in a molecule, the more prone the oil is to oxidation.

The degree of unsaturation of a fat or oil is typically measured as an iodine number (IN), which is calculated based on the reaction of the oil with iodine.

The IN is defined as the mass of iodine (in grams) absorbed by 100 grams of the oil or fat.

### AIM

To determine the iodine number of samples of commercial oils

### VARIABLES

Independent variable: ________________________________________

Dependent variable: ________________________________________

Controlled variables: ________________________________________

### METHOD

0.200 g samples of the commercial oils were dissolved separately in a mixture of cyclohexane and acetic acid. 0.420 g $I_2$ was then dissolved in each solution, which is an excess amount of $I_2$.

After the mixtures had been left in the dark for a fixed period of time to allow reaction to occur, the excess unreacted iodine was determined by a redox titration with standard 0.100 M sodium thiosulfate solution, which reacts according to the equation:

$$I_2(aq) + 2S_2O_3^{2-}(aq) \rightarrow S_4O_6^{2-}(aq) + 2I^-(aq)$$

Because starch turns blue in the presence of iodine, it was used as an indicator for this titration.

The procedure was performed three times for each commercial oil, using the same masses of oil and $I_2$. The results for one of the oils, safflower oil, are shown in Table 1.

## PRACTICAL ACTIVITY 19

### RESULTS

**Table 1** Results for safflower oil

| Titration | Initial burette reading (mL) | Final burette reading (mL) | Titre (mL) |
|---|---|---|---|
| 1 | 0.22 | 8.88 | |
| 2 | 10.02 | 18.70 | |
| 3 | 21.04 | 29.74 | |

### CALCULATIONS

| Question | Calculation and discussion |
|---|---|
| **1** Find the average of three concordant titres of sodium thiosulfate solution. | |
| **2** Using the concentration of the standard sodium thiosulfate solution (0.001 M), calculate the number of moles of sodium thiosulfate in the average titre. | |
| **3** Deduce the number of moles of excess $I_2$ present in each flask before the titration was commenced. | |
| **4** Calculate the number of moles of $I_2$ in the 0.420 g that was initially added to the reaction flask | |
| **5** Subtract the answers to step 4 and step 3 to find the number of moles of $I_2$ that reacted with the oil. | |
| **6** Calculate the mass of $I_2$ that reacted with the 0.200 g sample of oil. | |
| **7** Calculate the mass of $I_2$ that would react with 100 g of oil. | |
| **8** State the iodine number of the safflower oil. | |

### DISCUSSION

**1** Identify any systematic and random errors that might have occurred during this experiment.

systematic errors:

______________________________

random errors:

______________________________

**2** Suppose the burette was washed with water immediately before performing the titration. What effect is this likely to have on the measured value of the titre and on the calculated iodine number? Explain.

______________________________

______________________________

______________________________


ISBN 978 0 6557 0027 2

## PRACTICAL ACTIVITY 19

3 Linoleic acid ($M = 281$ g mol$^{-1}$) is a fatty acid that reacts with glycerol to form a triglyceride. The formula of linoleic acid is:

$$CH_3(CH_2)_4CH{=}CHCH_2CH{=}CH(CH_2)_7COOH$$

**a** Write the semi-structural formula of the product formed when linoleic acid reacts with iodine.

**b** What is the name given to this general type of reaction? ______________________

**c** Draw the semi-structural formula of the triglyceride formed from linoleic acid.

**d** Calculate the IN of linoleic acid.

**e** 1.40 g of a different fatty acid, called linolenic acid, reacts with 3.75 g of $I_2$. Calculate the number of double bonds in linolenic acid, given $M$(linolenic acid) = 278 g mol$^{-1}$.

## FURTHER DATA ANALYSIS (OPTIONAL)

In this data analysis, you will explore the relationship between IN and the degree of unsaturation of oils.

Vegetable oils contain mixtures of fatty acids, some with no double bonds, others with one, two or three double bonds.

Table 2 lists the major fatty acid composition of a range of vegetable oils.

**Table 2** The major fatty acid composition of a range of vegetable oils (percentages do not add up to 100% because there are other components in the mixture)

| Vegetable oil | % saturated fatty acids | % monounsaturated fatty acids | % with two double bonds | % with three double bonds |
|---|---|---|---|---|
| grapeseed | 11 | 19 | 64 | 0 |
| linseed | 18 | 19 | 15 | 57 |
| peanut | 10 | 54 | 24 | 0 |
| rapeseed | 7 | 60 | 19 | 8 |
| sesame | 15 | 40 | 40 | 0 |
| sunflower | 11 | 31 | 55 | 0 |
| walnut | 10 | 17 | 57 | 12 |

## PRACTICAL ACTIVITY 19

**1** To compare the degree of unsaturation of different vegetable oils, a 'double bond equivalent' can be calculated from the percentage composition of each oil. The double bond equivalent is calculated for each vegetable oil by adding the product of the percentage composition by the number of double bonds.

e.g. Double bond equivalent for grapeseed oil = $(11 \times 0) + (19 \times 1) + (64 \times 2) + (0 \times 3) = 147$

Calculate and enter the double bond equivalent values for each oil in Table 3. The first one has been entered for you from the calculation above.

**Table 3** Double bond equivalent for each vegetable oil

| Vegetable oil | Double bond equivalent |
|---|---|
| grapeseed | 147 |
| linseed | |
| peanut | |
| rapeseed | |
| sesame | |
| sunflower | |
| walnut | |

**2** Table 4 lists the average iodine number for each of the vegetable oils.

**Table 4** Average iodine number (IN) for the vegetable oils

| Vegetable oil | IN |
|---|---|
| grapeseed | 130 |
| linseed | 180 |
| peanut | 93 |
| rapeseed | 105 |
| sesame | 108 |
| sunflower | 130 |
| walnut | 145 |

Plot a graph of iodine number vs double bond equivalents.

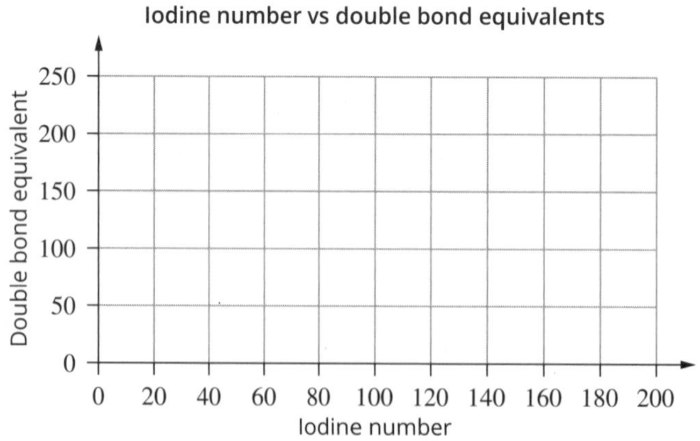


**3** Discuss the shape of this graph.



ISBN 978 0 6557 0027 2

4 Is this outcome expected? Explain your answer.

5 From your value for the iodine number of safflower oil calculated earlier, what is the double bond equivalent of this oil predicted from this graph?

6 If the fatty acid composition of safflower oil is 6% palmitic acid (saturated), 2% stearic acid (saturated), 13% oleic acid (monounsaturated) and 76% linoleic acid (two double bonds), calculate its double bond equivalent.

## CONCLUSION

# PRACTICAL ACTIVITY 20

Controlled experiment

# Analysis of ascorbic acid in vitamin C tablets

## SUGGESTED DURATION

- 80 minutes

## INTRODUCTION

The human body needs vitamin C (ascorbic acid) for a number of reasons; for example, it plays an essential role in the formation of the protein that connects cells. Vitamin C is not stored in body fat as some other vitamins are, so we need daily supplies of the vitamin in our diets. It is present in high concentrations in various fruits and vegetables.

In this experiment, a solution of vitamin C is analysed by titration with iodine solution. Since iodine is a volatile solid, it is difficult to weigh samples accurately to make up a standard solution. Furthermore, iodine solution tends to decrease in concentration if left to stand. Therefore, it is desirable to standardise an iodine solution just before use; this is done by titrating iodine solution with a freshly prepared, standard solution of sodium thiosulfate ($Na_2S_2O_3$). Starch indicator turns dark blue in the presence of excess iodine, so addition of a small volume of starch solution can help you see the endpoint of the titration. The half-equations for the reaction are:

$$I_2(aq) + 2e^- \rightarrow 2I^-(aq)$$

$$2S_2O_3^{2-}(aq) \rightarrow S_4O_6^{2-}(aq) + 2e^-$$

As well as ascorbic acid, many vitamin C tablets contain a 'binder' of starch. Depending on the brand, the tablets may also contain a food colour, a flavouring and a sweetener. Ascorbic acid ($C_6H_4O_2(OH)_4$) is quite stable in the solid form but, in the form of an aqueous solution, readily undergoes oxidation:

$$C_6H_4O_2(OH)_4(aq) \rightarrow C_6H_4O_4(OH)_2(aq) + 2H^+(aq) + 2e^-$$

It is this oxidation reaction that occurs when vitamin C tablets are dissolved in water and the resulting solution is titrated against a solution of iodine. Once again, starch indicator can be used to ascertain the end point.

## MATERIALS

- 3 vitamin C tablets
- 60 mL standard sodium thiosulfate solution, $Na_2S_2O_3(aq)$ (approximately 0.1 M)
- 200 mL of 0.05 M iodine solution, $I_2(aq)$
- starch indicator
- 150 mL deionised water
- 5 × 250 mL conical flasks
- 20 mL pipette
- pipette filler
- 10 mL measuring cylinder
- 100 mL measuring cylinder
- stirring rod
- burette and stand
- small funnel
- white tile
- weighing bottle or watch glass
- electronic balance
- safety glasses

## AIM

To determine the ascorbic acid content of vitamin C tablets

| PRE-LAB SAFETY INFORMATION | | |
|---|---|---|
| **Material or procedure** | **Hazard** | **Control** |
| sodium thiosulfate solution, $Na_2S_2O_3(aq)$ | • may cause irritation to the skin, eyes and respiratory system<br>• harmful if swallowed<br>• avoid contact | Avoid contact and breathing vapour. Wear safety glasses and a laboratory coat. |
| iodine solution, $I_2(aq)$ | • stains bench tops and clothing | Avoid contact. Wear safety glasses and a laboratory coat. |
| Complete and indicate that you have understood the information in the safety table.<br><br>Name (print): ________<br><br>I understand the safety information (signature): ________ | | |


ISBN 978 0 6557 0027 2

## METHOD

### PART A • STANDARDISATION OF IODINE SOLUTION

1 ▪ Record the concentration of a standard sodium thiosulfate solution in Table 1.
2 ▪ Use a pipette to transfer a 20.00 mL aliquot of the standard solution to a 250 mL conical flask. Add 1–2 mL starch indicator solution.
3 ▪ Fill a burette with iodine solution. Record the initial burette reading in Table 2.
4 ▪ Titrate the aliquot of sodium thiosulfate solution with the iodine solution. The end point occurs when the first permanent tinge of blue appears in the flask. Record the final burette reading.
5 ▪ Repeat steps 2–4 until three concordant titres have been obtained.

### PART B • DETERMINATION OF THE ASCORBIC ACID CONTENT OF VITAMIN C TABLETS

1 ▪ Weigh a vitamin C tablet and record its mass and other details in Table 3.
2 ▪ Transfer the tablet to a 250 mL conical flask. Add about 50 mL of deionised water and stir, crushing the tablet with a stirring rod until as much dissolves as possible. Add 1–2 mL starch indicator solution.
3 ▪ Refill the burette used in Part A with the iodine solution and record the burette reading in Table 4.
4 ▪ Titrate the vitamin C solution with the iodine solution until the point where the solution turns permanently blue. Record the final burette reading.
5 ▪ Repeat the procedure with a second and, if time permits, a third vitamin C tablet.

## RESULTS

### PART A

| Table 1 | |
|---|---|
| concentration of standard $Na_2S_2O_3$ solution (M) | |

Table 2

| Titration | Initial burette reading (mL) | Final burette reading (mL) | Titre (mL) |
|---|---|---|---|
| 1 | | | |
| 2 | | | |
| 3 | | | |

### PART B

| Table 3 | |
|---|---|
| brand of vitamin C tablets | |
| available vitamin C content specified by manufacturer | |
| mass of vitamin C tablet (g): tablet 1 | |
| mass of vitamin C tablet (g): tablet 2 | |

Table 4

| Titration | Initial burette reading (mL) | Final burette reading (mL) | Titre (mL) |
|---|---|---|---|
| tablet 1 | | | |
| tablet 2 | | | |

PRACTICAL ACTIVITY 20

DISCUSSION

## PART A • CALCULATIONS FOR THE STANDARDISATION OF THE IODINE SOLUTION

1 Write an ionic equation for the overall reaction that occurs during the titration of sodium thiosulfate solution with iodine solution.

2 Calculate the amount of sodium thiosulfate, in mol, in each 20.00 mL aliquot.

3 Calculate the average of the three concordant titres of iodine solution.

4 Find the amount, in mol, of iodine present in the average titre.

5 Calculate the concentration of the iodine solution in mol $L^{-1}$.

## PART B • CALCULATIONS FOR THE ASCORBIC ACID CONTENT OF VITAMIN C TABLETS

1 Write an ionic equation for the overall reaction that occurs during the titration of ascorbic acid solution with iodine solution.

2 Calculate the amount of iodine, in mol, in the average titre and hence the amount of ascorbic acid, in mol, in a tablet.

3 Calculate the mass of ascorbic acid in a vitamin C tablet.

4 How does your result compare with the manufacturer's specifications? If there is a discrepancy between your result and the manufacturer's claim, suggest why it might have arisen.

5 Pure ascorbic acid is a white, crystalline compound. Suggest why manufacturers of vitamin C tablets choose to give the tablets a yellow or orange colour.

6 Why is solid iodine not considered to be a good primary standard?


ISBN 978 0 6557 0027 2

7 New bottles of vitamin C tablets often contain a small packet of desiccating agent (a chemical that absorbs moisture from the air). Explain why this helps keep the vitamin C tablets in good condition.

8 Indicate any possible random and systematic errors of this analysis. Discuss how to improve the precision and accuracy of the results.

9 State which liquid is used to rinse each of the pieces of equipment listed. Indicate your reason for rinsing with this solution or liquid.

pipette:

burette:

standard flask:

conical flask (titrating flask):

## CONCLUSION

State the results of your determination of ascorbic acid in vitamin C tablets.

# PRACTICAL ACTIVITY 21

Controlled experiment

# Chromatography of a vegetable extract—extraction of natural plant compounds

## SUGGESTED DURATION

- 80 minutes

## INTRODUCTION

This experiment uses thin-layer chromatography (TLC) to separate the components in a green vegetable. A plate of silica or alumina is used as the stationary phase. The separation of the components in a mixture by chromatography depends on their attraction to the stationary surface and their solubility in the mobile phase. The effectiveness of two different solvents for this purpose are compared.

Once separated on the chromatogram, the silica powder containing each component can be scraped off separately and, using a suitable solvent, the component can be extracted from the solid silica powder.

These days many natural products are separated and extracted by HPLC (high-performance liquid chromatography), which is a more sophisticated type of chromatography than TLC. Often the desired component is present in the plant sample at low concentration, and HPLC is a more accurate and precise technique.

## AIM

To separate the naturally occurring pigments in a green vegetable by thin-layer chromatography, and to compare the effectiveness of two solvents for this purpose

| PRE-LAB SAFETY INFORMATION | | |
|---|---|---|
| **Material used** | **Hazard** | **Control** |
| acetone, $CH_3COCH_3(l)$, cyclohexane, $C_6H_{12}(l)$ | • irritating to the eyes, skin and respiratory system | Work in a fume hood if possible or in a well-ventilated area; keep away from flames and avoid breathing vapours.<br>Wear safety glasses and a laboratory coat. |
| ethanol, $CH_3CH_2OH(l)$ | • flammable | Keep away from flames and avoid breathing vapours. Wear safety glasses and a laboratory coat. |
| Please indicate that you have understood the information in the safety table.<br>Name (print):<br>I understand the safety information (signature): | | |

## MATERIALS

- finely chopped portion of spinach or silverbeet leaves (frozen spinach works well)
- 12 mL acetone (propanone), $CH_3COCH_3(l)$
- 2 mL cyclohexane, $C_6H_{12}(l)$
- 25 mL ethanol, $CH_3CH_2OH(l)$
- fine sand
- 600 mL beaker
- 10 mL measuring cylinder
- mortar and pestle
- 4 cm × 6.5 cm thin-layer chromatography plate*
- pencil and ruler
- spatula
- stirring rod
- glass capillary (melting point tube)
- safety glasses
- hair dryer (optional)

*Chromatography paper may be used instead of a thin-layer chromatography plate, with a 9 : 1 mixture of pentane and acetone as the solvent. An unfolded paper clip pushed through one end of the paper can be used to support the paper in the beaker.

ISBN 978 0 6557 0027 2

## PRACTICAL ACTIVITY 21

### METHOD

1. Use a pencil to draw a line across a thin-layer chromatography plate, 1 cm from one narrow end. This is known as the origin.
2. Extract the pigment from a spinach or silverbeet leaf by grinding the chopped leaf with about 10 mL of acetone (propanone), using a mortar and pestle. Addition of about half a spatula of fine sand will aid the crushing process.
3. In a fume hood, pour the acetone solution into a 100 mL beaker. Dip a capillary into the acetone and use the solution that rises into the capillary to draw a thin line along the pencil line on the chromatography plate.
4. Allow the line to dry, or blow it dry with a hair dryer; repeat the procedure until a thin, deep-green line is formed.
5. Use a measuring cylinder to pour 2 mL of acetone and 2 mL of cyclohexane into a 100 mL beaker. Use a stirring rod to mix the two liquids. (They will not dissolve in each other completely.)
6. Place the chromatography plate in the beaker so that the bottom of it is in the solvent and the plate rests against the rim of the beaker. The top of the solvent must be below the green line on the plate. Invert a 600 mL beaker over the top of the smaller beaker.
7. Allow the solvent to rise up the plate to a height of 5.5–6 cm.
8. Remove the plate, allow it to dry and use a pencil to mark the level the solvent has reached (this is called the solvent front) and the position of each component. (The colour of some components fades after several hours.)
9. Repeat steps 1–8 using a mixture of 1.5 mL ethanol, 1.5 mL acetone and 1.5 mL cyclohexane in the beaker in step 5.

### RESULTS

For each major component in the vegetable extract, describe its colour, measure its distance from the origin and record this information in Table 1. Also calculate and record its $R_f$ value in the table.

**Table 1**

| Cyclohexane–acetone<br>Distance of solvent front from origin: ________ cm | | | Ethanol–acetone–cyclohexane<br>Distance of solvent front from origin: ________ cm | | |
|---|---|---|---|---|---|
| Colour of band | Distance from origin (cm) | $R_f$ value calculation | Colour of band | Distance from origin (cm) | $R_f$ value calculation |
| | | | | | |
| | | | | | |
| | | | | | |
| | | | | | |

### DISCUSSION

The following information about likely components in plant extracts may be useful.

| Component | Colour |
|---|---|
| chlorophyll a | blue-green |
| chlorophyll b | yellow-green |
| xanthophyll | orange-yellow |
| β-carotene | yellow |

1 a What is the independent variable? ____________________

b What is the dependent variable? ____________________

c What are the controlled variables? ____________________

2 Why should the line of the mixture in step 4 of the Method be as thin as possible and concentrated?

____________________

____________________

## PRACTICAL ACTIVITY 21

**3** Why must the line where the extract is placed on the plate be above the level of the solvent at the beginning of the experiment?

---

**4** Explain in detail how the components of the plant material separate as they move up the plate.

---

---

**5** Why is the site of the origin marked with a pencil and not a pen?

---

**6** From your determination the $R_f$ value of each component in your chromatogram, and using the data about likely components given above, state the $R_f$ value for each component in the green pigment mixture.

---

---

---

**7** Discuss the differences between the cyclohexane–acetone chromatogram and the ethanol–acetone-cyclohexane chromatogram. Which solvent is more polar and which was more effective at separating the components?

---

---

---

**8** Why do different solvents produce different chromatograms?

---

---

**9** The orange colour of carrots and pumpkins is due to the presence of carotene. How could you determine whether one of the compounds found in the vegetable extract analysed in this experiment is carotene?

---

---

**10** Compare the SDS for cyclohexane and ethanol. What principles of green chemistry are being applied if an industry uses ethanol instead of cyclohexane to extract a component from plant material?

---

---

### CONCLUSION

Summarise the use of TLC as a simple method of extraction for natural plant products that may be used in medicines.

---

---

---


ISBN 978 0 6557 0027 2

# PRACTICAL ACTIVITY 22

Controlled experiment

# Action of catalysts—peroxide and enzymes

## SUGGESTED DURATION

- 50 minutes

## INTRODUCTION

In this experiment, students investigate catalysis by both inorganic compounds and an enzyme. Hydrogen peroxide slowly decomposes, forming water and oxygen gas. The decomposition process is very slow, so hydrogen peroxide may be safely stored in sealed containers. The reaction is represented by the equation:

$$2H_2O_2(aq) \rightarrow 2H_2O(l) + O_2(g)$$

A catalyst such as manganese dioxide, or the enzyme present in potato, will speed up this process so that it may be readily observed. The enzyme in potato is known as catalase and its function is to destroy the hydrogen peroxide produced during reactions in cells and thus prevent the accumulation of this substance, which is toxic. A single catalase molecule can decompose five million hydrogen peroxide molecules in 1 second!

Enzymes are proteins that catalyse specific biochemical reactions. Their mode of action is related to their shape. Heating may permanently alter the shape of an enzyme, preventing it from performing its function. When this occurs, an enzyme is said to have been denatured.

## AIM

- To demonstrate the action of an inorganic catalyst and an enzyme
- To show that an enzyme present in potato and liver acts as a catalyst and demonstrate that this enzyme may be denatured by heat

| PRE-LAB SAFETY INFORMATION | | |
|---|---|---|
| **Material used** | **Hazard** | **Control** |
| hydrogen peroxide, $H_2O_2(aq)$ | • strong oxidising agent | Avoid contact with skin and eyes. |
| manganese dioxide, $MnO_2(s)$ | • irritates skin and eyes<br>• harmful if inhaled | Wear safety glasses and lab coat. |
| Please indicate that you have understood the information in the safety table.<br>Name (print):<br>I understand the safety information (signature): | | |

## MATERIALS

- 70 mL of 20 volume hydrogen peroxide solution, $H_2O_2(aq)$
- manganese dioxide, $MnO_2(s)$
- half a potato, finely chopped
- liver, roughly chopped
- ice
- bottle of concentrated liquid detergent
- 7 test tubes, 25 mm × 180 mm
- test-tube rack
- 4 × 250 mL beakers (for water baths)
- 10 mL measuring cylinder
- spatula
- 2 Bunsen burners, tripod stands and gauze mats
- 2 bench mats
- 4 thermometers, –10 to 110°C
- knife
- lab coat
- safety glasses

ISBN 978 0 6557 0027 2

## METHOD

## PART A • ACTION OF CATALYSTS ON HYDROGEN PEROXIDE

1 ▪ Place 10 mL of hydrogen peroxide solution in each of three test tubes, labelled 1, 2 and 3. Add 2–3 drops of detergent to each test tube.

2 ▪ To test tube 1 add a small amount of manganese dioxide. Record your observations in Table 1.

3 ▪ Add a small amount of finely chopped potato to test tube 2 and add chopped liver to test tube 3. Record your observations in Table 1.

## PART B • EFFECT OF TEMPERATURE ON ENZYME ACTIVITY

1 ▪ Set up four water baths at temperatures of about 0°C (ice–water), 20°C, 37°C (body temperature) and 70°C. Measure and record the exact temperature of each bath in Table 2.

2 ▪ Insert finely chopped potato to a depth of 2 cm in each of the four test tubes. Place 2–3 drops of detergent in each test tube. Place a test tube in each of the water baths. Allow the test tubes to remain in the baths for 5 minutes.

3 ▪ Add 10 mL of hydrogen peroxide solution to each test tube and leave the test tubes in the water baths for a further 5 minutes.

4 ▪ Measure and record the height of the foam in each test tube in Table 2.

### Variables

Independent variable: ______________________

Dependent variable: ______________________

Controlled variables: ______________________

## RESULTS

**Table 1** Observations

| Test tube 1: $H_2O_2$ solution + manganese dioxide | Test tube 2: $H_2O_2$ solution + potato | Test tube 3: $H_2O_2$ solution + liver |
|---|---|---|
| | | |

**Table 2** $H_2O_2$ solution + potato at different temperatures

| Water bath temperature (°C) | Exact temperature (°C) | Height of foam column (mm) |
|---|---|---|
| 0 (ice–water) | | |
| 20 | | |
| 37 (body temperature) | | |
| 70 | | |

## DISCUSSION

1 Discuss the different effects, in each case, of adding manganese dioxide, liver or potato to the $H_2O_2$ solution.

______________________

______________________

2 Explain why the potato and liver are chopped before use in this experiment.

______________________

 ISBN 978 0 6557 0027 2

3 Draw a graph of foam height against temperature. The height of the foam is a measure of the rate of oxygen production during the decomposition of hydrogen peroxide.

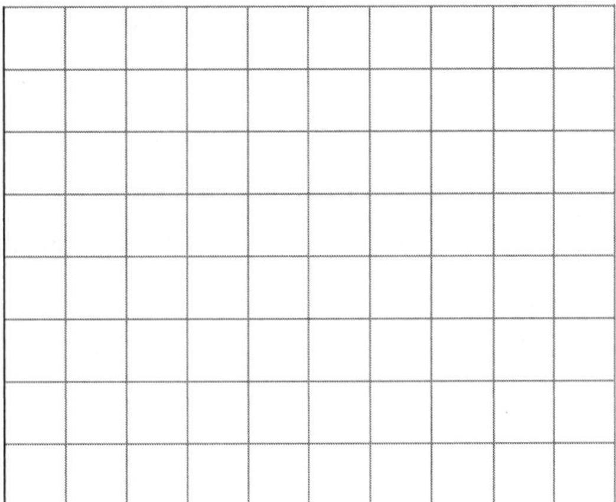

4 Explain the shape of the following sections of your graph in terms of the rate of reaction and enzyme structure.

a 0–37°C: ______

b 37–70°C: ______

5 If you were to repeat this experiment, identify how you could improve the precision of your results. Explain your answer.

## CONCLUSION

# EXAM QUESTIONS

## Multiple-choice questions

**Question 1** VCE Chemistry 2017 (A) 15

Which one of the following is a correct statement about the denaturation of a protein?

**A.** Denaturation is characterised by the release of peptides.

**B.** Alcohol denatures proteins by disrupting the hydrogen bonding.

**C.** Denaturation involves disruption of all bonds in the tertiary structure.

**D.** The primary and secondary structures are disrupted when denaturation occurs.

**Question 2** VCE Chemistry 2017 (A) 17

Shown below is the infrared spectrum of an organic compound.

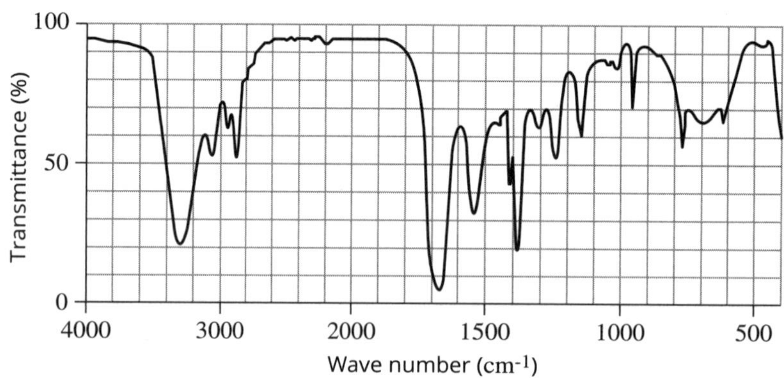


The organic compound that produces this spectrum is an

**A.** aldehyde.

**B.** alcohol.

**C.** amide.

**D.** ester.

 ISBN 978 0 6557 0027 2

# EXAM QUESTIONS

*Use the following information to answer Questions 3 and 4.*

The mass of caffeine in a particular coffee drink was determined by high-performance liquid chromatography (HPLC).

The calibration curve produced from running standard solutions of caffeine through an HPLC column is shown below.

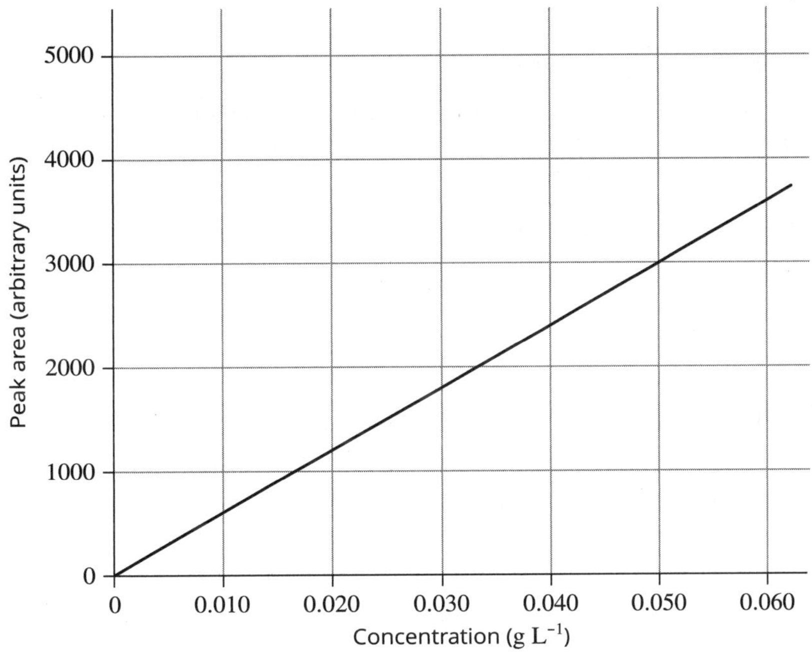


A 5.0 mL aliquot of the coffee drink was diluted to 50.0 mL with deionised water. A sample of the diluted coffee drink was run through the HPLC column under identical conditions to those used to obtain the calibration curve. The peak area obtained for this diluted sample was 2400 arbitrary units.

**Question 3** VCE Chemistry 2017 (A) 21

The HPLC column used has a non-polar stationary phase.

The most suitable solvent for determining the concentration of caffeine in the sample is

**A.** carbon tetrachloride, $CCl_4$.

**B.** methanol, $CH_3OH$.

**C.** octanol, $C_8H_{17}OH$.

**D.** hexane, $C_6H_{14}$.

**Question 4** VCE Chemistry 2017 (A) 22

The mass of caffeine, in grams, in 350 mL of the undiluted coffee drink is closest to

**A.** 0.014

**B.** 0.070

**C.** 0.14

**D.** 0.40

# EXAM QUESTIONS

**Question 5** VCE Chemistry 2018 (A) 19

Which one of the following molecules contains a chiral carbon?

**A.** $CH_2CHCH_2CH_3$

**B.** $CH_2FCH_2CH_2Cl$

**C.** $CH_3CHOHCH_2CH_3$

**D.** $CH_3CH_2CFClCH_2CH_3$

*Use the following information to answer Questions 6 and 7.*

A clear, colourless liquid extract of the rhubarb plant was analysed for the concentration of oxalic acid, $H_2C_2O_4$, by direct titration with a recently standardised and acidified potassium permanganate solution, $KMnO_4(aq)$. The balanced equation for this titration is shown below.

$$2MnO_4^-(aq) + 5C_2O_4^{2-}(aq) + 16H^+(aq) \rightarrow 2Mn^{2+}(aq) + 10CO_2(g) + 8H_2O(l)$$

purple colourless colourless

The steps in the titration were as follows:

Step 1—A 20.00 mL aliquot of the rhubarb extract was placed in a 200 mL conical flask.

Step 2—The burette was filled with acidified 0.0200 M $KMnO_4$ solution.

Step 3—The acidified 0.0200 M $KMnO_4$ solution was titrated into the rhubarb extract in the conical flask. The titration was considered to have reached the end point when the solution in the conical flask showed a permanent change in colour to pink. The volume of the titre was recorded.

Step 4—The titration was repeated until three concordant results were obtained.

The average of the concordant titres was 21.7 mL.

**Question 6** VCE Chemistry 2018 (A) 17

The concentration of $H_2C_2O_4$ in the rhubarb extract is closest to

**A.** $5.43 \times 10^{-2}$ M

**B.** $5.00 \times 10^{-2}$ M

**C.** $2.17 \times 10^{-2}$ M

**D.** $7.40 \times 10^{-4}$ M

**Question 7** VCE Chemistry 2018 (A) 18

Which of the following rinses is least likely to affect the accuracy of the results?

| | Item | Rinse solution |
|---|---|---|
| **A.** | burette | distilled water |
| **B.** | burette | rhubarb extract |
| **C.** | pipette | $KMnO_4(aq)$ |
| **D.** | conical flask | distilled water |

**Question 8** VCE Chemistry 2019 (A) 27

An organic compound has a molar mass of 88 g mol$^{-1}$. The $^{13}C$ NMR spectrum of the organic compound shows four distinct peaks. The organic compound is most likely

**A.** butan-1-ol.

**B.** 2-methylbutan-1-ol.

**C.** 2-methylbutan-2-ol.

**D.** 2,2-dimethylpropan-1-ol.

 ISBN 978 0 6557 0027 2

# EXAM QUESTIONS

### Question 9 VCE Chemistry 2020 (A) 29

Which of the following combinations of bonds can be broken during the breakdown of a protein that a person has eaten?

**A.** covalent bonds in the secondary structure and hydrogen bonds in the primary structure

**B.** covalent bonds in the tertiary structure and hydrogen bonds in the secondary structure

**C.** covalent bonds in the secondary structure and hydrogen bonds in the tertiary structure

**D.** covalent bonds in the quaternary structure and hydrogen bonds in the primary structure

### Question 10 VCE Chemistry 2017 (A) 10

Which one of the following structures represents a zwitterion of a 2-amino acid?

**A.**

$$\begin{array}{ccccc} & & \mathrm{O} & & \\ & & \| & & \\ & \mathrm{CH_2} & - \mathrm{C} - & \mathrm{NH_2} & \\ & | & & & \\ \mathrm{H_2N} - & \mathrm{CH} & - \mathrm{COO^-} & & \end{array}$$

**B.**

$$\begin{array}{ccc} \mathrm{CH_3} - & \mathrm{CH} & - \mathrm{CH_3} \\ & | & \\ \mathrm{H_3N^+} - & \mathrm{CH} & - \mathrm{COOH} \end{array}$$

**C.**

$$\begin{array}{ccc} & \mathrm{CH_2} & - \mathrm{CH_2} - \mathrm{COO^-} \\ & | & \\ \mathrm{H_3N^+} - & \mathrm{CH} & - \mathrm{COO^-} \end{array}$$

**D.**

$$\begin{array}{ccc} & \mathrm{CH_2} & - \mathrm{OH} \\ & | & \\ \mathrm{H_3N^+} - & \mathrm{CH} & - \mathrm{COO^-} \end{array}$$

## Short-answer questions

### Question 1 (9 marks) VCE Chemistry 2018 (B) 5

Researchers have developed a technique that allows high-performance liquid chromatography (HPLC) to simultaneously determine the concentration of sugar, organic acids and alcohols in fermentation products and food samples. The table below shows the retention times for some common organic molecules using this technique.

| Retention times for some organic molecules | |
|---|---|
| **Organic molecule** | **Retention time (min)** |
| maltose | 12.50 |
| lactose | 12.70 |
| glucose | 14.45 |
| galactose | 15.35 |
| succinic acid | 18.25 |
| glycerol | 20.33 |
| ethanol | 30.63 |

# EXAM QUESTIONS

Below is a chromatogram of four organic molecules.

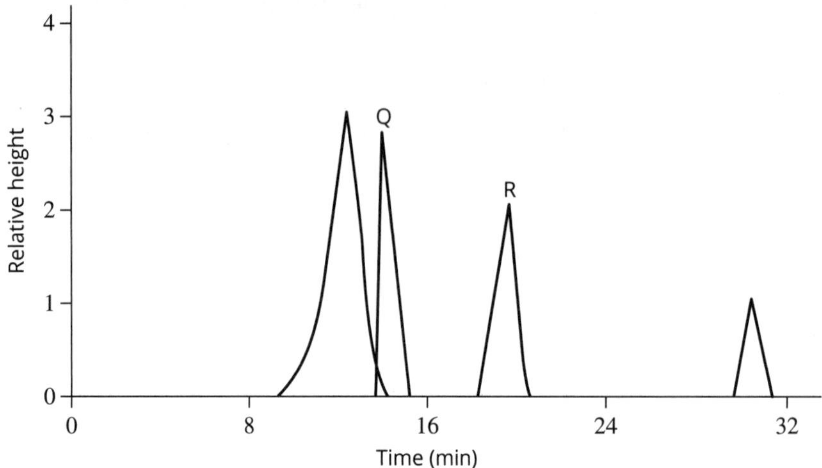


**a.** Identify the substances responsible for each of the peaks Q and R. 2 marks

**b.** Two substances have different retention times under identical conditions using the same HPLC equipment. Explain this difference in retention times. 3 marks

**c.** *Miscanthus floridulus* is a type of grass that is approximately 37% cellulose by mass. *M. floridulus* is being researched as a feedstock for bioethanol, $CH_3CH_2OH$, production. The cellulose in this grass can be used to produce $CH_3CH_2OH$, as summarised in the flow chart below. In each step of this process, there is incomplete conversion and waste products are formed.

cellulose $(C_6H_{11}O_5)_x$ —enzymatic hydrolysis→ glucose $xC_6H_{12}O_6$

$C_6H_{12}O_6$ —yeast fermentation→ $CH_3CH_2OH + CO_2$

HPLC can be used to determine the concentration of fermentation products at different points in the process.

 ISBN 978 0 6557 0027 2

**i.** Researchers use percentage by mass of $CH_3CH_2OH$ produced to make comparisons between different grasses as feedstocks. Using HPLC, a researcher determined that 144 L of $CH_3CH_2OH$ was produced from 1000 kg of *M. floridulus*.

Calculate the percentage by mass of $CH_3CH_2OH$ produced from the mass of cellulose in *M. floridulus*, assuming the density of $CH_3CH_2OH$ is 0.79 g $mL^{-1}$. 3 marks

**ii.** Calculate the amount of energy, in kilojoules, that would be produced if complete combustion of 144 L of $CH_3CH_2OH$ occurred. 1 mark

**Question 2** (9 marks) VCE Chemistry 2019 (B) 8

An unknown organic compound contains carbon, hydrogen and oxygen.

It is known that:

- the compound does not contain carbon-to-carbon double bonds (C=C)
- the molecular ion peak is found at a mass-to-charge ratio (*m/z*) of 74
- the $^{13}C$ NMR has three distinct peaks.

**a.** A small peak in the mass spectrum can be identified at *m/z* = 75.

Explain the presence of this peak. 1 mark

**b. i.** Use the information provided to give **two** possible molecular formulas for this compound. 2 marks

**ii.** The $^{1}H$ NMR spectrum of the compound shows three sets of peaks with a peak area ratio of 3 : 2 : 1.

What does this information tell you about the structure of the compound and its molecular formula? Justify your answer by referring to the information given about the peaks in the $^{1}H$ NMR spectrum. 2 marks

ISBN 978 0 6557 0027 2

# EXAM QUESTIONS

**c.** There are many structural isomers of this compound.

Draw the structural formulas of **two** possible isomers. 2 marks

**d.** The infrared (IR) spectrum of the compound is shown below.

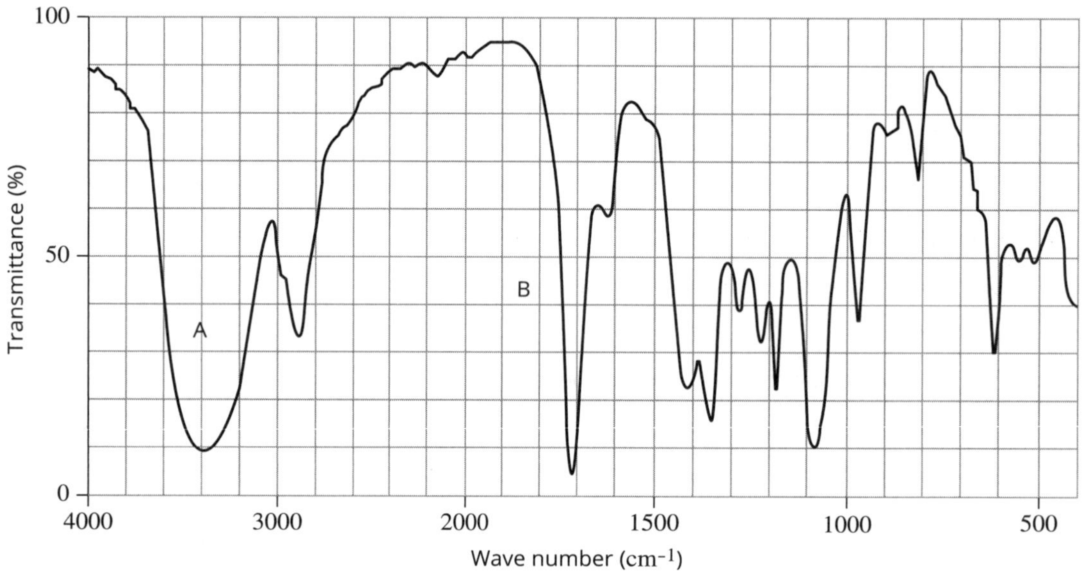


**i.** Identify the functional groups responsible for the absorption peaks labelled A and B in the IR spectrum. 1 mark

A: ______________________________

B: ______________________________

**ii.** Using the $^{1}H$ NMR information given in part b.ii and the IR spectrum provided above, draw the structural formula of the compound. 1 mark

 ISBN 978 0 6557 0027 2

# EXAM QUESTIONS

**Question 3** (9 marks) VCE Chemistry 2021 (B) 6

A reaction pathway beginning with 1-bromopentane is shown below.

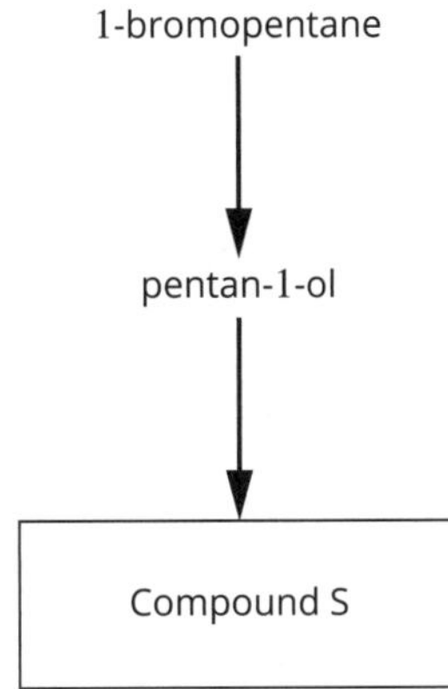


**a.** Draw the structural formula for an isomer of 1-bromopentane that contains a chiral carbon and circle this chiral carbon. 2 marks

**b. i.** Write a balanced equation for the reaction that will produce pentan-1-ol from 1-bromopentane and a sodium salt. 2 marks

**ii.** Calculate the atom economy in the production of pentan-1-ol from 1-bromopentane and a sodium salt. 3 marks

**c.** Pentan-1-ol is fully oxidised to compound S. Write the IUPAC name of compound S in the box in the diagram above. 1 mark

**d.** In an alternative reaction pathway, pentanamide can be formed from 1-bromopentane. Draw the skeletal formula for pentanamide. 1 mark

**Question 4** (8 marks) VCE Chemistry 2021 (B) 7

Two students are given a homework assignment that involves analysing a set of spectra and identifying an unknown compound. The unknown compound is one of the molecules shown below.

| | |
|---|---|
| P | Q |
| R | S |
| T | |

The $^{13}C$ NMR spectrum of the unknown compound is shown below.

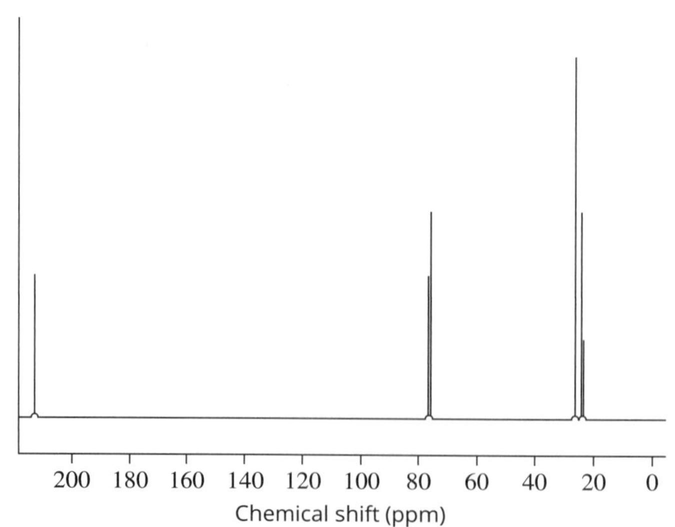


**a.** Based on the number of peaks in the $^{13}C$ NMR spectrum, which compound—P, Q, R, S or T—could be eliminated as the unknown compound? 1 mark

 ISBN 978 0 6557 0027 2

# EXAM QUESTIONS

**b.** The infrared (IR) spectrum of the unknown compound is shown below.

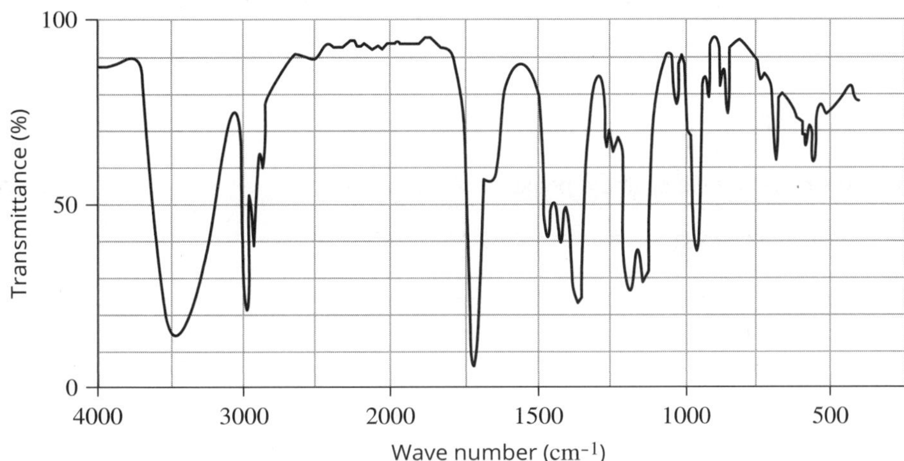


Identify which of the five compounds can be eliminated on the basis of the IR spectrum. Justify your answer using data from the IR spectrum. 3 marks

**c.** The mass spectrum of the unknown compound is shown below.

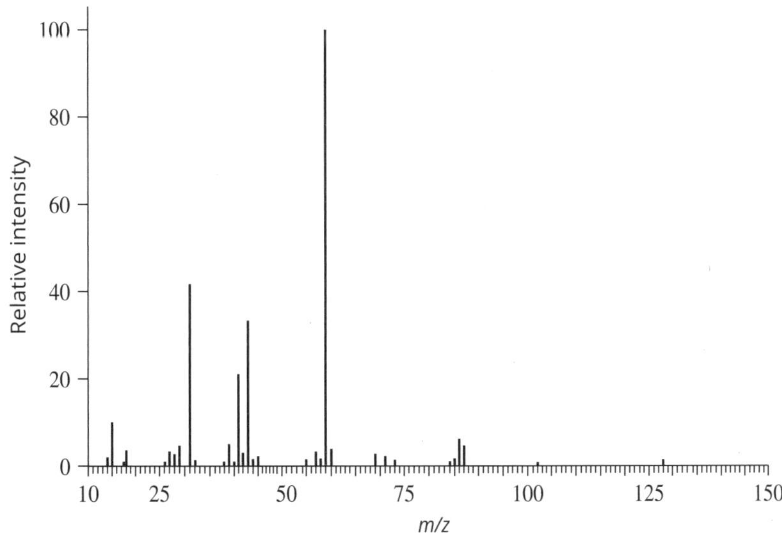


**i.** Write the chemical formula of the species that produces a peak at $m/z = 43$. 1 mark

ii. Define *m/z* as used in mass spectroscopy. 1 mark

iii. Explain why one molecule can produce multiple peaks on a mass spectrum. 2 marks

**Question 5** (7 marks) VCE Chemistry 2020 (B) 10

Analytical chemistry deals with methods for determining the chemical composition of samples of matter. A qualitative method yields information about the identity of atomic or molecular species or the functional groups in the sample ...

Analytical methods are often classified as being either classical or instrumental.

Source: DA Skoog, FJ Holler and SR Crouch, *Principles of Instrumental Analysis*, 6th edition, Thomson Brooks/Cole, Belmont (CA), 2007, p. 1

Classical methods include qualitative analysis, such as treating a compound with reagents to observe any reaction, and quantitative methods, such as volumetric analysis, where the amount of a compound is determined by its reaction with a standard reagent.

Instrumental methods include a variety of spectroscopy, such as IR spectroscopy and NMR spectroscopy.

**a.** Explain how the classical methods of analytical chemistry can be used to determine information about alcohols. In your answer, refer to:

- qualitative analysis and how it can be used to determine whether a compound is an alcohol and, if it is, the type of alcohol
- quantitative analysis. 3 marks

 ISBN 978 0 6557 0027 2

**b.** $C_3H_6O$ can exist as a ketone or as a primary alcohol.

Explain how the principles of IR spectroscopy and $^1H$ NMR spectroscopy lead to different spectra for the ketone and primary alcohol isomers of $C_3H_6O$, which can then be used to differentiate between the two molecules. 4 marks

ISBN 978 0 6557 0027 2

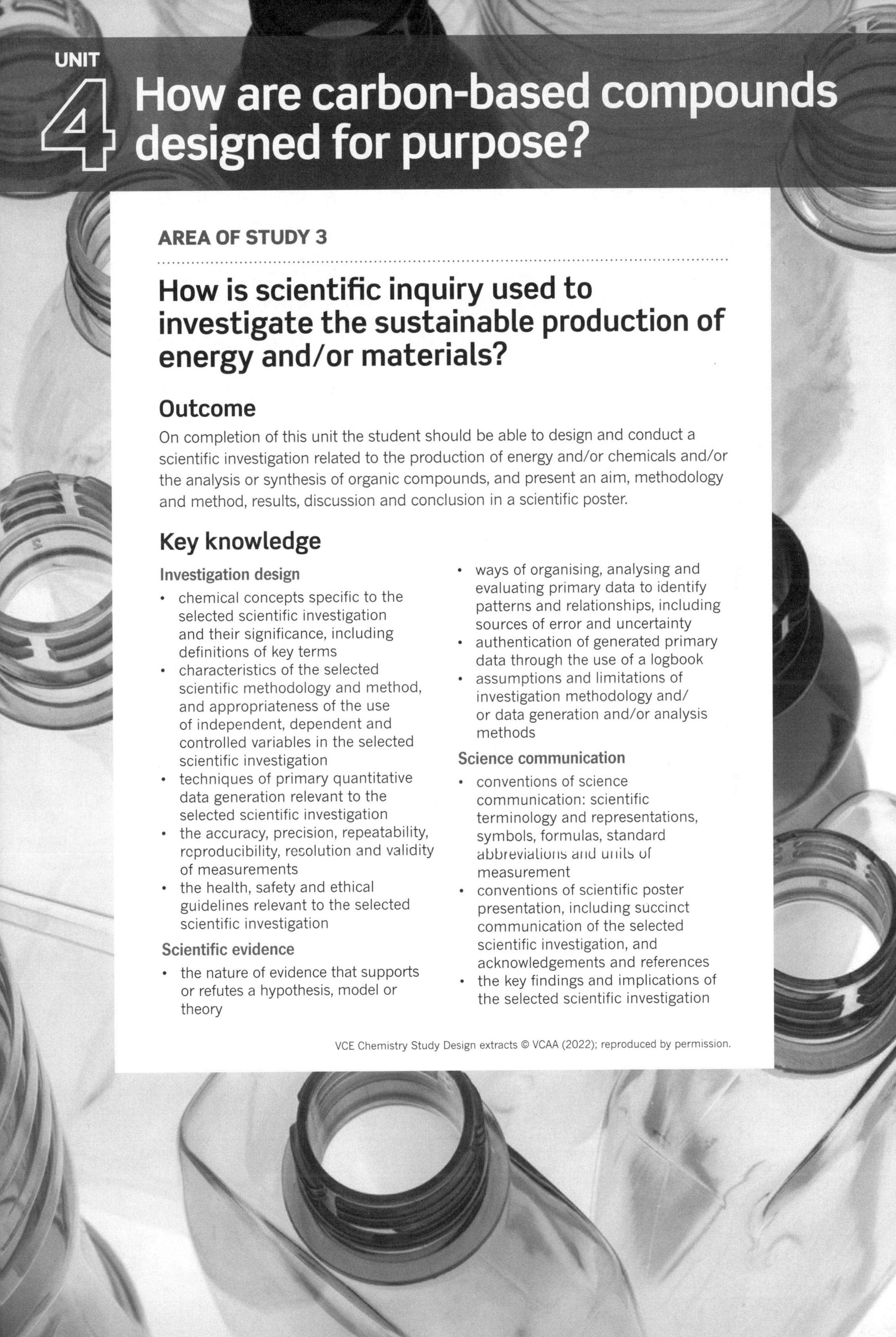

UNIT 4

# How are carbon-based compounds designed for purpose?

## AREA OF STUDY 3

## How is scientific inquiry used to investigate the sustainable production of energy and/or materials?

### Outcome

On completion of this unit the student should be able to design and conduct a scientific investigation related to the production of energy and/or chemicals and/or the analysis or synthesis of organic compounds, and present an aim, methodology and method, results, discussion and conclusion in a scientific poster.

### Key knowledge

**Investigation design**

- chemical concepts specific to the selected scientific investigation and their significance, including definitions of key terms
- characteristics of the selected scientific methodology and method, and appropriateness of the use of independent, dependent and controlled variables in the selected scientific investigation
- techniques of primary quantitative data generation relevant to the selected scientific investigation
- the accuracy, precision, repeatability, reproducibility, resolution and validity of measurements
- the health, safety and ethical guidelines relevant to the selected scientific investigation

**Scientific evidence**

- the nature of evidence that supports or refutes a hypothesis, model or theory
- ways of organising, analysing and evaluating primary data to identify patterns and relationships, including sources of error and uncertainty
- authentication of generated primary data through the use of a logbook
- assumptions and limitations of investigation methodology and/or data generation and/or analysis methods

**Science communication**

- conventions of science communication: scientific terminology and representations, symbols, formulas, standard abbreviations and units of measurement
- conventions of scientific poster presentation, including succinct communication of the selected scientific investigation, and acknowledgements and references
- the key findings and implications of the selected scientific investigation

VCE Chemistry Study Design extracts © VCAA (2022); reproduced by permission.

# Sample scientific investigation: Exploring the optimum conditions for electroplating

## STUDENT-DESIGNED SCIENTIFIC INVESTIGATION

You will design and conduct a scientific investigation related to the production of energy and/or chemicals and/or the analysis or synthesis of organic compounds, and present an aim, methodology and methods, results, discussion and a conclusion in a scientific poster.

The investigation draws on the knowledge and related key science skills developed across Units 3 and 4 and is undertaken in the laboratory.

## ASSESSMENT FOR OUTCOME 3

The design, analysis and findings of your scientific investigation will be presented as a structured scientific poster and your logbook entries. The poster should not exceed 600 words.

## SUGGESTED DURATION

A minimum of 10 hours of class time should be devoted to undertaking and communicating findings related to Area of Study 3. As in Areas of Study 1 and 2, your teacher will be your primary guide. To assist you, the key steps to follow, relevant poster section and suggested time allocation are set out below.

The sample investigation on the following page steps you through a controlled experiment, conducted following the designing and planning phase of the investigation.

## USING THIS GUIDE

There is scope for developing an investigation on a range of topics relevant to Units 3 and 4 and your teacher will direct you. The following sample investigation is drawn from Unit 3 Area of Study 2, with a focus on the key knowledge and skills related to the application of electrolysis. The investigation explores the optimum conditions for electroplating and some of the factors that affect the quality and quantity of the deposit formed in an electroplating cell.

In your student-designed investigation, you will need to carefully consider how to monitor the variable under investigation to achieve valid and reliable results. Safety considerations and a risk assessment must be considered in the planning process and you also need to address social and ethical issues. The sample will guide you when you are planning, conducting and reporting on your results. All details of your investigation must be recorded in your logbook as it provides evidence of your own work.

Refer to the Toolkit at the beginning of this book for more detailed information on designing and conducting scientific investigations, and presenting a scientific report.

The online resources for this Unit 4 Area of Study 3 sample investigation provide a guide to completing your logbook and writing your scientific report in the format of a scientific poster.

| Scientific investigation section | Key step | Relevant poster section(s) | Suggested time allocation |
|---|---|---|---|
| Designing and planning your investigation | Step 1: Developing your research question, aim and hypothesis | Title<br>Introduction | 60–120 minutes |
| | Step 2: Determining appropriate methodology and methods | Methodology and methods | 60–120 minutes |
| Conducting your investigation and recording and presenting data | Step 3: Conducting your investigation to generate primary data | | 120–240 minutes |
| | Step 4: Recording, organising and presenting your data | Results | 120–180 minutes |
| Discussing your investigation and drawing evidence-based conclusions | Step 5: Analysing and evaluating your data | Discussion<br>Conclusion | 60–120 minutes |
| | Step 6: Referencing | References and acknowledgements | 30 minutes |
| Reporting on your investigation | Step 7: Preparing your scientific poster | All sections | 60–180 minutes |


ISBN 978 0 6557 0027 2

## INTRODUCTION

### BACKGROUND

Using a simple copper-plating cell, it is quite difficult to obtain a bright, shiny coating of copper metal that adheres well to the cathode. Instead, the copper deposit is likely to be dark brown due to irregularities in its thickness. Industrial electroplating involves the careful control of conditions and of electrolyte composition. In this investigation, some of the factors that affect the quality and quantity of the deposit formed in an electroplating cell are explored.

### AIM

To design and perform experiments to determine the best conditions for electroplating, considering the significance of one or more of the following factors:

- concentration of the copper(II) electrolyte solution
- temperature of the electrolyte
- current
- surface area of the cathode immersed in the electrolyte.

### HYPOTHESIS

You will develop a hypothesis such as those in the following list.

- The longer the time that current flows through the cell, the greater the mass of copper coated at the cathode.
- For a particular time interval, the greater the current that flows through the cell, the greater the amount of copper deposited.
- The amount of copper deposited increases as the concentration of the electrolyte increases.
- As the temperature of the cell increases, the smaller the amount of copper deposited.

## METHODOLOGY AND METHODS

The methodology is a brief description of the general approach taken to investigate the research question or hypothesis and the reasons this approach is taken. The methodology can be described as the rationale behind your investigative methods. The methodology for this scientific investigation is a controlled experiment. Your method is the series of steps followed to generate the data during the investigation.

### METHODS

Ensure the steps in the method are clear and written in the third person. State the variable to be considered and how you will control the other variables. The use of a labelled diagram of the cell will be helpful.

Your earlier practical activities will help you to determine a list of chemicals and any materials needed.

## RESULTS

The results will be qualitative and quantitative and could include:

- presentation of the results in tables and graphs to illustrate trends and summarise relationships
- photographs to record the experimental set-up and results
- all observations
- any difficulties that occurred as well as all the positive outcomes
- errors and uncertainties.

Everything should be included in your logbook, which is your essential authentication document.

## DISCUSSION

The discussion is a critical section where you will analyse, interpret and evaluate the primary data collected. You will also need to evaluate the methodologies, methods and the impact of any limitations, and suggest improvements for later, similar, investigations. Your results should be linked to specific chemical concepts, the aim and the hypothesis.

## CONCLUSION

This is a clear, concise statement linking the question (title), aim and hypothesis to the results of the investigation.

## REFERENCES AND ACKNOWLEDGEMENTS

Remember to reference and acknowledge all secondary sources used in the course of planning and conducting your investigation. This may include:

- your logbook—primary data, conversations, research etc.
- material supplied by your teacher and the school
- internet research and sources.

It is important to provide bibliographical data according to accepted protocols, such as the APA style. For example, for books:

Commons, P. (2023). *Heinemann Chemistry 2 Skills and Assessment*. Pearson Australia.

# Periodic table

**Key:**
12 (ATOMIC NUMBER) | **Mg** (SYMBOL) | 24.3 (RELATIVE ATOMIC MASS; ( ) indicates most stable isotope) | 1090 (BOILING POINT °C) | 650 (MELTING POINT °C) | 1.3 (ELECTRONEGATIVITY) | $[Ne]3s^2$ (ELECTRON STRUCTURE) | Magnesium

Cell format: atomic number, **symbol**, relative atomic mass; boiling point °C, melting point °C, electronegativity; electron structure, name.

s BLOCK: groups 1–2; d BLOCK: groups 3–12; p BLOCK: groups 13–18

| Periods | Group 1 | 2 | 3 | 4 | 5 | 6 | 7 | 8 | 9 | 10 | 11 | 12 | 13 | 14 | 15 | 16 | 17 | 18 |
|---|---|---|---|---|---|---|---|---|---|---|---|---|---|---|---|---|---|---|
| 1 | | | | | | | | | 1 **H** 1.0<br>-252.9, -259, 2.2<br>$1s^1$ Hydrogen | | | | | | | | | 2 **He** 4.0<br>-268.9, -272, –<br>$1s^2$ Helium |
| 2 | 3 **Li** 6.9<br>1342, 180, 1.0<br>$1s^22s^1$ Lithium | 4 **Be** 9.0<br>2468, 1287, 1.6<br>$1s^22s^2$ Beryllium | | | | | | | | | | | 5 **B** 10.8<br>4000, 2077, 2.0<br>$1s^22s^22p^1$ Boron | 6 **C** 12.0<br>4827, 3500, 2.6<br>$1s^22s^22p^2$ Carbon | 7 **N** 14.0<br>-195.8, -210, 3.0<br>$1s^22s^22p^3$ Nitrogen | 8 **O** 16.0<br>-183, -219, 3.4<br>$1s^22s^22p^4$ Oxygen | 9 **F** 19.0<br>-188.1, -220, 4.0<br>$1s^22s^22p^5$ Fluorine | 10 **Ne** 20.2<br>-246, -249, –<br>$1s^22s^22p^6$ Neon |
| 3 | 11 **Na** 23.0<br>882.9, 97.8, 0.9<br>$[Ne]3s^1$ Sodium | 12 **Mg** 24.3<br>1090, 650, 1.3<br>$[Ne]3s^2$ Magnesium | | | | | | | | | | | 13 **Al** 27.0<br>2519, 660.3, 1.6<br>$[Ne]3s^23p^1$ Aluminium | 14 **Si** 28.1<br>3265, 1414, 1.9<br>$[Ne]3s^23p^2$ Silicon | 15 **P** 31.0<br>280.5, 44.2, 2.2<br>$[Ne]3s^23p^3$ Phosphorus | 16 **S** 32.1<br>444.6, 115.2, 2.6<br>$[Ne]3s^23p^4$ Sulfur | 17 **Cl** 35.5<br>-34, -102, 3.2<br>$[Ne]3s^23p^5$ Chlorine | 18 **Ar** 39.9<br>-185.8, -189, –<br>$[Ne]3s^23p^6$ Argon |
| 4 | 19 **K** 39.1<br>758.8, 63.4, 0.8<br>$[Ar]4s^1$ Potassium | 20 **Ca** 40.1<br>1484, 842, 1.0<br>$[Ar]4s^2$ Calcium | 21 **Sc** 45.0<br>2836, 1541, 1.4<br>$[Ar]3d^14s^2$ Scandium | 22 **Ti** 47.9<br>3287, 1670, 1.5<br>$[Ar]3d^24s^2$ Titanium | 23 **V** 50.9<br>3407, 1910, 1.6<br>$[Ar]3d^34s^2$ Vanadium | 24 **Cr** 52.0<br>2671, 1907, 1.7<br>$[Ar]3d^54s^1$ Chromium | 25 **Mn** 54.9<br>2061, 1246, 1.6<br>$[Ar]3d^54s^2$ Manganese | 26 **Fe** 55.8<br>2861, 1538, 1.8<br>$[Ar]3d^64s^2$ Iron | 27 **Co** 58.9<br>2927, 1495, 1.9<br>$[Ar]3d^74s^2$ Cobalt | 28 **Ni** 58.7<br>2913, 1455, 1.9<br>$[Ar]3d^84s^2$ Nickel | 29 **Cu** 63.5<br>2560, 1085, 1.9<br>$[Ar]3d^{10}4s^1$ Copper | 30 **Zn** 65.4<br>907, 419.5, 1.6<br>$[Ar]3d^{10}4s^2$ Zinc | 31 **Ga** 69.7<br>2229, 27.8, 1.8<br>$[Ar]3d^{10}4s^24p^1$ Gallium | 32 **Ge** 72.6<br>2833, 938.2, 2.0<br>$[Ar]3d^{10}4s^24p^2$ Germanium | 33 **As** 74.9<br>613, 816.8, 2.2<br>$[Ar]3d^{10}4s^24p^3$ Arsenic | 34 **Se** 79.0<br>684.8, 220.8, 2.6<br>$[Ar]3d^{10}4s^24p^4$ Selenium | 35 **Br** 79.9<br>58.8, -7.1, 3.0<br>$[Ar]3d^{10}4s^24p^5$ Bromine | 36 **Kr** 83.8<br>-153.4, -157, –<br>$[Ar]3d^{10}4s^24p^6$ Krypton |
| 5 | 37 **Rb** 85.5<br>687.8, 39, 0.8<br>$[Kr]5s^1$ Rubidium | 38 **Sr** 87.6<br>1377, 769, 1.0<br>$[Kr]5s^2$ Strontium | 39 **Y** 88.9<br>3345, 1522, 1.2<br>$[Kr]4d^15s^2$ Yttrium | 40 **Zr** 91.2<br>4406, 1854, 1.3<br>$[Kr]4d^25s^2$ Zirconium | 41 **Nb** 92.9<br>4741, 2477, 1.6<br>$[Kr]4d^45s^1$ Niobium | 42 **Mo** 96.0<br>4639, 2622, 2.2<br>$[Kr]4d^55s^1$ Molybdenum | 43 **Tc** (98)<br>4262, 2157, 2.1<br>$[Kr]4d^55s^2$ Technetium | 44 **Ru** 101.1<br>4147, 2333, 2.2<br>$[Kr]4d^75s^1$ Ruthenium | 45 **Rh** 102.9<br>3695, 1963, 2.3<br>$[Kr]4d^85s^1$ Rhodium | 46 **Pd** 106.4<br>2963, 1555, 2.2<br>$[Kr]4d^{10}5s^0$ Palladium | 47 **Ag** 107.9<br>2162, 961.8, 1.9<br>$[Kr]4d^{10}5s^1$ Silver | 48 **Cd** 112.4<br>766.8, 321.1, 1.7<br>$[Kr]4d^{10}5s^2$ Cadmium | 49 **In** 114.8<br>2072, 156.6, 1.8<br>$[Kr]4d^{10}5s^25p^1$ Indium | 50 **Sn** 118.7<br>2602, 231.9, 2.0<br>$[Kr]4d^{10}5s^25p^2$ Tin | 51 **Sb** 121.8<br>1587, 630.6, 2.0<br>$[Kr]4d^{10}5s^25p^3$ Antimony | 52 **Te** 127.6<br>987.8, 449.5, 2.1<br>$[Kr]4d^{10}5s^25p^4$ Tellurium | 53 **I** 126.9<br>184.4, 113.7, 2.7<br>$[Kr]4d^{10}5s^25p^5$ Iodine | 54 **Xe** 131.3<br>-108.1, -112, 2.6<br>$[Kr]4d^{10}5s^25p^6$ Xenon |
| 6 | 55 **Cs** 132.9<br>670.8, 28.5, 0.8<br>$[Xe]6s^1$ Caesium | 56 **Ba** 137.3<br>1845, 727, 0.9<br>$[Xe]6s^2$ Barium | 57 **La** 138.9<br>2464, 920, 1.1<br>$[Xe]5d^16s^2$ Lanthanum | 72 **Hf** 178.5<br>4600, 2233, 1.3<br>$[Xe]4f^{14}5d^26s^2$ Hafnium | 73 **Ta** 180.9<br>5455, 3017, 1.5<br>$[Xe]4f^{14}5d^36s^2$ Tantalum | 74 **W** 183.9<br>5555, 3414, 1.7<br>$[Xe]4f^{14}5d^46s^2$ Tungsten | 75 **Re** 186.2<br>5596, 3454, 1.9<br>$[Xe]4f^{14}5d^56s^2$ Rhenium | 76 **Os** 190.2<br>5008, 3033, 2.2<br>$[Xe]4f^{14}5d^66s^2$ Osmium | 77 **Ir** 192.2<br>4428, 2446, 2.2<br>$[Xe]4f^{14}5d^76s^2$ Iridium | 78 **Pt** 195.1<br>3825, 1768, 2.2<br>$[Xe]4f^{14}5d^96s^1$ Platinum | 79 **Au** 197.0<br>2836, 1064, 2.4<br>$[Xe]4f^{14}5d^{10}6s^1$ Gold | 80 **Hg** 200.6<br>356.6, -38.8, 1.9<br>$[Xe]4f^{14}5d^{10}6s^2$ Mercury | 81 **Tl** 204.4<br>1473, 303.8, 1.8<br>$[Xe]4f^{14}5d^{10}6s^26p^1$ Thallium | 82 **Pb** 207.2<br>1749, 327.5, 1.8<br>$[Xe]4f^{14}5d^{10}6s^26p^2$ Lead | 83 **Bi** 209.0<br>1564, 271.4, 1.9<br>$[Xe]4f^{14}5d^{10}6s^26p^3$ Bismuth | 84 **Po** (210)<br>962, 253.8, 2.0<br>$[Xe]4f^{14}5d^{10}6s^26p^4$ Polonium | 85 **At** (210)<br>366.8, 301.8, 2.2<br>$[Xe]4f^{14}5d^{10}6s^26p^5$ Astatine | 86 **Rn** (222)<br>-61.8, -71.2, –<br>$[Xe]4f^{14}5d^{10}6s^26p^6$ Radon |
| 7 | 87 **Fr** (223)<br>676.8, 27, 0.7<br>$[Rn]7s^1$ Francium | 88 **Ra** (226)<br>1140, 699.8, 0.9<br>$[Rn]7s^2$ Radium | 89 **Ac** (227)<br>3200, 1050, 1.1<br>$[Rn]6d^17s^2$ Actinium | 104 **Rf** (261)<br>–, –, –<br>$[Rn]5f^{14}6d^27s^2$ Rutherfordium | 105 **Db** (262)<br>–, –, –<br>$[Rn]5f^{14}6d^37s^2$ Dubnium | 106 **Sg** (266)<br>–, –, –<br>$[Rn]5f^{14}6d^47s^2$ Seaborgium | 107 **Bh** (264)<br>–, –, –<br>$[Rn]5f^{14}6d^57s^2$ Bohrium | 108 **Hs** (267)<br>–, –, –<br>$[Rn]5f^{14}6d^67s^2$ Hassium | 109 **Mt** (268)<br>–, –, –<br>$[Rn]5f^{14}6d^77s^2$ Meitnerium | 110 **Ds** (271)<br>–, –, –<br>$[Rn]5f^{14}6d^87s^2$ Darmstadtium | 111 **Rg** (272)<br>–, –, –<br>$[Rn]5f^{14}6d^97s^2$ Roentgenium | 112 **Cn** (285)<br>–, –, –<br>$[Rn]5f^{14}6d^{10}7s^2$ Copernicium | 113 **Nh** (284)<br>–, –, –<br>$[Rn]5f^{14}6d^{10}7s^27p^1$ Nihonium | 114 **Fl** (289)<br>–, –, –<br>$[Rn]5f^{14}6d^{10}7s^27p^2$ Flerovium | 115 **Mc** (289)<br>–, –, –<br>$[Rn]5f^{14}6d^{10}7s^27p^3$ Moscovium | 116 **Lv** (292)<br>–, –, –<br>$[Rn]5f^{14}6d^{10}7s^27p^4$ Livermorium | 117 **Ts** (294)<br>–, –, –<br>$[Rn]5f^{14}6d^{10}7s^27p^5$ Tennessine | 118 **Og** (294)<br>–, –, –<br>$[Rn]5f^{14}6d^{10}7s^27p^6$ Oganesson |

f BLOCK

| 58 **Ce** 140.1<br>3443, 799, 1.1<br>$[Xe]4f^25d^06s^2$ Cerium | 59 **Pr** 140.9<br>3520, 931, 1.1<br>$[Xe]4f^35d^06s^2$ Praseodymium | 60 **Nd** 144.2<br>3074, 1016, 1.1<br>$[Xe]4f^45d^06s^2$ Neodymium | 61 **Pm** (145)<br>3000, 1042, –<br>$[Xe]4f^55d^06s^2$ Promethium | 62 **Sm** 150.4<br>1794, 1072, 1.1<br>$[Xe]4f^65d^06s^2$ Samarium | 63 **Eu** 152.0<br>1794, 1072, –<br>$[Xe]4f^75d^06s^2$ Europium | 64 **Gd** 157.3<br>3273, 1313, 1.2<br>$[Xe]4f^75d^16s^2$ Gadolinium | 65 **Tb** 158.9<br>3230, 1359, –<br>$[Xe]4f^95d^06s^2$ Terbium | 66 **Dy** 162.5<br>2567, 1412, 1.2<br>$[Xe]4f^{10}5d^06s^2$ Dysprosium | 67 **Ho** 164.9<br>2700, 1472, 1.2<br>$[Xe]4f^{11}5d^06s^2$ Holmium | 68 **Er** 167.3<br>2868, 1529, 1.2<br>$[Xe]4f^{12}5d^06s^2$ Erbium | 69 **Tm** 168.9<br>1950, 1545, 1.3<br>$[Xe]4f^{13}5d^06s^2$ Thulium | 70 **Yb** 173.1<br>1196, 824, –<br>$[Xe]4f^{14}5d^06s^2$ Ytterbium | 71 **Lu** 175.0<br>3402, 1663, 1.0<br>$[Xe]4f^{14}5d^16s^2$ Lutetium |
|---|---|---|---|---|---|---|---|---|---|---|---|---|---|
| 90 **Th** 232.0<br>4875, 1750, 1.3<br>$[Rn]5f^06d^27s^2$ Thorium | 91 **Pa** 231.0<br>–, 1572, 1.5<br>$[Rn]5f^26d^17s^2$ Protactinium | 92 **U** 238.0<br>4131, 1135, 1.7<br>$[Rn]5f^36d^17s^2$ Uranium | 93 **Np** (237)<br>3902, 644, 1.3<br>$[Rn]5f^46d^17s^2$ Neptunium | 94 **Pu** (244)<br>3228, 640, 1.3<br>$[Rn]5f^66d^07s^2$ Plutonium | 95 **Am** (243)<br>2011, 1176, –<br>$[Rn]5f^76d^07s^2$ Americium | 96 **Cm** (247)<br>3110, 1340, –<br>$[Rn]5f^76d^17s^2$ Curium | 97 **Bk** (247)<br>–, 986, –<br>$[Rn]5f^96d^07s^2$ Berkelium | 98 **Cf** (251)<br>–, 900, –<br>$[Rn]5f^{10}6d^07s^2$ Californium | 99 **Es** (252)<br>–, –, –<br>$[Rn]5f^{11}6d^07s^2$ Einsteinium | 100 **Fm** (257)<br>–, –, –<br>$[Rn]5f^{12}6d^07s^2$ Fermium | 101 **Md** (258)<br>–, 827, –<br>$[Rn]5f^{13}6d^07s^2$ Mendelevium | 102 **No** (259)<br>–, –, –<br>$[Rn]5f^{14}6d^07s^2$ Nobelium | 103 **Lr** (262)<br>–, –, –<br>$[Rn]5f^{14}6d^17s^2$ Lawrencium |



ISBN 978 0 6557 0027 2

# APPENDIX 2 Electrochemical series

| Reaction | Standard electrode potential ($E°$) in volts at 25°C |
|---|---|
| $F_2(g) + 2e^- \rightleftharpoons 2F^-(aq)$ | +2.87 |
| $H_2O_2(aq) + 2H^+(aq) + 2e^- \rightleftharpoons 2H_2O(l)$ | +1.77 |
| $Au^+(aq) + e^- \rightleftharpoons Au(s)$ | +1.68 |
| $Cl_2(g) + 2e^- \rightleftharpoons 2Cl^-(aq)$ | +1.36 |
| $O_2(g) + 4H^+(aq) + 4e^- \rightleftharpoons 2H_2O(1)$ | +1.23 |
| $Br_2(l) + 2e^- \rightleftharpoons 2Br^-(aq)$ | +1.09 |
| $Ag^+(aq) + e^- \rightleftharpoons Ag(s)$ | +0.80 |
| $Fe^{3+}(aq) + e^- \rightleftharpoons Fe^{2+}(aq)$ | +0.77 |
| $O_2(g) + 2H^+(aq) + 2e^- \rightleftharpoons H_2O_2(aq)$ | +0.68 |
| $I_2(s) + 2e^- \rightleftharpoons 2I^-(aq)$ | +0.54 |
| $O_2(g) + 2H_2O(l) + 4e^- \rightleftharpoons 4OH^-(aq)$ | +0.40 |
| $Cu^{2+}(aq) + 2e^- \rightleftharpoons Cu(s)$ | +0.34 |
| $Sn^{4+}(aq) + 2e^- \rightleftharpoons Sn^{2+}(aq)$ | +0.15 |
| $S(s) + 2H^+(aq) + 2e^- \rightleftharpoons H_2S(g)$ | +0.14 |
| $2H^+(aq) + 2e^- \rightleftharpoons H_2(g)$ | 0.00 |
| $Pb^{2+}(aq) + 2e^- \rightleftharpoons Pb(s)$ | –0.13 |
| $Sn^{2+}(aq) + 2e^- \rightleftharpoons Sn(s)$ | –0.14 |
| $Ni^{2+}(aq) + 2e^- \rightleftharpoons Ni(s)$ | –0.25 |
| $Co^{2+}(aq) + 2e^- \rightleftharpoons Co(s)$ | –0.28 |
| $Cd^{2+}(aq) + 2e^- \rightleftharpoons Cd(s)$ | –0.40 |
| $Fe^{2+}(aq) + 2e^- \rightleftharpoons Fe(s)$ | –0.44 |
| $Zn^{2+}(aq) + 2e^- \rightleftharpoons Zn(s)$ | –0.76 |
| $2H_2O(l) + 2e^- \rightleftharpoons H_2(g) + 2OH^-(aq)$ | –0.83 |
| $Mn^{2+}(aq) + 2e^- \rightleftharpoons Mn(s)$ | –1.18 |
| $Al^{3+}(aq) + 3e^- \rightleftharpoons Al(s)$ | –1.66 |
| $Mg^{2+}(aq) + 2e^- \rightleftharpoons Mg(s)$ | –2.37 |
| $Na^+(aq) + e^- \rightleftharpoons Na(s)$ | –2.71 |
| $Ca^{2+}(aq) + 2e^- \rightleftharpoons Ca(s)$ | –2.87 |
| $K^+(aq) + e^- \rightleftharpoons K(s)$ | –2.93 |
| $Li^+(aq) + e^- \rightleftharpoons Li(s)$ | –3.04 |

Extract from VCE Chemistry's Written examination: Data Book December 2020. © VCAA (2020), reproduced by permission.